Enabling Smart Urban Services with GPS Trajectory Data

Chao Chen • Daqing Zhang • Yasha Wang •
Hongyu Huang

Enabling Smart Urban Services with GPS Trajectory Data

Springer

Chao Chen
College of Computer Science
Chongqing University
Chongqing, China

Daqing Zhang
School of Electronics Engineering
and Computer Science
Peking University
Beijing, China

Yasha Wang
School of Electronics Engineering
and Computer Science
Peking University
Beijing, China

Hongyu Huang
College of Computer Science
Chongqing University
Chongqing, China

ISBN 978-981-16-0180-4 ISBN 978-981-16-0178-1 (eBook)
https://doi.org/10.1007/978-981-16-0178-1

This Springer imprint is published by the registered company Springer Nature Singapore Pte Ltd.
The registered company address is: 152 Beach Road, #21-01/04 Gateway East, Singapore 189721, Singapore

Preface

With the proliferation of GPS devices and wireless communication technologies in daily life, recent years have witnessed an increasing number of GPS-equipped vehicles, reporting their real-time moving locations to data centers continuously. On the other hand, due to convenience and speed, vehicles have become one of the most common transportation ways to move around the city. As a result, trajectory data that records where and when people move is now gathered and is readily available on a large scale, providing us a time-evolving view to understand how city transports from a data-driven perspective. Taxis' movements, compared to other vehicles, are more complicated. Obviously, taxis' trajectories contain richer information because their movements are affected not only by taxi drivers, but also by passengers, urban planners, etc. For instance, in a trip generation cycle, the heading destination is required by the passenger, while the taxi driver can make decisions on which route to deliver, and where to hunt for new passengers after arrivals. Whether taxis can move smoothly depends on the management efficiency of urban infrastructures by urban planners to a great degree. Generally, to provide various smart urban services for different beneficial parties, the very basic is to understand or model the individual and group/community human behaviors. In more detail, knowing how an individual behaves under various contexts can make the service provision more personalized; on the other hand, knowing how a group/community of people moves is able to understand the time-dependent traffic dynamics. What is more, their differences may correspond to some abnormal (e.g., detouring) or outstanding behaviors (e.g., efficient service strategy) of an individual. All kinds of behaviors and their differences are embedded and hidden behind the trajectory data, calling for new data mining and machine learning models. In one word, due to its richness and uniqueness, the research on mining trajectory data of moving taxis has attracted extensive attention from numerous scholars in the past decade.

In this monograph, we present our latest achievements on mining taxi GPS trajectory data to enable a number of smart urban services, and to bring us one step closer to the vision of smart mobility, concerning interests from the most-involved parties including taxi drivers, passengers to the little-involved parties like

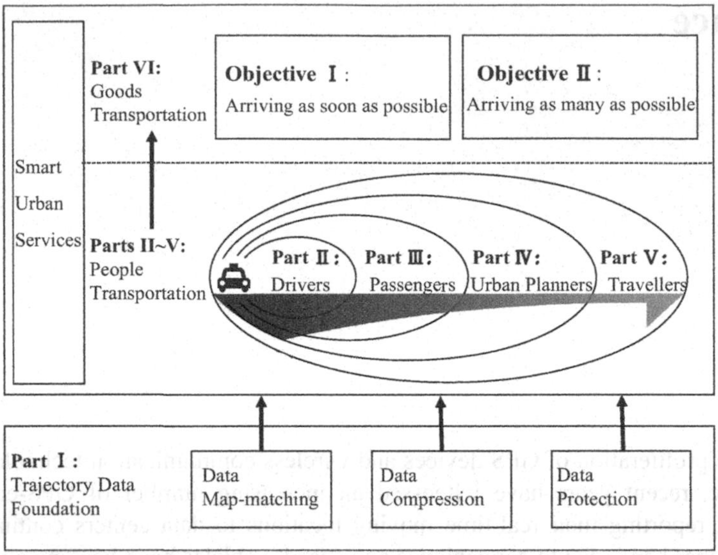
Part VI:
Goods
Transportation
Objective Ⅰ:
Arriving as soon as possible
Objective Ⅱ:
Arriving as many as possible
Smart
Urban
Services
Parts II~V:
People
Transportation
Part Ⅱ:
Drivers
Part Ⅲ:
Passengers
Part Ⅳ:
Urban Planners
Part Ⅴ:
Travellers
Part Ⅰ:
Trajectory Data
Foundation
Data
Map-matching
Data
Compression
Data
Protection

Fig. 1 The structure of main parts in this monograph

urban planners and travellers. The monograph is organized as follows, as shown in Fig. 1. Firstly, we focus on some fundamental issues in trajectory data mining and analytics, including data map-matching, data compression, and data protection. Secondly, driven by the real needs and the most common concerns of each involved party, we formulate each problem mathematically and propose novel data mining or machine learning methods to solve it. Multiple human behavior-related data including check-in data, POI data are also coupled and combined to gain more complete and detailed understanding about human mobility when necessary. Extensive evaluations with real-world datasets are also provided to demonstrate the effectiveness and efficiency of using trajectory data. Lastly, we discuss the most important scientific problems and open issues in mining GPS trajectory data. In summary, this monograph gradually introduces background, related research, new definitions and technologies for solving problems that beginners will encounter when entering this newly exciting research field.

This monograph differs from existing literatures in the following aspects. On the one hand, we focus on analyzing the GPS trajectory data particularly since such data is adequately representative of human behaviors contributed by multiple kinds of urban residents. Besides, we also intend to cover a wider and more complete range, from trajectory data map-matching, data compression, and data protection to data-driven smart urban services. We group the smart urban services into four categories according to the beneficial party, i.e., drivers, passengers, urban planners as well as travellers (as shown in the middle part of Fig. 1). Beyond people transportation, we also extend smart urban services to goods transportation by introducing a novel idea of Crowdshipping, i.e., recruiting taxis to make package deliveries on the basis of real-time information. Since people and goods are two essential components of smart cities and their transportation shares the sufficient overlap (e.g., the same city infrastructure), we believe such extension is logical and essential.

The main contents of this monograph include seven parts. Below is a brief introduction to the topics that will be covered in each part:

Part I: Trajectory Data Foundations (Chaps. 1–3): It is essential to investigate some fundamental issues of GPS trajectory data before using it. To ensure the trajectory data that can reflect the true moving path of an interested taxi along the roads, the most intuitive way is to report the time-stamped locations as frequently as possible. However, such intuitive solution is not feasible and reliable due to overheads caused by data communication, storage, and processing. In real-life scenarios, GPS-equipped taxis only report their locations occasionally, e.g., once per 1 or 2 min. Although such naïve method can alleviate above-mentioned overheads, the collected data is usually in low quality which hinders further data mining tasks eventually. In more detail, on the one hand, the reported GPS locations are contaminated by errors from both GPS devices and digital road network data; on the other hand, the distance gap between two consecutive reported locations can be quite big especially when moving fast, generating a large degree of uncertainty in-between. To bridge the gap between the reported location sequence and the true driving path and to minimize the data that needs to be uploaded while well preserving the spatial information from the algorithm perspective, online trajectory data map-matching (Chap. 1) and data compression (Chap. 2) at the side of data generators (i.e., vehicles) are widely recognized as promising solutions. Online trajectory map-matching aims to reveal the true positions of moving taxis to facilitate data mining and analysis tasks; however, it inevitably results in the potential privacy leakage. Hence, trajectory data protection (Chap. 3) before publishing is also critical, in which the objective of balancing data usability and drivers' privacy should be achieved. In this part, we mainly focus on developing online algorithms (i.e., map-matching, data compression, data protection), aiming at alleviating issues regarding data accuracy, data volume, and data privacy near the data generators before sending them to data centers. In general, developing online algorithms is more challenging due to the incomplete information and the fast response requirement. Online algorithms can be easily extended and applied to offline working cases by simply specifying a time window including the whole trajectory. In this regard, we believe the investigation of online algorithms can be more important and useful in practice.

Part II: Enabling Smart Urban Services for Drivers (Chaps. 4 and 5): The most concern of taxi drivers is their income. The strategies adopted by taxi drivers play a direct influence on the amount of time and distance the taxi is occupied or vacant, resulting in the income difference. Fortunately, taxi driving and service behaviors, including the acceleration and deceleration behaviors, where they pick up the passengers and how they find and deliver the next passengers are implicitly conveyed in the GPS trajectory data. Such data provides us with a unique opportunity to understand and learn good taxi service strategies in various situations. Hence, in Chap. 4, we aim to *earn more* by understanding efficient taxi service strategies for drivers, i.e., increase their income by learning efficient taxi service strategies from others. In more detail, with the trajectory data, we discover the most differentiated service strategies between high-income and average-income drivers, from perspectives of passenger-searching, passenger-delivery, and service area preference. Moreover, the widely deployed OBD-II (On-Board-Diagnostics) sensors in cars can link the collected GPS trajectory data and the resulted fuel consumption, offering rich sources for investigating the inherent relationship between the personalized driving behaviors and the fuel economy. On this basis, in Chap. 5, we propose a two-phase framework named **GreenPlanner** to *save more* by taking the most-fuel-efficient driving route, i.e., increase their income by planning the most fuel-efficient driving route according to the personal driving style and the real-time traffic conditions.

Part III: Enabling Smart Urban Services for Passengers (Chaps. 6 and 7): Compared to taxi drivers, GPS trajectory data contains less information regarding passengers. To be more specific, only trajectories left when delivering passengers are relevant. Thus, in this part, first of all, we enable smart urban services for passengers based on the passenger-carrying trajectory data. Such trajectory data can tell that how passengers are delivered explicitly, i.e., the routes that taxi drivers choose. For passengers, different routes may correspond different distance and traffic jam duration, resulting in the difference of taxi fares that needed to pay. Hence, in Chap. 6, we aim to detect the anomalous passenger-delivering trajectories that can inform and alert passengers in time to prevent from being overcharged. In more detail, we propose an isolation tree-based anomalous trajectory (**iBOAT**) detection algorithm that can identify the anomalous trajectory parts correctly on the fly. Understanding why people take taxis (i.e., trip purposes) is quite essential in recommending services after drop-offs. Unfortunately, passengers' trajectories after being dropped off cannot be recorded by GPS-equipped taxis. Hence, in the second, to fill this gap, in Chap. 7, coupling with the complementary spatial and temporal context around the drop-off point provided by Foursquare check-in data, we propose a probabilistic two-phase framework called **TripImputor** that can make real-time taxi trip purpose imputation correctly.

Part IV: Enabling Smart Urban Services for Urban Planners (Chaps. 8 and 9): Trajectory data does not contain any related information regarding urban planners at first glance. However, as urban planners design, build, and manage urban infrastructures, they actually have a kind of relationship with the taxi movement. The quality of GPS data is vital in enabling a wide range of location-based applications and services. In addition to the quality degradation caused by GPS devices, the urban environment (e.g., structure, layout, buildings, tunnels) also contributes such degradation. One promising way is to deploy a number of dedicated localizers along roads to improve the positioning accuracy. The very basis is to evaluate the city-scale GPS Environment Friendliness (GEF) on the road level. GEF of the road should be low when GPS observations vary a lot when traversing the same road. Hence, in Chap. 8, we choose the trajectory data of buses which have fixed and known driving routes in advance and evaluate the GEF based on distributions of GPS observations. Compared to buses, taxis operate in a 24/7 manner, and they become dominate transportation mean during late night and early morning. In this context, the taxi trajectory data generated during that time can almost reflect the true travel demands. In Chap. 9, based on the travel demands, we propose a two-phase framework called **B-Planner** including bus station identification and route generation to plan night-through bus routes to deliver passengers aggregately in a more energy-efficient way.

Part V: Enabling Smart Urban Services for Travellers (Chaps. 10 and 11): The fifth part consists of two chapters, targeting at offering smart urban services to travellers. More travellers can be attracted if the related urban services are smarter, however, compared to above-mentioned parties, the use of GPS trajectory data is not that intuitive. In Chap. 10, we show that the typical time-dependent travel times between any two points can be inferred by mining historical GPS trajectory data. Coupled with other related data regarding user preference on venues, we build a personalized, interactive, and traffic-aware trip planning service for travellers. In Chap. 11, rather than recommending the fastest or shortest driving routes between two points for travellers, we develop a route planner that can trade-off the quality of sightseeing along the driving route and the travel time budget.

Part VI: Enabling Smart Urban Services for Goods Transportation (Chaps. 12 and 13): Beyond the citywide people transportation, the sixth part concerns goods transportation. We overview the concept and framework of Crowdshipping by using taxis as package hitchhikers, discuss the research challenges and potential solutions. In Chaps. 12 and 13, we respectively define two different concrete objectives, i.e., sending packages as soon as possible and sending them by the deadline. We further elaborate the algorithm design details for the two objectives with extensive evaluations.

Part VII: Open Issues, Future Directions, and Conclusions (Chap. 14): In the last part, we will discuss the some of the open issues and future research directions in enabling smart urban services with GPS trajectory data, followed by a conclusion to the whole monograph.

We hope this monograph provides a clear and concise picture of state-of-the-art GPS trajectory data mining or even spatial-temporal data mining approaches to the researchers in this field.

Chongqing, China — Chao Chen
Beijing, China — Daqing Zhang
Beijing, China — Yasha Wang
Chongqing, China — Hongyu Huang
December 2020

Acknowledgments

We would like to express our appreciation to the many colleagues and students who had worked with us and contributed directly or indirectly to the monograph. During the past 10 years, we have received financial supports continuously from European and Chinese governments, including the National Key Research and Development Project of China under Grant 2017YFB1002000, the National Natural Science Foundation of China under Grant 61872050, and so forth. A special thanks goes to our collaborator, Prof. Shijian Li, from Zhejing University, who shared the valuable taxi GPS trajectory data, which is the solid foundation for all the outcomes. We also would like to thank Prof. Bin Guo, Zhiwen Yu, and Zhu Wang at School of Computer Science of Northwestern Polytechnical University, China; Prof. Shijian Li, Gang Pan, Zhaohui Wu at School of Computer Science of Zhejiang University, China; Prof. Zhi-Hua Zhou at School of Artificial Intelligence of Nanjing University, China; Dr. Nan Li at Microsoft; and Prof. Bin Li at School of Computer Science of Fudan University, China. We also want to express our thanks to all the members of Big Data and Smart Computing group of Chongqing University, China for their valuable discussions, insights, and constructive suggestions. Thanks also goes to my student who help proofread the draft of this book carefully and completely, including Linli Jiang, Chengwu Liao, Xingchen Wang, Jie Zhao, Liping Gao, Chenxi Liu, and Xuansu Gao. We would like to thank the staff at Springer, Mr. Nick Zhu, Ms. Riya Rathore, and Ms. Sagaai Vilma Paul for their kind help throughout the publication and preparation processes of the monograph.

Contents

Part I
Foundations

Chapter 1
Trajectory Data Map-matching

1.1 Introduction

The wide deployment of GPS devices in vehicles has made the collection of moving trajectories more easily. There is still a huge gap between the raw data and the wide spectrum of smart urban services although the pervasiveness and easy availability of such data [1–3]. The problems of *sparseness* and *uncertainty* commonly exist in the GPS trajectory data since the GPS devices equipped in the moving objects (e.g., vehicles) only report the locations occasionally to save energy and communication costs. To make matter worse, location observations are imprecise and erroneous caused by both measurement errors from GPS devices and inaccuracy from the digital maps. The objective of map-matching is to find the true positions and driving paths since vehicles can only move with the 2D space constrained by the road network [4–6]. There are two categories for the map-matching algorithms, i.e., online and offline. In general, the online map-matching algorithms are desirable since the real-time intelligent transportation services such as taxi dispatch, vehicle monitoring, transportation resources allocation can be appropriately supported.

Generally speaking, there are mainly three steps for the online map-matching procedure, namely, finding the true position, inferring and refining the true driving paths. A number of map-matching algorithms have been proposed, but it is still challenging to balance the two conflicting objectives, i.e., the matching quality and the computation time. In this chapter, to achieve the two objectives simultaneously, a novel three-stage online map-matching algorithm, named Spatial-Direction-Matching (SD-Matching for short in the rest of presentation) has been proposed. In this new algorithm, it smartly utilizes an *important yet not fully-explored* dimension information (i.e., *heading direction*) contained in the collected GPS trajectory

---

Part of this chapter is based on a previous work: C. Chen, Y. Ding, X. Xie and S. Zhang, "A three-stage online map-matching algorithm by fully using vehicle heading direction," vol. 9, no. 5, pp. 1623–1633.

C. Chen et al., *Enabling Smart Urban Services with GPS Trajectory Data*,
https://doi.org/10.1007/978-981-16-0178-1_1

data at all three stages. In more detail, at the stage of candidate edge identification for a given GPS point, the heading direction is used to enhance the probability computation. At the stage of filling the distance gap between two consecutive GPS points, the heading direction is adopted as a cost-effective guider to accelerate the shortest path finding by narrowing down the searching space. At the stage of refining path for a sequence of GPS points, with the topology of the road network, the heading direction is finally utilized to refine the driving path by picking the one with the maximum likelihood. To sum up, we mainly make the following contributions in this chapter.

- We target at a fundamental and preliminary problem for many trajectory data mining tasks, that is the online map-matching problem. We propose to fully explore the usability of the heading direction at all three stages to trade-off the quality and the computational efficiency of the online map-matching. Furthermore, we adopt a number of effective techniques including simple indexing, tree building and pruning in implementing SD-Matching algorithm under the real and mobile environment.
- Based on a large-scale trajectory and road network datasets collected from the city of Beijing, China, we design a number of experimental plans to verify the performance of the proposed SD-Matching algorithm in terms of matching accuracy and running time. To calculate the matching quality correctly, we also invite volunteers to manually label the ground truth driving path for each vehicle trajectory in the dataset. The final evaluation results demonstrate that the SD-Matching algorithm outperforms while compared to the state-of-arts.

1.2 Related Work

To the best of our knowledge, there are four survey papers on map-matching algorithms [6–9] in literature, among them three of them focus on the taxonomy of map-matching algorithms from different perspectives, while the rest one concentrates on the real applications of map-matching algorithms. In earlier related work, the authors mainly aimed to identify “true” positions for each individual GPS point [10], which can address the measurement errors only. Previously, before finding the true position for a given GPS point, a set of candidate edges within the error range is firstly identified. The probability of true position on each edge in the set is different. For instance, a common method to obtain such probability is the geometric analysis, which is based on the geometric information including the distance of ‘point-to-curve’ or ‘point-to-shape’ [6]. Among the candidate edges, the one with the highest probability value would be picked as the true edge that the given GPS point locates. Hence, the key issue is how to compute the probability accurately. Focusing on the heading direction, previous studies have demonstrated its usefulness in improving the probability computation accuracy [11]. However, the heading direction is not collected *directly* but is inferred by the difference between the current GPS observation to its closet one, which actually refers to the *moving direction*. Such heading direction may be not that effective when the sampling time interval is too big. To make matter worse, it cannot be obtained when dealing with the task of online map-matching, since the next position is unknown. Fortunately, the instantaneous heading direction at the sampling time and location can be collected in current GPS-enabled vehicles, which has been successfully integrated in pinpointing the true position more correctly [12]. Inspired by the previous conclusions, similar to the previous studies, we also intend to explore how to utilize the directly collected heading direction to enhance the probability computation performance.

More research work focus on filling the distance gap for a sequence of GPS points by reconstructing the travelling path during recent years. More specifically, the two edges that the two consecutive GPS points locate at are generally disconnected in the road network. The distance gap can be quite large when the speed is high or the sampling time interval is big. In this regard, this sort of work aims at addressing the issues of uncertainty and sparseness [13]. Additional information including road network topology (e.g., link connectivity and contiguity), road attributes (e.g., speed limits) and motion laws are also used in order to reconstruct the true driving path [14, 15]. As for the matching algorithms, some probabilistic algorithms such as Hidden Markov Model-based and more advanced algorithms are developed and proposed to improve the matching performance [16, 17]. To reconstruct the driving path for a given sequence of GPS points reliably, the very common operation is to find the driving path between every two consecutive GPS points within the sequence, followed by the path refining globally. However, such common operation is usually resource-hungry and time-consuming. On one hand, the core algorithm within the common operation is the most likely path finding (e.g., the shortest path) such as Dijkstra and A* algorithms [18]. On the other hand, there are many OD pairs needing the shortest path finding, since each GPS point corresponds a set of candidate edges [19]. As a result, the running time is extremely long and becomes one of the bottlenecks for online map-matching algorithms. Hence, in this chapter, we propose metaheuristic to explore the usability of the direct collected heading direction in both path-finding and path-refining, with the objective of overcoming the running time bottleneck in the computation of shortest path-finding.

1.3 Main Concepts

Definition 1.1 (Road Network). A road network is a graph $G(N, E)$, consisting of an edge set E and a node set N. Each element in E is a directed edge e_i that is associated with two nodes n_h and n_t (i.e., the head and tail nodes). Each element n in N is an intersection of a pair of longitude and latitude coordinates, representing its spatial location.

Definition 1.2 (Edge Direction). Each edge e_i has two edge directions, $d(n_h, n_t)$ and $d(n_t, n_h)$ respectively. $d(n_h, n_t)$ is defined as the edge direction from node n_h to node

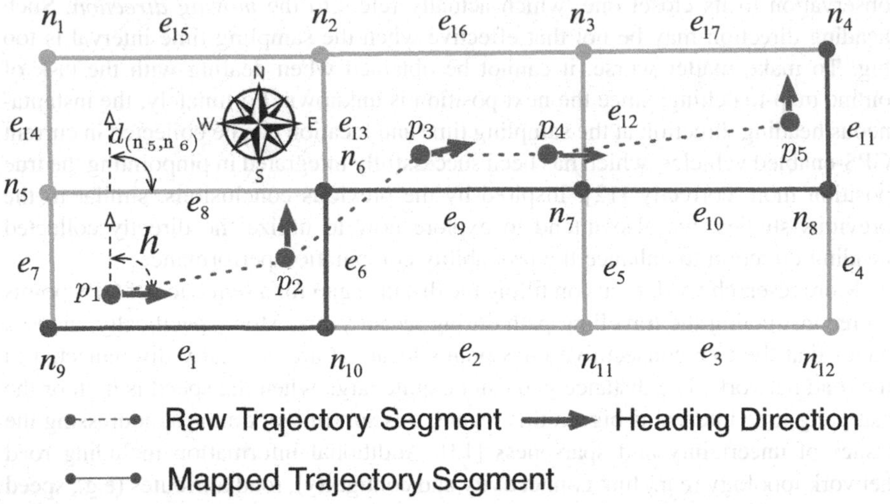

n1
e15
n2
e16
n3
e17
n4
N
W
E
S
e14
d(n 5,n 6)
e13
p3
p4
e12
e11
p5
n6
n5
e8
e9
n7
e10
n8
h
e7
e6
e5
e4
p1
p2
n9
e1
n10
e2
n11
e3
n12
Raw Trajectory Segment
Heading Direction
Mapped Trajectory Segment

Fig. 1.1 Illustration of main concepts used in this chapter

n_t, which can be easily calculated based on their longitudes and latitudes, using north direction as a basis. For instance, $d(n_5, n_6)$ in Fig. 1.1 is the edge direction of e_8 from n_5 to n_6.

Definition 1.3 (Heading Direction). Heading direction h ($0° \leq h < 360°$) refers to the direction of the vehicle heads at a sampling position, using north direction as a basis again. This information can be accessible directly from the trajectory data. For example, the vehicle's heading direction at point p_1 is h_1, as shown in Fig. 1.1.

Definition 1.4 (GPS Point). A GPS point p_i records the spatio-temporal information of a vehicle, which consists of a timestamp, a geospatial coordinate, an instantaneous speed, and a heading direction, denoted by $p_i = (t_i, lat_i, lon_i, v_i, h_i)$.

Definition 1.5 (Raw Trajectory Segment). A raw trajectory segment τ is a sequence of GPS points, represented by $\tau = \langle p_i, p_{i+1}, \cdots, p_{i+l} \rangle$. Parameter l controls the length of the segment. A raw trajectory stream T consists of an unbounded set of raw trajectory segments τ, denoted by $T = \langle \tau_1, \tau_2, \cdots \rangle$.

Definition 1.6 (Mapped Trajectory Segment). Given a raw trajectory segment, its mapped segment τ_m is the real path that the vehicle travels in the road network. τ_m is denoted by $\langle e_i, e_{i+1}, \cdots, e_n \rangle$, where each pair of two consecutive edges are connected. A mapped trajectory stream $T_m = \langle \tau_{m1}, \tau_{m2}, \cdots \rangle$ is an unbounded set of mapped trajectory segments τ_m.

1.4 The SD-Matching Algorithm

1.4.1 Algorithm Overview

The overview of the SD-Matching algorithm is shown in Fig. 1.2. As GPS trajectory data coming point by point, data splitting is first applied to divide the data stream into small equal-length trajectory segment (i.e., raw trajectory segment), which is the basic data unit for processing. For each raw trajectory data, the online map-matching algorithm handles it at three stages, by integrating the road network data and heading direction comprehensively. At the first stage, when computing the likelihood of each GPS point belonging to a given nearby edge, both the spatial and direction probabilities are weighted. At the second stage, the heading direction is used in both narrowing down the search space and accelerating the path-finding. At the third stage, to identify the true driving path for the given raw trajectory segment, both the heading direction and the road network topology are considered.

1.4.2 *Identifying Top-k Candidate Edges for a Given GPS Point*

For a given GPS point, we first determine a set of candidate edges within the circle of radius. Note that each point usually has several candidate edges, especially in the area with dense road network. Intuitively, to determine such a set, we need to compute all distances from that point to all edges in the road network. To accelerate the process, we adopt a simple grid indexing mechanism. In more detail, the whole

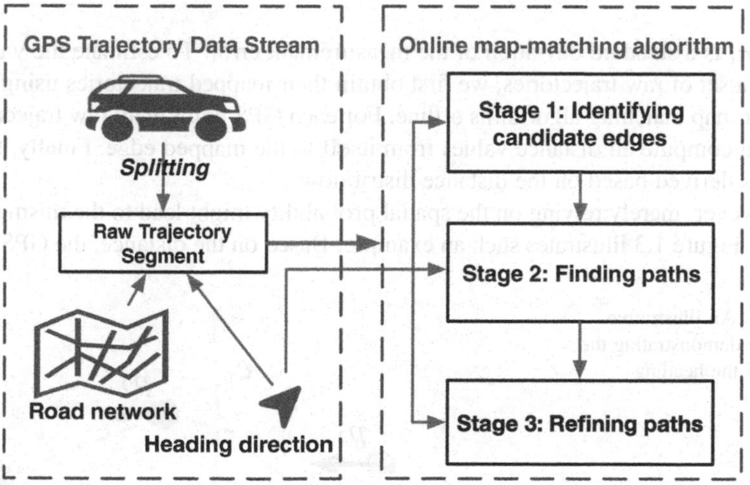
GPS Trajectory Data Stream
Splitting
Raw Trajectory Segment
Road network
Heading direction
Online map-matching algorithm
Stage 1: Identifying candidate edges
Stage 2: Finding paths
Stage 3: Refining paths

Fig. 1.2 Overview of the proposed SD-Matching algorithm

city is divided into a number of small equal-sized grid cells, inside which nodes and edges can be easily known. It is also easy to determine the corresponding grid cell. In this manner, to identify the set of candidate edges, it only needs to compute the distances from the given GPS point to edges in its neighboring grid cells. To find the truly correctly-matched edge for the given GPS point, it is necessary to further reduce the number of candidate edges. Here, for each candidate edge, we use the spatial probability and direction probability together to measure the probability of obtaining a correctly-matched result for the given GPS point, then identify the top-k candidate edges for further processing, shown as follows.

1.4.2.1 Spatial Probability

The spatial probability is defined as the *likelihood* that a GPS point p_i matches a candidate edge e_j based on the distance between p_i and e_j. The distance $\left(H_{e_j}^{p_i}\right)$ from p_i to e_j is the minimal one among $dist(p_i, c_1)$, $dist(p_i, c_2)$, $dist(p_i, c_3)$, where function *dist* returns the distance between two points, and c_1 and c_2 are the two endpoints of edge e_j. The point c_3 is the vertical projection point on edge e_j of p_i. Note that we manually set $dist(p_i, c_3)$ to $+\infty$ if c_3 is out of the range of e_j.

Generally speaking, the noise of location measurement can be reasonably modeled as a standard-normal distribution of the distance $H_{e_j}^{p_i}$. Such distribution indicates that how likely p_i can be matched to the candidate edge e_j. Thus, the spatial probability can be calculated based on Eq. (1.1).

$$G_1\left(H_{e_j}^{p_i}\right) = \frac{1}{\sqrt{2\pi}\sigma_1} e^{-\frac{\left(H_{e_j}^{p_i}\right)^2}{2\sigma_1^2}} \quad (1.1)$$

where σ_1 is a standard deviation of the measurement error. To estimate the value of σ_1, for a set of raw trajectories, we first obtain their mapped trajectories using other popular map matching algorithms offline. For each GPS point in all raw trajectories, we then compute all distance values from itself to the mapped edge. Finally, the σ_1 value is derived based on the distance distribution.

However, merely relying on the spatial probability might lead to the mismatched results. Figure 1.3 illustrates such an example. Based on the distance, the GPS point

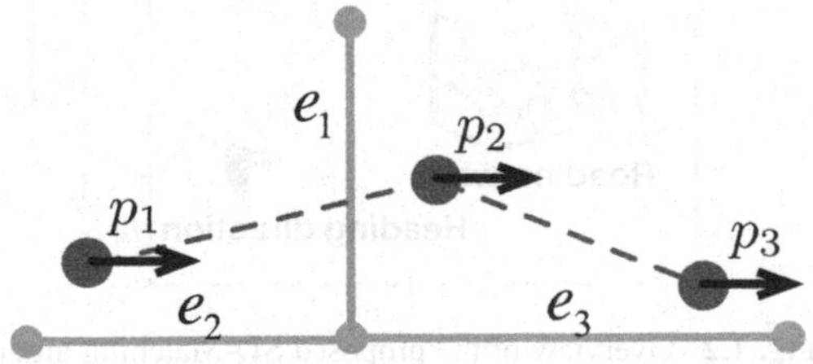

e1
p2
p1
p3
e2
e3

Fig. 1.3 An illustrative example demonstrating the utility of the heading direction

p_2 should be mapped to edge e_1, which is incorrect in the real case. As can be seen, the vehicle heading direction at the point p_2 is much closer to the edge direction of e_3, which motivates us to bring in the heading direction information to avoid mismatching. More specifically, *direction probability* is introduced, detailed as follows.

1.4.2.2 Direction Probability

The direction probability is defined as the likelihood that a GPS point p_i matches a candidate edge e_j based on the angle difference between the vehicle's heading direction h_i at the point p_i and the edge direction of e_j. The angle difference $\left(A_{e_j}^{p_i}\right)$ between p_i and e_j is computed as $\min\{|h_i - d_{th}|, |h_i - d_{ht}|\}$. Similar to the spatial probability, the direction probability can be calculated according to Eq. (1.2).

$$G_2\left(A_{e_j}^{p_i}\right) = \frac{1}{\sqrt{2\pi}\sigma_2} e^{-\frac{\left(A_{e_j}^{p_i}\right)^2}{2\sigma_2^2}} \tag{1.2}$$

where σ_2 is a standard deviation of the direction measurement error, which can be estimated in the similar way to σ_1. Combining the probability G_1 and G_2, we define the comprehensive probability G as the geometrical mean of the spatial probability and the direction probability, as shown in Eq. (1.3).

$$G = \sqrt{G_1\left(H_{e_j}^{p_i}\right) \times G_2\left(A_{e_j}^{p_i}\right)} \tag{1.3}$$

As mentioned previously, the GPS point may have a large number of candidate edges. Each candidate edge has a value of comprehensive probability G. We can identify k candidate edges with top probability values (i.e., top-k candidate edges) for each GPS point. Note that the number of candidate edges (i.e., the parameter k) can impact at the accuracy and time cost of the proposed map matching algorithm, which will be evaluated in Sect. 1.5.

1.4.3 *Finding Paths Between Two Consecutive GPS Points*

The candidate edges of two consecutive GPS points p_i and p_{i+1} are generally unconnected in the road network, since the distance between them may be far. Thus, we need to find potential paths between two consecutive GPS points. For each point p_i, it has k candidate edges. It is easy to understand that for a pair of two consecutive points, it would have k^2 possible paths in total. In previous work, such one path is commonly discovered by A* or Dijkstra algorithm, which is time-consuming.

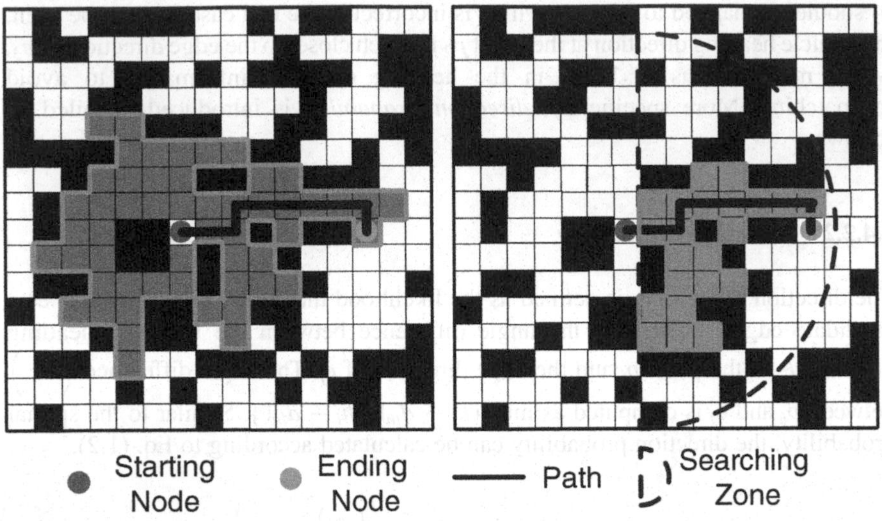
Starting Node
Ending Node
Path
Searching Zone

Fig. 1.4 An illustrative example of searching zone comparison with and without the consideration of the heading direction

Moreover, the computation time issue is even more critical in our case, since all k^2 possible paths are needed to be found. To alleviate the issue, we propose to *fully use the heading direction* to (1) *narrow down the searching zone* and (2) *serve as a cost-effective guider* in finding paths.

Before going to the details of the proposed algorithm, we provide an illustrative example to gain an overall idea of the effectiveness by using the heading direction information. Figure 1.4 (left) shows the procedure of finding the path from a given starting node to the ending node based on the A* algorithm. The big rectangle is the potential searching zone, in which black cells are physical barriers that cannot be travelled and white cells can be travelled by vehicles. Before returning the path, A* algorithm needs to search all gray cells (within the regions bounded by the solid blue lines). As a comparison, Fig. 1.4 (right) shows the searching zone and together with the area that needed to be covered (gray cells) before finding the final path from the starting node to the ending node with our proposed algorithm. As can be observed, both searching zone and area of gray cells are much smaller, thus the procedure should be more efficient.

The proposed path-finding algorithm contains two steps, i.e., determining the potential searching zone and finding paths within the potential searching zone, respectively.

Step 1: Determining the potential searching zone. The potential searching zone is a sector (e.g., the gray area in Fig. 1.5), which can be well-determined as follows: (1) the apex of this sector is the point p_i; (2) the radius $r_{\max}$ of the sector is calculated by $\max\left(v_{p_i}, v_{p_{i+1}}\right) \times \Delta t + c$, where v_{p_i} and $v_{p_{i+1}}$ are the speed values of the vehicle at pints p_i and p_{i+1} respectively; c is constant (we set $c = 50$ m) and

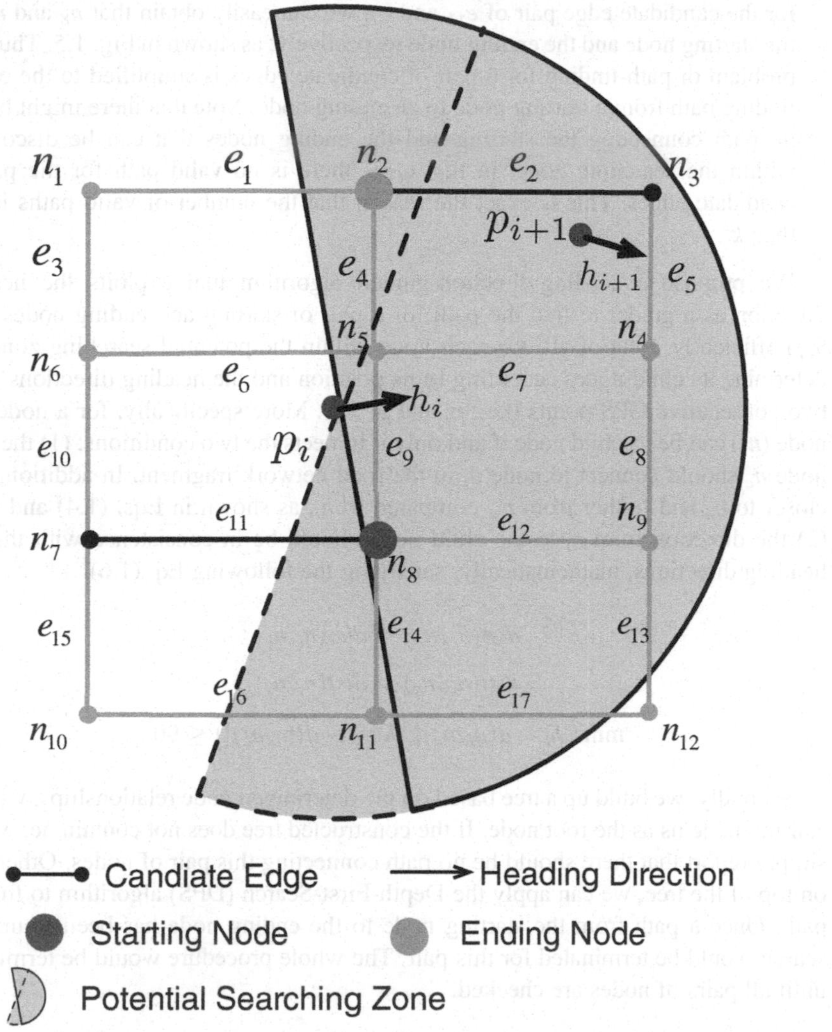
n_1
e_1
n_2
e_2
n_3
p_{i+1}
e_3
e_4
h_{i+1}
e_5
n_5
n_4
n_6
e_6
e_7
h_i
p_i
e_{10}
e_9
e_8
e_{11}
e_{12}
n_9
n_7
n_8
e_{15}
e_{14}
e_{13}
e_{16}
e_{17}
n_{10}
n_{11}
n_{12}
Candiate Edge
Heading Direction
Starting Node
Ending Node
Potential Searching Zone

Fig. 1.5 An illustrative example of the proposed two-step path-finding

Δt is the sampling time interval. Note that r_{max} indicates the maximum driving distance of the vehicle from point p_i to p_{i+1}; (3) the sector is composed of two semicircles. Each diameter of semicircles are perpendicular to the vehicle's heading direction h_i and h_{i+1} at point p_i and p_{i+1} (solid and dashed semicircles in Fig. 1.5), respectively.

Step 2: Finding paths within the potential searching zone. For the given GPS points p_i and p_{i+1}, we need to find k^2 possible paths corresponding to k^2 pairs of candidate edges within the potential searching zone (returned by Step 1). For a pair of candidate edges, according to the heading directions at p_i and p_{i+1}, it is easy to determine the starting node and the ending node of the path. For instance, for the candidate edge pair of e_{11} and e_2, we can easily obtain that n_8 and n_2 are the starting node and the ending node respectively, as shown in Fig. 1.5. Thus, the problem of path-finding for a pair of candidate edges is simplified to the one of finding path from a starting node to an ending node. Note that there might be just no path connecting the starting and the ending nodes that can be discovered within the searching zone. In this case, there is no valid path for the pair of candidate edges. This is exact the reason that the number of valid paths is less than k^2.

We propose a heading-direction-guided algorithm that exploits the heading direction as a guider to find the path for a pair of starting and ending nodes ($\langle n_s, n_e \rangle$) efficiently. First of all, for each node within the potential searching zone, we determine its child nodes according to its position and the heading directions at the two consecutive GPS points (i.e., p_i and p_{i+1}). More specifically, for a node n_i, a node (n_c) can be its child node if and only if it meets the two conditions: (1) the child node n_c should connect to node n_i in the road network fragment. In addition, n_c is closer to n_e and farther from n_s, compared to n_i, as shown in Eqs. (1.4) and (1.5); (2) the direction from n_i to the child node should be in consistence with the two heading directions, mathematically, satisfying the following Eq. (1.6).

$$dist(n_c, n_e) < dist(n_i, n_e) \tag{1.4}$$

$$dist(n_c, n_s) > dist(n_i, n_s) \tag{1.5}$$

$$\min\left(|h_i - d(n_i, n_c)|, |h_{i+1} - d(n_i, n_c)|\right) \leq 90^{\circ} \tag{1.6}$$

Secondly, we build up a tree based on the determined node relationships, with the starting node ns as the root node. If the constructed tree does not contain ne, we can simply report that there should be no path connecting this pair of nodes. Otherwise, on top of the tree, we can apply the Depth-First-Search (DFS) algorithm to find the path. Once a path from the starting node to the ending node has been found, the search would be terminated for this pair. The whole procedure would be terminated until all pairs of nodes are checked.

1.4.4 *Refining Paths for a Given Raw Trajectory Segment*

Between two consecutive GPS points, with the previous two steps, we could obtain at most k^2 possible paths. Thus, for a raw trajectory segment with l GPS points, ideally there would be totally $k^2(l-1)$ possible paths connecting them. However, the actual number is much smaller since (1) there are a much smaller number of paths ($\ll k^2$) for two consecutive GPS points (i.e., there would be no path returning for some node pairs); (2) not all paths of different consecutive GPS points can be connected in the road network, and path refining based on the topology of the road network is necessary. Moreover, among all possible paths, each of them has different

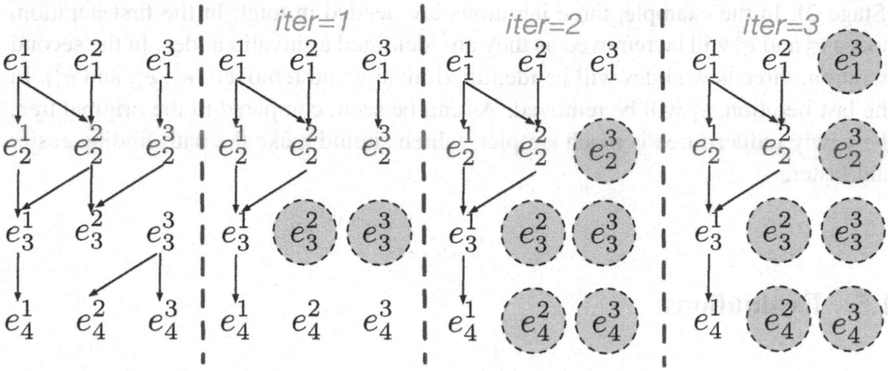
iter=1
iter=2
iter=3

Fig. 1.6 Illustrative example on graph pruning strategy

probabilities to be the real mapped trajectory, due to that each candidate edge has different probabilities to be the correctly mapped one for the GPS point. Thus, we further refine the path based on the possibility of a path (τ_{mi}) being the real mapped trajectory. The possibility is estimated by summarizing all the comprehensive probabilities of GPS points being correctly mapped to the candidate edge, as shown in Eq. (1.7).

$$P_{\tau_{mi}} = \sum_{i=1}^{l} G\left(p_i, e_j^{p_i}\right) \tag{1.7}$$

where $G\left(p_i, e_j^{p_i}\right)$ refers to the comprehensive probability of p_i being correctly mapped to edge e_j, as defined in Eq. (1.3). Note that one possible path for a given raw trajectory segment $\langle p_1, p_2, \cdots, p_\delta \rangle$ can be denoted by $\tau_{mi} = \left\langle e_j^{p_1} \rightarrow e_j^{p_2} \cdots \rightarrow e_j^{p_\delta} \right\rangle$, where $e_j^{p_i}$ refers to the jth candidate edge of p_i, $e_j^{p_i} \rightarrow e_j^{p_{i+1}}$ refers to the path from $e_j^{p_i}$ to $e_j^{p_{i+1}}$.

In the implementation of path-refining, we first construct a directed tree in which each node refers to a candidate edge for a GPS point. The directed edge in the tree refers to the path discovered at Stage 2 that connects two candidate edges. On top of the built tree, we apply a simple tree pruning strategy to get a much tighter one by removing invalid nodes iteratively. A node is claimed invalid if its in-degree or out-degree equals 0, except for the starting and ending nodes.

To make our idea clear, we use an illustrative example, as shown in Fig. 1.6. For better demonstration, we set l and k to be 4 and 3, respectively. The depth of the tree is exactly equal to the size of the trajectory segment (i.e., l). The first level corresponds to the first GPS point while the last level corresponds to the last GPS point. The number of nodes in each level is identical, which just equals to the number of identified candidate edges (i.e., k). There does (not) exist an arrow if a path is (not) returned for a pair of candidate edges belonging to two consecutive GPS points

(Stage 2). In the example, three iterations are needed in total. In the first iteration, nodes e_3^2 and e_3^3 will be removed as they are identified as invalid nodes. In the second iteration, three new nodes will be identified invalid and removed (e_2^3, e_4^2, and e_4^3). In the last iteration, e_1^3 will be removed. As can be seen, compared to the original tree, the newly reduced tree is much simpler, which should make the path-finding easier and faster.

1.5 Evaluations

In this section, we conduct extensive experiments to demonstrate the effectiveness and efficiency of the proposed SD-Matching algorithm, by comparing it to the state-of-art algorithms in terms of effectiveness and efficiency.

1.5.1 Data Preparation

We mainly use three datasets from the city of Beijing, China, in the evaluations. The first dataset is the road network that is extracted from OpenStreetMap, which collects data via a crowdsourcing manner. The second dataset is the raw GPS trajectory data generated by 50 taxis on 15th September, 2015, which contains over 50,000 GPS points. The sampling rate for all taxis is identical and constant, with a value of around 6 s. The third dataset is a human-labelled trajectory data. To be more specific, for each set of trajectory data generated by one taxi, we first manually split it into a number of trajectory segments containing equal-sized GPS points. The number of trajectory segments is determined by its length (l) and the number of GPS points in each set of trajectory data. We set $l = 5, 10, 15, 25, 30$ in our experiments and get different numbers of trajectory segments, with the details shown in Table 1.1. Then, for each trajectory segment, with the digital map, we recruit volunteers to identify its corresponding mapped trajectory segment in the road network manually. The human-labelled mapped trajectory segments can be viewed as the ground truth for the raw trajectory segments.

Table 1.1 Number of raw trajectory segments under different ls

l	5	10	15	20	25	30
# trajectory segments	10,366	5048	3456	2645	2569	1824

1.5.2 Evaluation Criteria

To evaluate SD-Matching, we propose two metrics. The first one is the mean accuracy, which is defined as the ratio of the number of correct matched GPS points to the number of total GPS points to be matched. The other one is the time cost, which refers to the computation time that needed in mapping a raw trajectory segment. For the time cost, to gain a statistical sense, we further use both average time cost and the maximum time cost when evaluating a set of raw trajectory segments. The average time cost is important to evaluate the overall performance of different map-matching algorithms, while the maximum time cost corresponds to the longest time delay of SD-Matching algorithm.

1.5.3 Baseline Methods

To show the superior performance, we compare SD-Matching with the start-of-art map-matching algorithms, i.e., the ST-Matching algorithm and the S-Matching. Note that the ST-Matching algorithm is the most popular competitor in the research of map-matching [15, 20]. It differs from our proposed algorithm in the following two main aspects:

1. For a given GPS point, it identifies the candidate edges based on the spatial and temporal information, while our method considers both the spatial and heading direction information collectively. When calculating the probability of a candidate edge of being the correctly-matched edge for the given GPS point, it uses both spatial and temporal information, while our method uses both spatial and heading direction information.
2. When searching the path between two consecutive GPS points, it adopts either Dijkstra (ST-Matching_v_1) or A* (ST-Matching_v_2) algorithm. Note that different versions of ST-Matching algorithm would generate the same map-matching accuracy, but it is expected that ST-Matching_v_2 is more efficient. As a comparison, we propose a novel path-finding algorithm that fully uses the heading direction information.

S-Matching differs from ST-Matching algorithm as it only leverages the spatial information (i.e., based on spatial probability only) in identifying the candidate edges. In summary, the most salient and distinct feature of SD-Matching algorithm is that it leverages heading direction information in different stages of map-matching.

1.5.4 Effectiveness Study

There are two important parameters that can influence the mean accuracy of SD-Matching algorithm. One is the length of the trajectory segments (i.e., l), the other is the number of identified candidate edges for each given GPS point, referred as k. We study the performance of SD-Matching algorithm on the mean accuracy under different ls and ks respectively. As a comparison, the results of the baseline methods are also provided.

1.5.4.1 Varying l

We report the results of mean accuracy for different algorithms under various lengths of trajectory segment (l) in Fig. 1.7a. We increase the length from 5 to 30, with an equal interval of 5. Note that we only show the result of one version of ST-Matching algorithm, since different versions only affect the time cost at mapping trajectories. To be more specific, A* algorithm needs less time than Dijkstra in path-finding, but they would return the same result. As can be observed, the mean accuracy for all three algorithms increases as the trajectory segment grows longer. This is probably because: with the increase of trajectory segment, more spatial context information among the GPS points can be used to improve the accuracy. *This is also the key reason why offline map-matching algorithms usually have a better accuracy performance than online ones since they can use the spatial information about the whole completed trajectory.* What is more, SD-Matching algorithm achieves slightly better mean accuracy for all lengths of trajectory segment than ST-Matching algorithm. They both perform significantly better than S-Matching algorithm for all lengths. In short, the performance on the mean accuracy of the proposed SD-Matching algorithm is the best, demonstrating the effectiveness of using heading direction information in map-matching. We fix $k = 6$ in this experiment.

1.5.4.2 Varying k

The results for all three algorithms under different number of identified candidate edges for each GPS point (i.e., k) are shown in Fig. 1.7b. As can be observed, the mean accuracy can be improved if we identify a bigger number of candidate edges (k) for each GPS point. *The reason behind is: the probability that a GPS point can be mapped to one of candidate mapped edges correctly gets bigger with k increases, leading to a higher mean accuracy.* However, the improvement on mean accuracy is quite limited when k increases from 3 to 6. Again, the proposed SD-Matching algorithm performs slightly better than that of ST-Matching algorithm, and they are both significantly better than that of S-Matching for all ks. We fix $l = 10$ in this experiment.

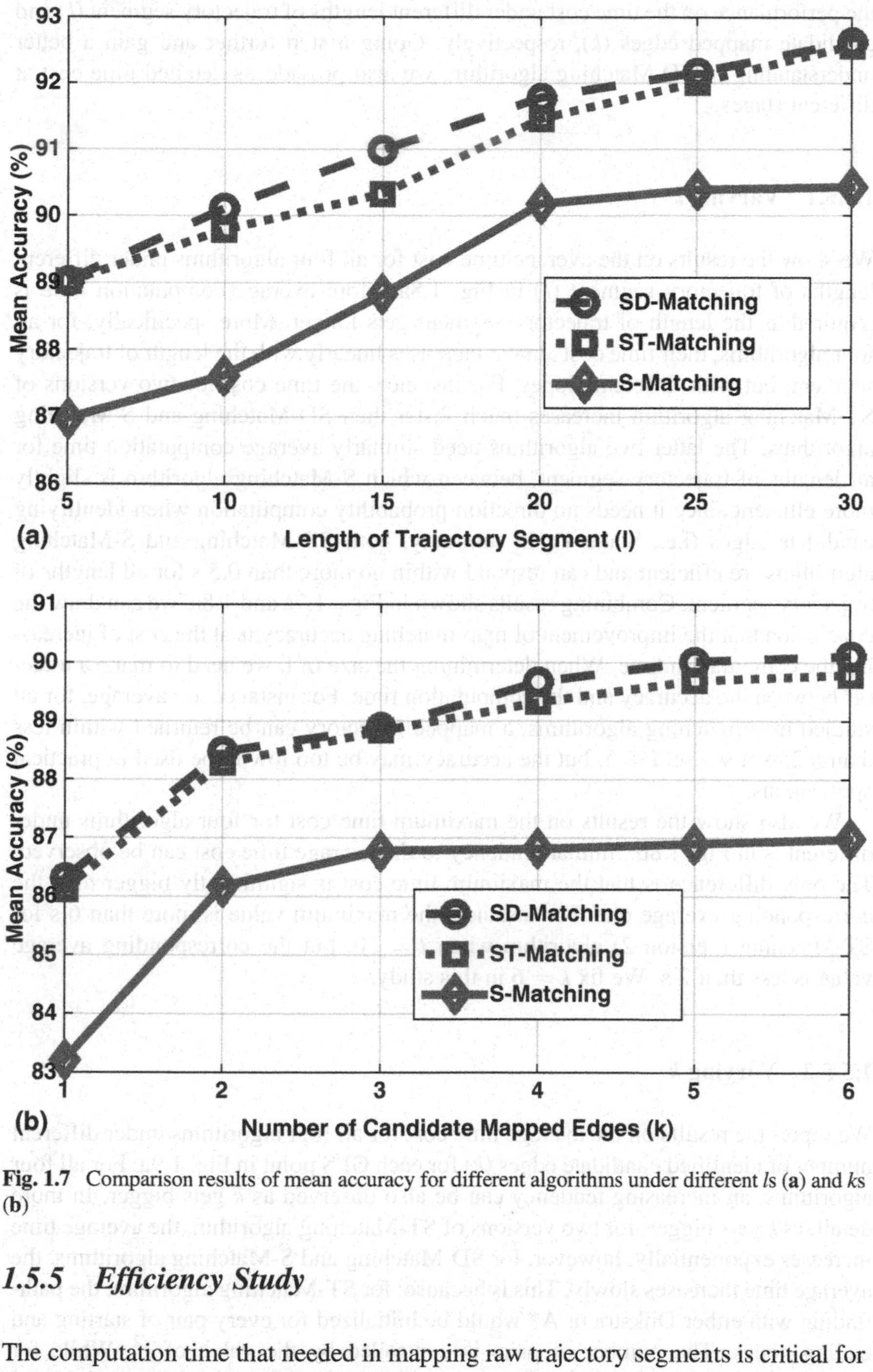

93
92
91
90
89
88
87
86
Mean Accuracy (%)
5
10
15
20
25
30
(a)
Length of Trajectory Segment (l)
SD-Matching
ST-Matching
S-Matching
91
90
89
88
87
86
85
84
83
Mean Accuracy (%)
1
2
3
4
5
6
(b)
Number of Candidate Mapped Edges (k)
SD-Matching
ST-Matching
S-Matching

Fig. 1.7 Comparison results of mean accuracy for different algorithms under different *l*s (**a**) and *k*s (**b**)

1.5.5 Efficiency Study

The computation time that needed in mapping raw trajectory segments is critical for online map-matching algorithms. Similar to the effectiveness study, we investigate the performance on the time cost under different lengths of trajectory segment (l) and candidate mapped edges (k), respectively. Going a step further and gain a better understanding of SD Matching algorithm, we also provide its detailed time cost at different stages.

1.5.5.1 Varying l

We show the results on the average time cost for all four algorithms under different lengths of trajectory segment (l) in Fig. 1.8a. More average computation time is required as the length of trajectory segment gets longer. More specifically, for all four algorithms, their time cost almost increases linearly with the length of trajectory segment, but with different slopes. For instance, the time cost for two versions of ST-Matching algorithm increases much faster than SD-Matching and S-Matching algorithms. The latter two algorithms need similarly average computation time for all lengths of trajectory segment, between which S-Matching algorithm is slightly more efficient since it needs no direction probability computation when identifying candidate edges (i.e., Stage 1). In summary, both SD-Matching and S-Matching algorithms are efficient and can respond within no more than 0.5 s for all lengths of trajectory segment. Combining results shown in Figs. 1.7a and 1.8a, we can draw the conclusion that the improvement of map-matching accuracy is at the cost of increasing the computation time. When determining the size of l, we need to make a trade-off between the accuracy and the computation time. For instance, on average, for all studied map-matching algorithms, a mapped trajectory can be returned within less than 0.25 s if we set $l = 5$, but the accuracy may be too low to be used in practical applications.

We also show the results on the maximum time cost for four algorithms under different ls in Fig. 1.8b. Similar tendency to the average time cost can be observed. The only difference is that the maximum time cost is significantly bigger than the corresponding average one. For instance, the maximum value is more than 6 s for ST-Matching (version 2) algorithm when $l = 30$, but the corresponding average value is less than 2 s. We fix $k = 6$ in this study.

1.5.5.2 Varying k

We report the results on the average time cost for all four algorithms under different number of identified candidate edges (k) for each GPS point in Fig. 1.9a. For all four algorithms, an increasing tendency can be also observed as k gets bigger. In more detail, as k gets bigger, for two versions of ST-Matching algorithm, the average time increases exponentially, however, for SD-Matching and S-Matching algorithms, the average time increases slowly. This is because: for ST-Matching algorithm, the path-finding with either Dijkstra or A* would be initialized for every pair of starting and ending nodes. The number of pairs is controlled by the value of k^2. While for SD-Matching and S-Matching algorithms, the path-finding only would be initiated

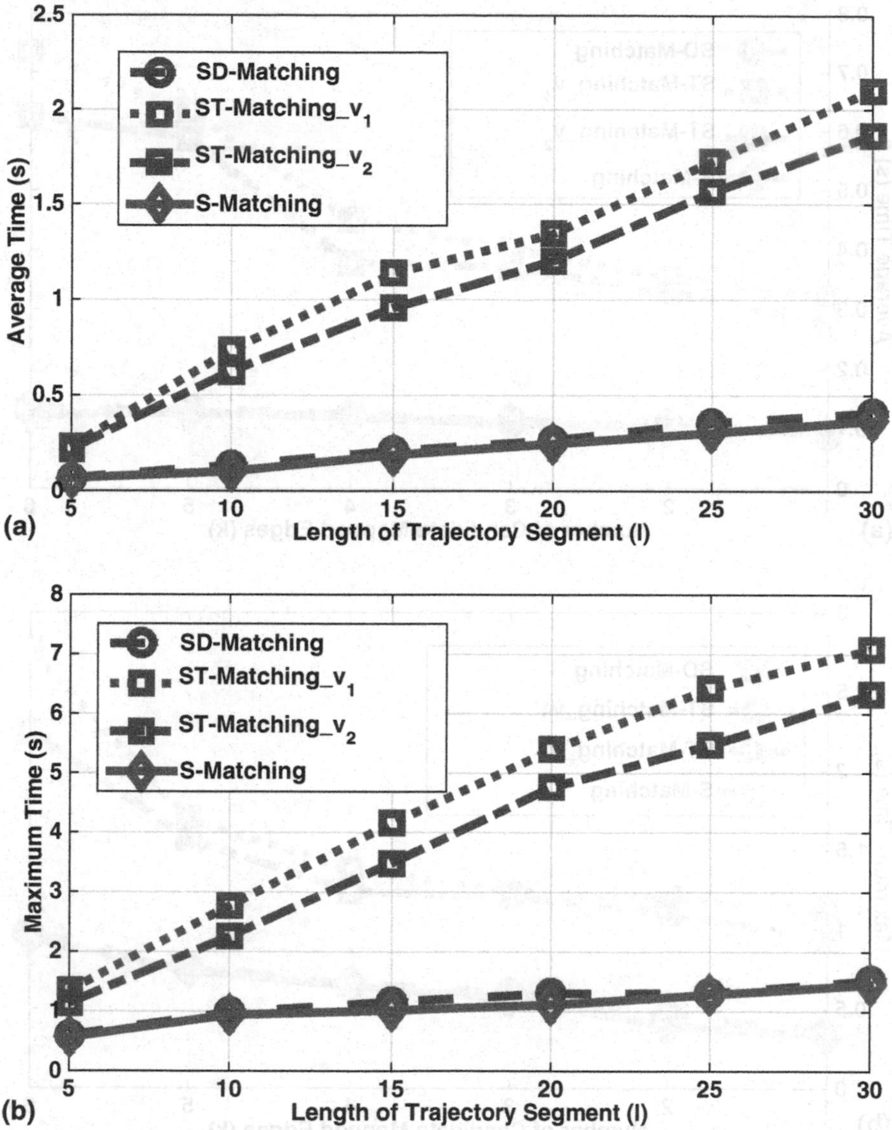

2.5
2
1.5
1
0.5
0
Average Time (s)
SD-Matching
ST-Matching_v1
ST-Matching_v2
S-Matching
5
10
15
20
25
30
(a)
Length of Trajectory Segment (l)
8
7
6
5
4
3
2
1
0
Maximum Time (s)
SD-Matching
ST-Matching_v1
ST-Matching_v2
S-Matching
5
10
15
20
25
30
(b)
Length of Trajectory Segment (l)

Fig. 1.8 Comparison results of the average (**a**) and the maximum (**b**) time cost for different algorithms under different *l*s

on top of the trees that contain both the starting and ending nodes. The number of such trees is greatly smaller than k^2 in real cases. Moreover, path-finding on top of the trees with DFS is also more efficient than Dijkstra and A*.

Similarly, we also show the results on the maximum time cost for all four algorithms under different *k*s in Fig. 1.9b. The maximum time cost is almost 4 times more than the corresponding average time cost. For SD-Matching and

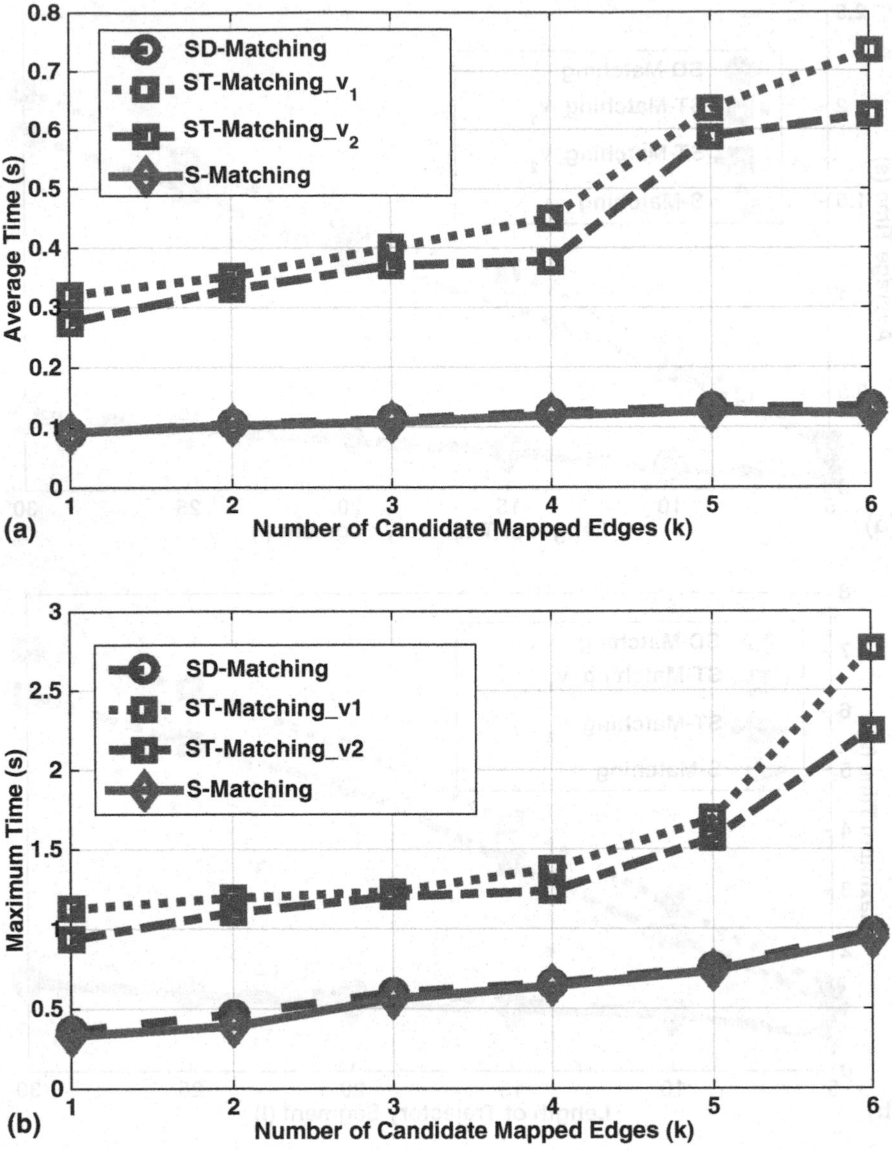
SD-Matching
ST-Matching_v1
ST-Matching_v2
S-Matching
Average Time (s)
Number of Candidate Mapped Edges (k)
(a)
Maximum Time (s)
(b)

Fig. 1.9 Comparison results of the average (**a**) and the maximum (**b**) time cost for different algorithms under different *k*s

S-Matching algorithms, compared to the average time cost, the maximum time cost increases more remarkably with k gets bigger. We fix $l = 10$ in this study.

1.5.5.3 Time Cost at Different Stages

We show the results of detailed time cost at different stages in the proposed SD-Matching algorithm under different *l*s and *k*s in Fig. 1.10. As can be observed from both charts, most of the computation time is consumed at the first two stages. Time cost at the third stage is marginal. The first stage is time-consuming mainly due

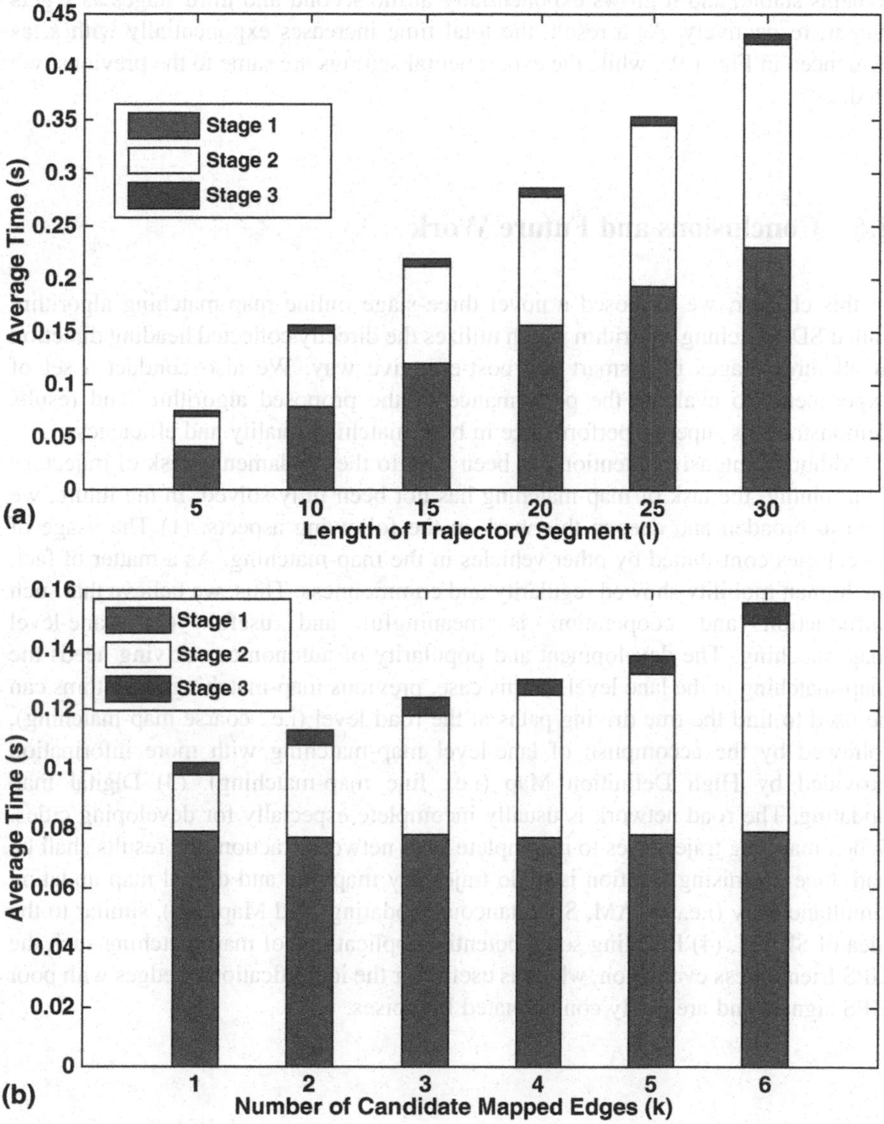

(a)
Average Time (s)
Length of Trajectory Segment (l)
Stage 1
Stage 2
Stage 3
0.45
0.4
0.35
0.3
0.25
0.2
0.15
0.1
0.05
0
5
10
15
20
25
30
(b)
Average Time (s)
Number of Candidate Mapped Edges (k)
0.16
0.14
0.12
0.1
0.08
0.06
0.04
0.02
1
2
3
4
5
6

Fig. 1.10 Detailed time cost at different stages in SD-Matching algorithm under different *l*s (**a**) and *k*s (**b**) to the following reason: in order to identify top-k candidate mapped edges, we need to first compute the distance from the given GPS points to all edges in the urban road network. However, this stage can be easily accelerated by implementing some indexing techniques, which will be investigated in the future work. As shown in Fig. 1.10a, with l gets longer, the time cost at the first two stages increases linearly, resulting in the total time cost also being increased nearly linearly, as shown in Fig. 1.8a. On the contrary, as shown in the Fig. 1.10b, the time cost at the first stage remains stable, and it grows exponentially at the second and third stages as k gets bigger, respectively. As a result, the total time increases exponentially with k, as evidenced in Fig. 1.9a, while the experimental settings are same to the previous two studies.

1.6 Conclusions and Future Work

In this chapter, we proposed a novel three-stage online map-matching algorithm called SD-Matching algorithm which utilizes the directly collected heading direction at all three stages in a smart and cost-effective way. We also conduct a set of experiments to evaluate the performance of the proposed algorithm, and results demonstrate its superior performance in both matching quality and efficiency.

Although intensive attention has been paid to the fundamental task of trajectory data mining, the task of map-matching has not been fully solved. In the future, we plan to broaden and deepen this work in the following aspects: (1) The usage of trajectories contributed by other vehicles in the map-matching. As a matter of fact, the human mobility showed regularity and commonness. Thus, we believe that such introduction and cooperation is meaningful and useful. (2) Lane-level map-matching. The development and popularity of autonomous driving needs the map-matching at the lane level. In this case, previous map-matching algorithms can be used to find the true driving paths at the road level (i.e., coarse map-matching), followed by the accomplish of lane-level map-matching with more information provided by High Definition Map (i.e., fine map-matching). (3) Digital map updating. The road network is usually incomplete especially for developing cities. When mapping trajectories to incomplete road network fraction, the results shall be bad. One promising solution is to do trajectory mapping and digital map updating simultaneously (i.e., SUAM, Simultaneous Updating And Mapping), similar to the idea of SLAM. (4) Enabling some potential applications of map-matching such the GPS friendliness evaluation, which is useful for the identification of edges with poor GPS signals and are easily contaminated by noises.

References

1. Chen C, Zhang D, Castro PS, Li N, Sun L, Li S, Wang Z. iBOAT: isolation-based online anomalous trajectory detection. IEEE Trans Intell Transp Syst. 2013;14(2):806–18.
2. Chen C, Wang Z, Guo B. The road to the Chinese smart city: progress, challenges, and future directions. IT Professional. 2016;18(1):14–7.
3. Chen C, Jiao S, Zhang S, Liu W, Feng L, Wang Y. TripImputor: real-time imputing taxi trip purpose leveraging multi-sourced urban data. IEEE Trans Intell Transp Syst. 2018;19 (10):3292–304.
4. Cho W, Choi E. A basis of spatial big data analysis with map-matching system. Clust Comput. 2017;20(3):2177–92.
5. Pink O, Hummel B. A statistical approach to map matching using road network geometry, topology and vehicular motion constraints. In: 2008 11th International IEEE conference on intelligent transportation systems, 2008. p. 862–7.
6. Quddus MA, Ochieng WY, Noland RB. Current map-matching algorithms for transport applications: state-of-the art and future research directions. Transp Res C Emerg Technol. 2007;15(5):312–28.
7. Hashemi M, Karimi HA. A critical review of real-time map-matching algorithms: Current issues and future directions. Comput Environ Urban Syst. 2014;48:153–65.
8. Kubicka M, Cela A, Mounier H, Niculescu S. Comparative study and application-oriented classification of vehicular map-matching methods. IEEE Intell Transp Syst Mag. 2018;10 (2):150–66.
9. Chao P, Xu Y, Hua W. A survey on map-matching algorithms. In: Proceedings of Australasian database conference, 2020. p. 121–33.
10. Li Y, Huang Q, Kerber M, Zhang L, Guibas L. Large-scale joint map matching of GPS traces. In: Proceedings of the 21st ACM SIGSPATIAL international conference on advances in geographic information systems, 2013. p. 214–23.
11. Greenfeld JS. Matching GPS observations to locations on a digital map. In: 81st Annual meeting of the transportation research board, 2002. p. 164–73.
12. Velaga NR, Quddus MA, Bristow AL. Developing an enhanced weight-based topological map-matching algorithm for intelligent transport systems. Transp Res C Emerg Technol. 2009;17(6):672–83.
13. Zheng K, Zheng Y, Xie X, Zhou X. Reducing uncertainty of low-sampling-rate trajectories. In: 2012 IEEE 28th international conference on data engineering, 2012. p. 1144–55.
14. Kang W, Li S, Chen W, Lei K, Wang T. Online map-matching algorithm using object motion laws. In: 2017 IEEE 3rd international conference on big data security on cloud (BigDataSecurity), IEEE international conference on high performance and smart computing (HPSC), and IEEE international conference on intelligent data and security (IDS), 2017. p. 249–54.
15. Lou Y, Zhang C, Zheng Y, Xie X, Wang W and Huang Y. Map-matching for low-sampling-rate GPS trajectories. In: Proceedings of the 17th ACM SIGSPATIAL international conference on advances in geographic information systems, 2009. p. 352–61.
16. Bierlaire M, Chen J, Newman J. A probabilistic map matching method for smartphone GPS data. Transp Res C Emerg Technol. 2013;26:78–98.
17. Goh CY, Dauwels J, Mitrovic N, Asif MT, Oran A, Jaillet P. Online map-matching based on Hidden Markov model for real-time traffic sensing applications. In: 2012 15th International IEEE conference on intelligent transportation systems, 2012. p. 776–81.
18. Newson P, Krumm J. Hidden Markov map matching through noise and sparseness. In: Proceedings of the 17th ACM SIGSPATIAL international conference on advances in geographic information systems, 2009. p. 336–43.

19. Wei H, Wang Y, Forman G, Zhu Y, Guan H. Fast Viterbi map matching with tunable weight functions. In: Proceedings of the 20th international conference on advances in geographic information systems, 2012. p. 613–6.
20. Yuan J, Zheng Y, Zhang C, Xie X, Sun G. An interactive-voting based map matching algorithm. In: 2010 11th International conference on mobile data management, 2010. p. 43–52.

Chapter 2
Trajectory Data Compression

2.1 Introduction

Due to the increasing of GPS-equipped public and private vehicles, trajectory data is accumulated and recorded rapidly, massively and abundantly. Trajectory data is sending to the data center continuously, causing communication overhead, storing and computing issues [1–3]. To make matter worse, massive data also makes computing tasks (e.g. data visualization, pattern mining) more complicated, resulting in worse performance. To ameliorate the above-mentioned issues, reducing the size of trajectory data before sending it to the data center is one of promising ways [4]. Intuitively, the easiest way is collecting the vehicle's locations less frequently. However, such method is problematic because the collected data may be too sparse to infer the detailed driving paths in-between, resulting in limited urban services [5]. On the contrary, trajectory compression based on online map-matching is a common but never well-addressed solution, which tries to address the issues from the perspective of seeking trajectory representation [6]. To be more specific, map-matching algorithm aligns raw trajectory onto the road network in advance. The mapped trajectory returned in terms of a sequence of connected edges in the road network is usually redundant since some edges in the trajectory can be recovered using the reserved edges. The objective of trajectory compression is to reduce such redundancy. However, there is an obvious conflict between the resource-hungry map-matching and trajectory-compressing tasks and the

---

Part of this chapter is based on two previous work:

C. Chen, Y. Ding, S. Zhang, Z. Wang and L. Feng, "TrajCompressor: An online map-matching-based trajectory compression framework leveraging vehicle heading direction and change," IEEE Transactions on Intelligent Transportation Systems, vol. 21, no. 5, pp. 2012–2028, Apr. 2019.

C. Chen, Y. Ding, Z. Wang, J. Zhao, B. Guo and D. Zhang, "VTracer: When online vehicle trajectory compression meets mobile edge computing," IEEE Systems Journal, vol. 14, no. 2, pp. 1635–1646, Aug. 2019.

C. Chen et al., *Enabling Smart Urban Services with GPS Trajectory Data*,
https://doi.org/10.1007/978-981-16-0178-1_2

resource-limited vehicle-mounted GPS devices. The vehicle-mounted GPS devices cannot undertake the heavy tasks themselves.

In this chapter, similar to the SD-Matching algorithm discussed in Chap. 1, we also intend to explore the heading direction information in online trajectory compression. To be more specific, we propose a heading-change-compression (HCC) algorithm, which only maintains edges whose heading change significantly, compared to the previously connected edges. To alleviate the issue caused by the limited computing capacity of GPS devices, we propose an idea of leveraging the smartphones of drivers as the local computing unit (i.e., cloudlet) to migrate the computing burdens, as inspired by the mobile edge computing [7]. In summary, the main contributions of this chapter are as follows.

- Based on the mapped trajectory, we propose a novel online trajectory compression algorithm that leverage the heading direction in a smart way. We also conduct extensive evaluations using real-life trajectory data generated by 595 taxis in a week and road network data in the city of Beijing to verify the efficiency and effectiveness of HCC. Experimental results show that HCC achieves a high compression ratio and can respond in real time.
- We develop a system called VTracer combining HCC with SD-Matching, which can save the cost on storage, communication significantly. The main noteworthy aspect of VTracer is that it brings the idea of mobile edge computing into a very preliminary but important step of trajectory data mining tasks (in the step of data collection). With VTracer, the data center can obtain a noise-free, more concise yet still completed trajectory representation for the moving vehicles. We also evaluate VTracer system in terms of the trajectory mapping quality, the compression ratio, memory, and energy consumptions, and the app size in real situations, using the 13-h real driving data generated in the city of Chongqing, China (equivalent to around 530 km in driving distance). Experimental results demonstrate the superior performance of VTracer.

2.2 Related Work

According to the utilization of the road network, trajectory compression algorithms are generally classified into two groups, i.e., line-simplification-based and map matching-based algorithms.

2.2.1 Line-Simplification-Based Compression

The GPS points of the raw trajectory can be divided into important and unimportant points in terms of their contribution to the trajectory shape. And the core idea of line-simplification-based algorithms is to maintain some important GPS points [8]. To evaluate the importance for a given GPS point, the distance from the point to the line segment composed by the end-points of the (sub-)trajectory is calculated. The GPS point is claimed important if the distance is bigger than the user-specified error bound. Therefore, the error bound value has a significant impact on the number of retained GPS points and the compression ratio. The DP and its variants are the most well-known representatives [8]. The importance of the GPS point is counted by considering the GPS location only, thus it can be wrongly evaluated due to the measurement errors. To alleviate such side-effect, map-matching-based algorithms are proposed by aligning the GPS points to the underlying road network before trajectory compression.

2.2.2 Map-Matching-Based Compression

The core idea of map-matching-based compression is mapping raw trajectory data onto the road network in advance [9, 10]. The utilization of road network further benefits the trajectory compression since it endows the trajectory with more correct and semantical information. Some recent work also introduces the trajectory prediction in the compression. Based on the historical trajectory data [11], will retain the trajectory which is not consistent with the results predicted by the Markov model, and discard that is consistent. In the overview of trajectory data mining [12], briefly summarize some popular trajectory compression algorithms. Here, we select three well-known compressors to review, i.e., MMTC [13], PRESS [3], CCF [1].

We know that in a road network there can be multiple paths between two points, and each path usually consists of a different number of edges, meaning that some paths just have more edges than others. In order to reduce the number of edges, MMTC algorithm makes use of the paths with fewer edges but with high similarity to replace the original sub-trajectory. In real life, most drivers prefer to select the shortest paths. Thus, PRESS utilizes the driving preference to accomplish trajectory compression. More specifically, PRESS replaces the shortest sub-path in the original trajectory by the pair of beginning and ending edges. However, the shortest path computation either causes high computation time (online) or storage space cost (offline), which greatly degrades the efficiency. Different from PRESS, CCF explores the usability of intersections in trajectory compression. It retains all out-edges at intersections to represent the compressed trajectory. Since the number of out-edges is much smaller than the edges in the mapped trajectory, CCF can save significant storage space. However, retaining all out-edges is still redundant and the trajectory can be further compressed. Going a step further, in this study, we propose a novel trajectory compression algorithm, which takes full advantage of heading changes at intersections and only retains out-edges with remarkable heading changes. The rationale behind is that the number of such out-edges only take a quite small fraction. In addition, HCC are spatial-lossless, while MMTC, PRESS and CCF are all spatial-loss.

2.3 Basic Concepts and System Overview

2.3.1 Basic Concepts

Definition 2.1 (Compressed Trajectory Segment). Given a mapped trajectory segment $\tau_m = \langle e_i, \cdots, e_n \rangle$, its compressed form is defined as $\{t_i, \tau_c\}$. t_i is the starting time of this mapped segment. τ_c is a subset of τ_m, denoted as $\tau_c = \langle e_i, \cdots e_m \rangle$, where $m \ll n$ and each pair of two consecutive edges is usually not connected in the road network. The compressed trajectory steam T_c is composed of an unbounded set of compressed trajectory segments τ_c and their starting times, denoted by $T_c = \langle \{t_1, \tau_{c1}\}, \{t_2, \tau_{c2}\}, \cdots \rangle$.

Definition 2.2 (Heading Change at Intersections). Heading change at intersections[2] is defined as the angle difference between the heading direction of the last GPS point (e.g., p_1) before the vehicle enters the intersection and the heading direction of the first GPS point (e.g., p_2) after leaving the intersection. According to the angle value, the heading change (0–360°) is divided into four categories, denoted as N (0–45°, 316–360°), E (46–135°), S (136–225°) and W (226–315°), as shown in Fig. 2.1. Heading changes belonging to E indicate that the vehicles ***go straight*** when passing intersections; heading changes belonging to other three categories mean that the vehicles ***make turns*** (e.g., left, right, or U) when passing intersections.

Definition 2.3 (In-Edge and Out-Edge). In-edge and out-edge refer to the edge that the vehicle enters at (e.g., e_1) and leaves from (e.g., e_2) an intersection respectively, as illustrated in Fig. 2.1. The out-edge is determined by the in-edge and the heading change of the vehicle at the intersection.

To ease the computation of heading changes, we can simply use the angle difference between the in-edge and out-edge directions along the vehicle moving direction to approximate the heading changes at the given intersection. In this way, heading changes of mapped trajectory segments can be also estimated, since mapped trajectory segments don't save heading direction information any more.

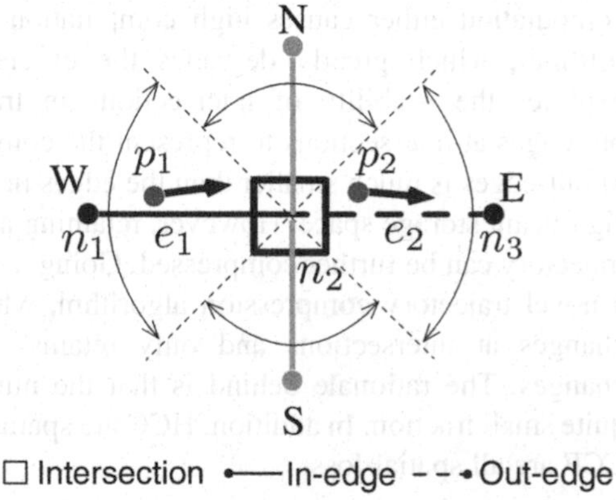
N
W
E
S
p_1
p_2
n_1
e_1
e_2
n_2
n_3
□ Intersection
In-edge
Out-edge

Fig. 2.1 Illustration of heading change, in- and out-edges

2.3.2 System Overview

The framework of VTracer consists of three major units, namely vehicle-mounted GPS devices, mobile phones and data center, as shown in Fig. 2.2. The vehicle-mounted GPS devices have two main functions. For one thing, the devices collect the vehicular locations (GPS points) at a constant rate, then transmit raw trajectory stream (an unbounded set of GPS points) into mobile phones via Bluetooth. For another, the devices receive the compressed data from mobile phones via Bluetooth, then send them to the data center via GPRS. Mobile phones comprise two components, i.e., Map Matcher and Compressor.

Map Matcher takes raw GPS trajectory stream as input and divides it into an unbounded set of raw trajectory segments whose length is controlled by user-specified parameter (i.e., l). Moreover, two consecutive segments share a common point, i.e., the first GPS point of the latter segment is also the last GPS point of the former segment. Afterwards, Map Matcher maps each raw trajectory segment onto a sequence of connected edges (i.e., the mapped trajectory segment) in the road network (i.e., map-matching). Then, for each mapped trajectory segment, Compressor compresses it and return the compressed representation to GPS devices. The data transmission between mobile phones and GPS devices is via Bluetooth, which is free of charge and has smaller latency. GPS devices receive the compressed trajectory segments and upload them to the data center via GPRS, which costs the third-party bandwidth. At last, the data center may decompress and manage the data for further applications, such as the common when-and-where query service.

Our developed system differs from the prior research mainly in the following aspects.

1. We collect the trajectory data densely to ease map matching and increase the accuracy. On the other hand, we also implement a more map matching algorithm which fully utilizes the under-explored heading direction information [14].
2. We offload the resource-hungry computing tasks, including online trajectory mapping and compression to the nearby smartphones of drivers without incurring extra communication cost, bringing the smart idea of mobile edge computing.

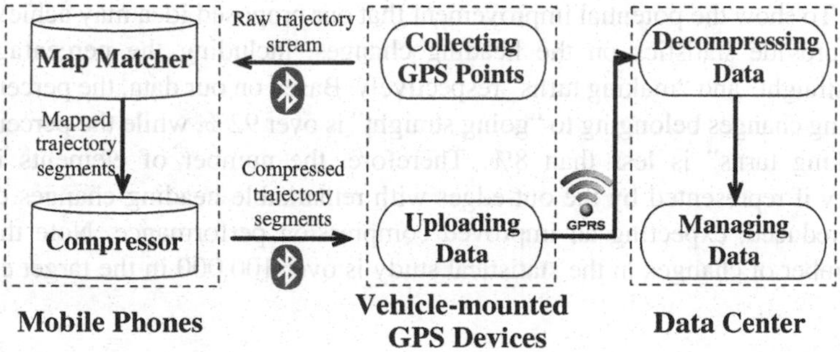
Map Matcher
Mapped trajectory segments
Compressor
Mobile Phones
Raw trajectory stream
Compressed trajectory segments
Collecting GPS Points
Uploading Data
Vehicle-mounted GPS Devices
GPRS
Decompressing Data
Managing Data
Data Center

Fig. 2.2 System overview of VTracer

3. The system can not only realize significant storage and communication savings, but also maintain a high-quality trajectory dataset in the data center for further applications.

2.4 HCC Algorithm

2.4.1 Motivation

We first use the following motivating example to illustrate the basic idea of our proposed trajectory compression algorithm.

Motivating Example: To recommend a unique route to drivers from the origin to the destination, instead of giving edge-by-edge guidance, GPS navigation systems usually only remind the drivers before changing driving directions at intersections (i.e., go straight, make left/right/U turn).

Driving is a prudent thing. Thus, navigation systems actually provide a more concise trajectory representation, i.e., using heading changes at intersections to avoid distracting drivers when recommending driving directions. Inspired by the idea [1], present a Clockwise Compression Framework (CCF), which represents trajectory as the sequence of out-edge at each by-passing intersection from the origin to the destination. More specially, CCF encodes the out-edge with a unique code based on the ID of the intersection and the heading change. Excluding the starting and ending edges, the number of elements in the representation is just the number of by-passing intersections, which is smaller than the number of edges, occupying less space. There is a clear daily experience that we prefer to "going straight" more often at intersections during driving, meaning that straight out-edges are repeated in the trajectory. In such a situation, we can discard out-edges with small heading change to further reduce trajectory size. Thus, on top of the CCF, *we argue that the representation can be further reduced if retaining out-edges with remarkable heading changes ("making turns") only*. In such manner, compared to CCF, it is expected the number of elements in the compressed trajectory can be even smaller, which motivates us to propose the trajectory compression algorithm based on the heading change. To show the potential improvement that our proposed idea may achieve, we further provide statistics on the heading changes, including the percentages of "going straight" and "making turns" respectively. Based on our data, the percentages of heading changes belonging to "going straight" is over 92%, while the percentages of "making turns" is less than 8%. Therefore, the number of elements in the trajectory if represented by the out-edges with remarkable heading changes can be greatly reduced, expecting an improved compression performance. Note that the total number of changes in the statistical study is over 100,000 in the target city.

2.4.2 Algorithm Details

As discussed, rather than retaining all out-edges in a mapped trajectory segment, the key idea of our proposed HCC is to retain the out-edges when the vehicle makes a turn (or U-turn). Algorithm 2.1 summarizes the whole procedure of the proposed HCC algorithm. Note that the mapped trajectory segment (τ_m) is the basic processing unit of HCC.

Algorithm 2.1. ***HCC***$(\tau_m, G(N,E))$

1: $\tau_c = e_1$;
2: **for** $i = \; = 2$ *to* $\mid \tau_m \mid \; - 1$ **do**
3: $\quad s =$ *identifyNode*(e_i, e_{i+1});
4: $\quad$ **if** *isIntersection*$(s, G(N,E))$ **then**
5: $\quad\quad$ **if** ~*isGoStraight*(e_i, e_{i+1}) **then**
6: $\quad\quad\quad \tau_c = \tau_c \cup e_{i+1}$;
7: $\quad\quad$ **end if**
8: $\quad$ **end if**
9: **end for**
10: $\tau_c = \tau_c \cup e_{|\tau_m|}$;

Line 1 in Algorithm 2.1 refers to the initialization of the compressed trajectory. More specially, HCC enrolls the first edge e_1 of the input trajectory into the compressed trajectory τ_c. Lines 2–6 refer to the loop operation. In the loop, HCC algorithm scans and checks the remaining edges of the input trajectory one by one before it reaches the last edge. In more detail, for each edge e_i in the input trajectory, HCC first identifies the node between e_i and its following edge e_{i+1} (Line 3), then judges whether it is an intersection node (Line 4). If yes, then it continues to identify the heading change of the vehicle at this intersection (Line 5). If the vehicle at the intersection is identified as NOT "going straight", we append the out-edge (i.e., e_{i+1}) in the previous obtained compressed trajectory τ_c (Line 6); otherwise, HCC just skips this edge and continues to process next one e_{i+1}. Finally, HCC enrolls the last edge $e_{|\tau_m|}$ into the compressed trajectory τ_c and terminates the whole procedure (Line 7). As can be observed, for any input mapped trajectory segment, HCC always retains the first and last edges in the compressed trajectory. The time complexity of HCC is $O(|\tau_m|)$, where $|\tau_m|$ is the number of edges in the input mapped trajectory τ_m.

However, HCC may result in trajectory ambiguity, because of the complexity of the road network and the coarse granularity of the defined heading change category. For example, different out-edges may be identified with a same heading change. We use a simple example to illustrate the issue of trajectory ambiguity, as shown in Fig. 2.3. There are two mapped trajectories with the same starting and ending edges, e.g., $\tau_{m1} = \langle e_1, e_2, e_3, e_4, e_8 \rangle$ and $\tau_{m2} = \langle e_1, e_5, e_6, e_7, e_8 \rangle$. According to our proposed HCC algorithm, the compressed trajectory for both trajectories are the same, i.e., $\langle e_1, e_8 \rangle$, since the heading change between e_1 and e_2, and the heading change between e_1 and e_5 are both identified as "going straight". In this case, both out-edges would be discarded, which is incorrect and should be avoided.

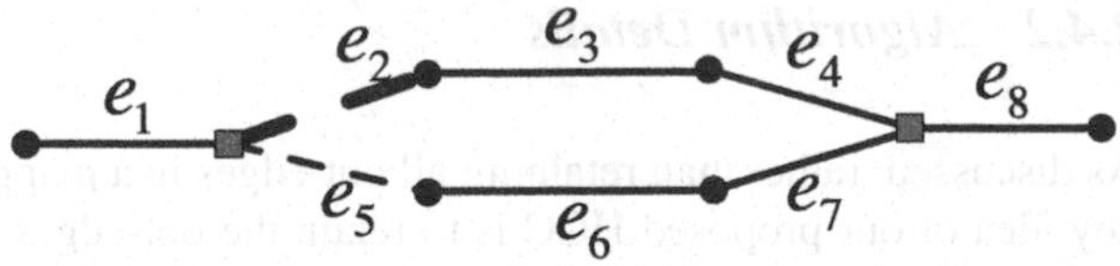
e_1
e_2
e_3
e_4
e_5
e_6
e_7
e_8

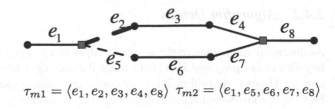

$\tau_{m1} = \langle e_1, e_2, e_3, e_4, e_8 \rangle$ $\tau_{m2} = \langle e_1, e_5, e_6, e_7, e_8 \rangle$

Fig. 2.3 Illustrative example of trajectory ambiguity

To address the issue, there is an intuitive idea that HCC still maintains the out-edge in the compressed trajectory even it is identified as "going straight", if the case that more than one out-edge at the intersection are identified as "going straight" category occurs (Case I for short in the rest of presentation). For instance, with the intuitive idea, the compressed trajectories for the two trajectories shown in Fig. 2.3 are $\tau_{c1} = \langle e_1, e_2, e_8 \rangle$ and $\tau_{c2} = \langle e_1, e_5, e_8 \rangle$, respectively. However, such a method increases the number of edges that have to be maintained. To resolve trajectory ambiguity without costing much storage space, we embed the idea of Frequent Edge Compression (FEC) into HCC algorithm, detailed as follows.

Frequent Edge Compression: Similar to the observation that drivers are inclined to choose the shortest path from the origin to the destination, drivers may prefer to select some out-edge than others when traversing the intersections. Inspired by the daily experience, trajectories containing out-edges travelled by drivers frequently can be further reduced. The frequent out-edges are no need to be maintained. With the idea of FEC, HCC is expected to achieve even better compression performance. Taking the two trajectories shown in Fig. 2.3 as the example again. Suppose that e_2 out-edge is more popular than e_5, thus we can discard e_2 for the trajectory τ_{m1} during compression. Bringing in the idea of FEC, the compressed trajectories for τ_{m1} and τ_{m2} are $\tau_{c1} = \langle e_1, e_8 \rangle$ and $\tau_{c2} = \langle e_1, e_5, e_8 \rangle$, respectively. The improvement is significant since Case I happens quite commonly in our trajectory data.

2.4.3 *Trajectory Decompression*

For a compressed trajectory, it is trivial to recover its original trajectory (i.e., the sequence of connected edges). Taking the example shown in Fig. 2.3 again, for the compressed trajectory $\tau_{c2} = \langle e_1, e_5, e_8 \rangle$, the first ($e_1$) and the last edge ($e_8$) corresponds to the starting and ending edge of the original trajectory respectively. The only remaining edge e_5 is the retained out-edge at the first intersection. There is no out-edge retained in the second intersection, which indicates that the vehicle generally goes through from e_5 to e_8. Hence, we can decompress the trajectory correctly. As a comparison, for the compressed trajectory $\tau_{c1} = \langle e_1, e_8 \rangle$, only the first and last edges are retained, which indicates that vehicles take the most common turn when travelling intersections in-between.

Although few semantics are embedded directly in the compressed trajectory, the trajectory decompression is capable of unveiling some common semantics by coupling with other multi-source urban datasets [15]. To name a few, the average speed of the trajectory segment can be computed by dividing the total driving distance (the sum of each edge) to the total time. The average speed is a clear indicator of the traffic state [16]. A staying state can be safely claimed if the starting and ending edge are unique. Moreover, with the point-of-interest and digital map data, the surrounding spatial context of the trajectory (e.g., street names, the number of lanes) can be inferred [15].

2.5 Evaluations of HCC

In this section, based on the GPS trajectory data collected from taxis in the real world, we conduct extensive experiments to evaluate HCC algorithm in terms of compression ratio and computation time under different lengths of trajectory segment. We also test the performance of HCC algorithm on the quality of response for the when-and-where query.

2.5.1 Baselines

We select three representative baselines to compare, i.e., PRESS [3], CCF [1] and DP algorithms [8], with their details as follows.

- With PRESS, the edge sequence between every pair of two consecutive edges in the compressed trajectory follows the shortest path exactly. Readers can refer to [3] for the details.
- From the starting edge to the ending edge, CCF only retains the out-edge ID and code (in the clock-wise order) at each intersection that the vehicle passes.
- DP aims at reducing the number of GPS points based on line-fitting. If the distance from the GPS point to the line segment is less than the error bound, then it will be discarded.

Remark PRESS and CCF are both based on the map matching and they are spatial-lossless, while DP is based on the raw trajectory segment (i.e., the sequence of GPS points) and spatial-lossy. Similar to HCC, for PRESS and CCF, the starting edge and ending edge of the trajectory segment are also always maintained. To accelerate the process of PRESS, the shortest paths for any two edges in the road network are usually pre-computed. However, it costs too much storage space (e.g., over 100 GB for Beijing City), which is intolerable in the mobile environment. Therefore, to make it comparable to HCC algorithm, we revise the original PRESS (i.e., the revised version) by adopting Dijkstra algorithm to achieve the online shortest path computation instead.

2.5.2 Experimental Setup

1. Data preparation: In the experiments, we prepare two different kinds of datasets from the city of Beijing, China. The basic dataset is the road network, which can be freely downloaded and extracted from OpenStreetMap. Totally, it contains 141,735 nodes and 157,479 edges. Moreover, OpenStreetMap also provides some additional attributes of edges, including number of lanes, number of traffic lights, speed limits and so on [17]. The second is the GPS trajectory data, which are generated by 595 taxis in one week (from 15th to 21st September, 2015). The sampling rate for all taxis is identical and constant, with a value of around 6 s. In terms of storage space, it takes up over 1.01 GB. Prior to compression, we used an SD-Matching algorithm to transform the raw trajectory data into a sequence of edges on the road network.
2. Evaluation metrics: For trajectory compression algorithms, the popular and well-known metric of compression ratio (*cr*) is used to quantify the compression effectiveness, which is defined as the ratio of the occupied storage space of the raw trajectory segments to the occupied storage space of the compressed trajectory segments. It is easy to understand that *cr* should be always bigger than 1. The compression performance is better if *cr* is bigger. We use the computation time cost at both trajectory mapping phase and compressing phase as the time cost when evaluating the efficiency for trajectory compression algorithms. Similarly, the average time is adopted to measure the overall efficiency.

There is an important user-specified parameter in HCC algorithm, i.e., the length of the trajectory segment (l). Thus, we study its effectiveness in terms of compression ratio and its efficiency in terms of time cost under different ls in the following. As a comparison, the results of the baseline algorithms are also shown.

2.5.3 Varying l

As can be seen from Fig. 2.4a, HCC, PRESS and CCF algorithms achieve much higher compression ratio than DP algorithm under all ls, demonstrating the superior performance of map-matching based trajectory compression algorithms. HCC algorithm achieves the compression ratio in-between among all three map-matching based algorithms, and PRESS obtains the best performance. The reason why HCC performance better than CCF is that: CCF retains the out-edge information at every intersection, while HCC only retains out-edges with significant heading changes, which usually take up a small fraction of all out-edges. One credible reason why PRESS performs the best may be due to that taxi drivers prefer to taking the shortest path in real cases when delivering passengers [18]. Under such circumstance, the edges between the origin and destination can be commonly discarded, resulting in the best compression ratio. One major drawback of PRESS algorithm is that the shortest path between every two edges in the whole road network should be

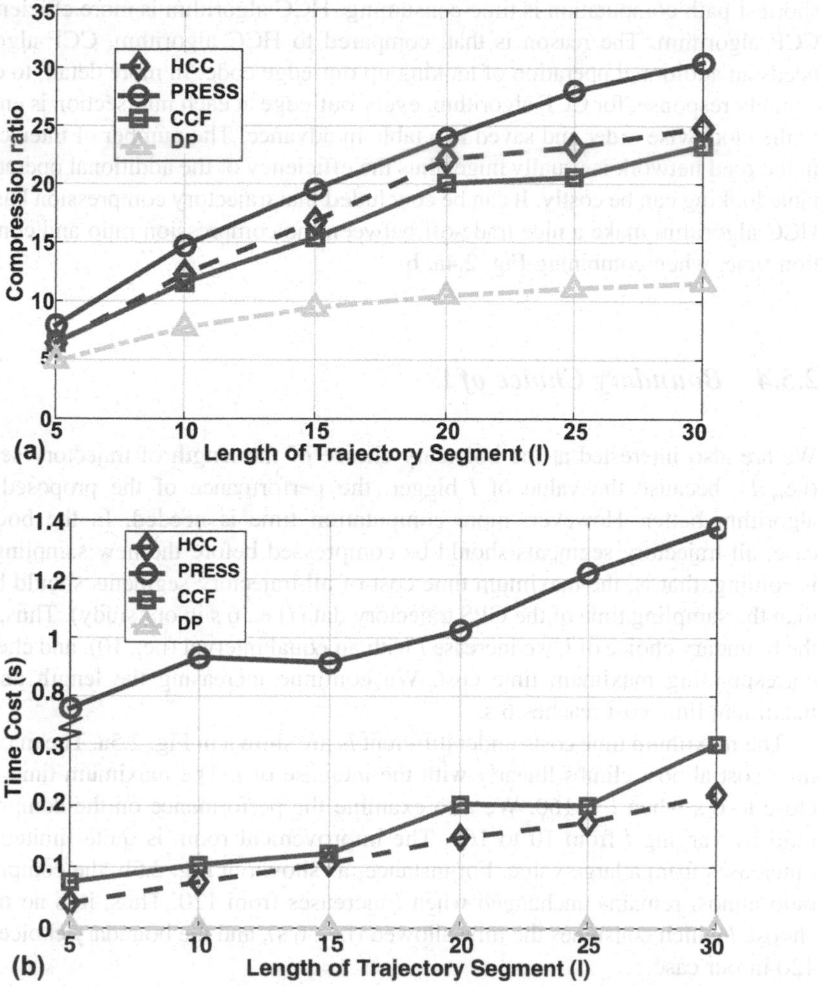
Compression ratio
HCC
PRESS
CCF
DP
35
30
25
20
15
10
5
0
5
10
15
20
25
30
(a)
Length of Trajectory Segment (l)
Time Cost (s)
1.4
1.2
1
0.8
0.3
0.2
0.1
0
(b)
Length of Trajectory Segment (l)

Fig. 2.4 Comparison results of compression ratio (**a**) and time cost (**b**) for different algorithms under different *l*s

pre-computed and stored in order to save the compression time, which is always a time-consuming process. To make matters worse, the process must be repeated once the road network updates that is quite common in developing cities, causing many sustainable maintenance issues.

We also show the results of the time cost for HCC under different ls, as well as the results of the baseline algorithms in Fig. 2.4b. For all four algorithms, more computation time is required as the length of trajectory segment gets longer. DP algorithm is the most time-efficient because it needs no map-matching. For the other three map-matching based algorithms, as it can be predicted, PRESS needs much more computation time under all ls than the other three algorithms, since the online shortest-path computation is time-consuming. HCC algorithm is more efficient than CCF algorithm. The reason is that, compared to HCC algorithm, CCF algorithm needs an additional operation of looking up out-edge code. In more detail, to ensure a timely response, for CCF algorithm, every out-edge at each intersection is encoded in the clockwise order and saved in a table in advance. The number of intersections in the road network is usually huge, thus the efficiency of the additional operation of table looking can be costly. It can be concluded that trajectory compression based on HCC algorithm make a nice trade-off between the compression ratio and computation time, when combining Fig. 2.4a, b.

2.5.4 *Boundary Choice of* l

We are also interested at the boundary choice of the length of trajectory segment (i.e., l), because the value of l bigger, the performance of the proposed HCC algorithm better. However, more computation time is needed. In the boundary case, all trajectory segments should be compressed before the new sampling GPS is coming, that is, the maximum time cost of all trajectory segments should be less than the sampling time of the GPS trajectory data (i.e., 6 s in our study). Thus, to get the boundary choice of l, we increase l with an equal interval (i.e., 10), and check the corresponding maximum time cost. We continue increasing the length until the maximum time cost reaches 6 s.

The maximum time costs under different ls are shown in Fig. 2.5a. The maximum time cost almost climbs linearly with the increase of l. The maximum time cost is close to 6 s when $l = 160$. We also examine the performance on the compression ratio by varying l from 10 to 160. The improvement room is quite limited when l increases from a large value. For instance, as shown in Fig. 2.5b, the compression ratio almost remains unchanged when l increases from 120. Thus, it is no need to choose l which consumes the time allowed (i.e., 6 s), and the boundary choice of l is 120 in our case.

2.5.5 *Quality of Response for When-and-Where Query*

To support location-based services (LBS), some common types of query are generally applied on the top of the compressed trajectory data directly, among which when-and-where is the most popular query. For such query, users are mainly concerned with where the vehicle located at what time. To ensure the quality of response, the estimated and the actual locations of the vehicle should be as close as possible. The results of mean error under different lengths of trajectory segment (l) are shown in Table 2.1. From the table, we can observe that the mean error increase with the trajectory segment gets longer, due to the reason the total travel distance of the vehicle becomes longer with l and the driving behavior is also more complex,

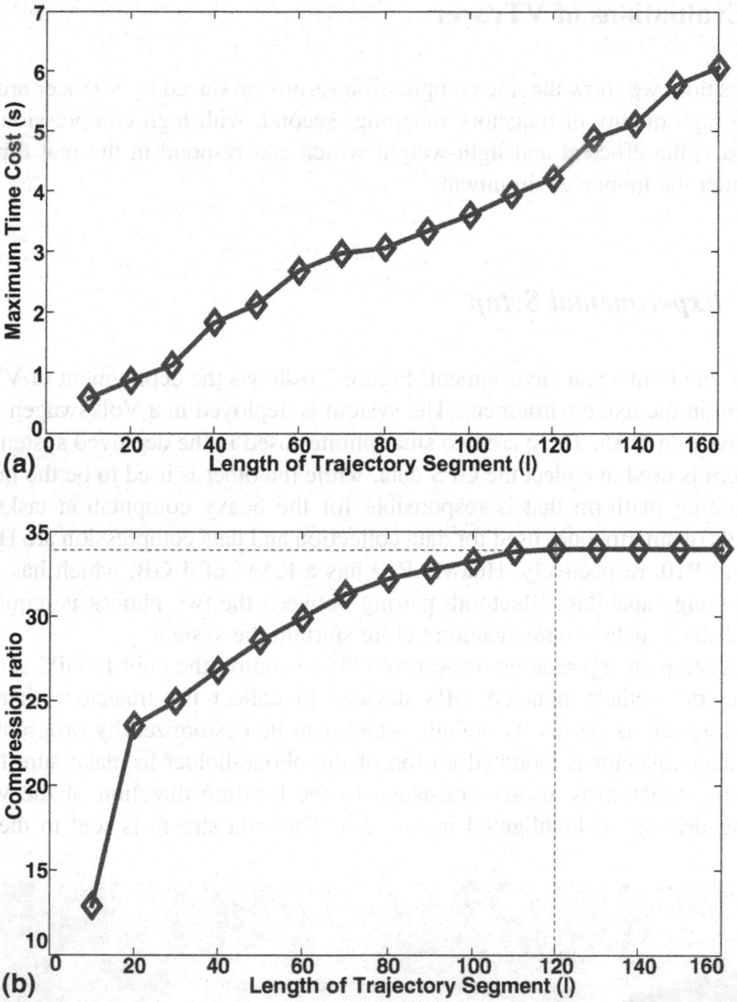

Maximum Time Cost (s)
(a)
Length of Trajectory Segment (l)
Compression ratio
(b)
Length of Trajectory Segment (l)

Fig. 2.5 Results of the boundary choice of *l*. (**a**) Maximum time costs under different *l*s. (**b**) Compression ratios under different *l*s

Table 2.1 Results of mean error in when-and-where query under different *l*s

*l*s	5	10	15	20	25	30
Mean error (m)	112.8	144.7	187.9	200.5	236.4	271.3

making it more difficult to estimate its position at the given time. However, compared to the results reported in [1, 3], our proposed algorithm still achieves a better query quality. For instance, the mean error of our method is less than 145 m when *l* equals 10, which can be usable for most of LBSs in real life. We fix $l = 10$ in this experiment.

2.6 Evaluations of VTracer

In this section, we show that the compression results produced by VTracer are: first, with the high quality of trajectory mapping, second, with high compression ratio, and third, quite efficient and light-weight which can respond in the real time and work under the mobile environment.

2.6.1 Experimental Setup

1. Deployment in a real environment: Figure 2.6 shows the deployment of VTracer system in the real environment. The system is deployed in a Volkswagen Passat Car made in 2015. There are two smartphones used in the deployed system. One of them is used to collect the GPS data, while the other is used to be the primary computing platform that is responsible for the heavy computation tasks. The models of smartphones used for data collection and data compression are Huawei P9 and P10, respectively. Huawei P10 has a RAM of 4 GB, which has strong computing capability. Bluetooth pairing between the two phones is required to establish the data communication before starting the system.
   We develop an app running on Android OS to control the built-in GPS sensor to mimic the vehicle-mounted GPS devices to collect the trajectory data. The sampling rate is set 6 s by default, which can be customized by programming. The data collector is mounted on top of the phone-holder to make sure that its heading direction is always consistent to the heading direction of the vehicle during driving, as highlighted in Fig. 2.6. The data stream is sent to the other

Smart phone
Built-in GPS

Fig. 2.6 Deployment of VTracer system in real environment smartphone via the established Bluetooth link. In the other smartphone, we also develop another app called *TrajCompressor* running on Android OS to compress the continuously received GPS data stream.

2. Data preparation: Two major datasets are prepared for the evaluation of VTracer in the city of Chongqing, China. The first one is the road network, which is downloaded from OpenStreetMap. Statistically, it totally contains 30,691 edges and 29,461 nodes. The second one is the raw GPS trajectory data containing around 7600 raw GPS points, which is collected from March 17th to April 21st, 2018. The accumulated driving distance of the trajectory dataset is around 513 km.
3. Baseline methods: We select two baselines to compare, i.e., PRESS and CCF. Moreover, we implement them in the same Android platform and develop the corresponding apps, namely PRESS and CCF.
4. Evaluation metrics: Two metrics are used to measure the effectiveness of VTracer. Specifically, the average matching accuracy (*acc*) used to quantify the quality of map-matching; and the compression ratio (*cr*) used to measure the performance of trajectory compression. We define *acc* as the difference between 1 and the ratio of the number of wrongly-matched GPS points to the total number of observed GPS points, as shown in Eq. (2.1).

$$acc = 1 - \frac{N_{wm}}{N_{total}} \tag{2.1}$$

where N_{wm} and N_{total} refer to the number of wrongly-matched GPS points and the total number of observed GPS points, respectively. A given GPS point is reported wrongly matched if it is not mapped to its true edge. It should be noted that we recruit three volunteers to manually label the "true edge" that each GPS point should locate by voting. Hence, it is challenging or even impossible to compute the value of map-matching accuracy "on-the-go" because the ground truth can only be known *ex post*.

In previous studies, *cr* is usually defined as the ratio between the disk space that the trajectory data occupy before and after applying data compression algorithms. However, as to our VTracer system, a data point is sent to the local computing center for processing immediately once produced. At the side of the computing center, it also never stores data in the hard disk. To make the computation of *cr* feasible, we approximate it using the following formula, as shown in Eq. (2.2).

$$cr = \frac{N_{total} \times c_1}{\lceil N_{total}/l \rceil \times c_2 + N_{edge} \times c_3} \tag{2.2}$$

where N_{total} refers to the number of total GPS points that the local computing center received and processed; $\lceil N_{total}/l \rceil$ gets the rounding up value of N_{total}/l, and it refers to the number of trajectory batches; N_{edge} is the number of retained edges in the compressed trajectory; c_1 is the value of constant bytes that a GPS point occupies, which is roughly estimated; similarly, c_2 and c_3 refer to the constant values of bytes that the time and an edge occupy, respectively. In our experiment, we set $c_1 = 40$ and $c_2 = c_3 = 9$. The compression performance is better if cr is bigger. Compared to acc, it is feasible to compute cr "on-the-go" since we can monitor the values of N_{total} and N_{edge} in real time. To be more specific, we simply set two counters to obtain the number of received GPS points and the number of retained edges during trajectory compression, which enables us to show and refresh the value of cr timely.

To measure how efficient that VTracer is, we count the total time cost at both phases including trajectory mapping and trajectory compression for each trajectory batch. Moreover, we use the average time value among all time costs for trajectory batches to quantify the overall efficiency of VTracer. In addition, the energy and memory that VTracer consumes are also used to reflect its efficiency.

2.6.2 Effectiveness Study

We are quite interested in the question, that is, *does VTracer perform consistently for all rides?* To address such issue, we intend to investigate the system performance for a sample of rides, in terms of the matching accuracy (acc) and the compression ratio (cr), respectively. Specifically, we first category all rides into five groups according to their driving distance, i.e., 0–10, 10–15, 15–20, 20–25, >25. Then, for all rides belonging to the same group, with VTracer, we obtain the average, minimum, and maximum values of the two performance indicators, as shown in Fig. 2.7. From results shown in two figures, we can draw the conclusion that the average performance is rather stable for rides with different driving distances. For instance, first, the accuracy of trajectory mapping is high and always above 95% for all rides; second, the compression ratio is also stable and above 15 for all rides. In addition, we can observe that, when the driving distance becomes longer, the range of both performance indicators become narrower, i.e., the system performs more stably. One possible explanation for such observation could be that, the vehicle may still stay within the same region (inner-region) with high confidence if driving shortly. Under this circumstance, the spatial context of the ride along the path may be very similar. Specifically, the road network along is either dense or sparse, which leads the performance to the extreme, i.e., either very poor or excellent. In contrast, the vehicle may cross different regions (inter-region) while driving far. In this case, the spatial context of the ride along the path may be averaged. Hence, to sum up, the performance can vary much more significantly for rides with smaller driving distance.

We also compare the compression ratio of VTracer to two baselines. The experiment results are shown in Table 2.2. As can be observed, VTracer achieves the compression ratio in-between among three systems, and PRESS obtains the best performance. The reason why VTracer performs better than CCF is that: CCF retains the out-edge information at every intersection, while VTracer only retains out-edges with significant heading changes, which usually take up a small fraction of all

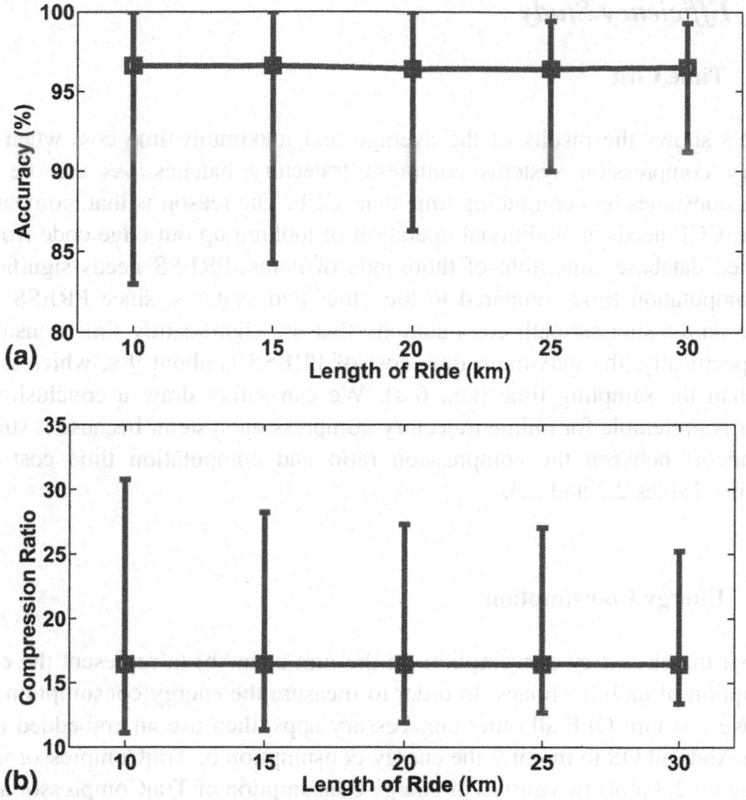
100
95
90
85
80
Accuracy (%)
(a)
10
15
20
25
30
Length of Ride (km)
35
30
25
20
15
10
Compression Ratio
(b)
10
15
20
25
30
Length of Ride (km)

Fig. 2.7 Results of *acc* and *cr* for rides belonging to different groups. (**a**) Accuracy of trajectory mapping. (**b**) Compression ratio of trajectory compression

Table 2.2 Compression ratios of different trajectory systems

System	VTracer	PRESS	CCF
cr	15.85	16.67	15.69

out-edges. One credible reason why PRESS performs the best may be due to that drivers prefer to taking the shortest path in real cases [19]. Under such circumstance, the edges between the origin and destination can be commonly discarded, leading to the best compression ratio.

2.6.3 Efficiency Study

2.6.3.1 Time Cost

Table 2.3 shows the results of the average and maximum time cost when three trajectory compression systems compress trajectory batches. As can be seen, VTracer consumes less computing time than CCF. The reason is that, compared to VTracer, CCF needs an additional operation of looking up out-edge code from the predefined database consisting of thousands of items. PRESS needs significantly more computation time compared to the other two systems, since PRESS needs multiple online shortest-path computations that are significantly time consuming. More specifically, the maximum time cost of PRESS is about 9 s, which is even larger than the sampling time (i.e., 6 s). We can safely draw a conclusion that VTracer is preferable for online trajectory compression system, because it strikes a nice tradeoff between the compression ratio and computation time cost when combining Tables 2.2 and 2.3.

2.6.3.2 Energy Consumption

We adopt the electricity consumption (in the unit of mAh) to represent the energy consumption of mobile phones. In order to measure the energy consumption accurately, we just turn OFF all other unnecessary apps, then use an embedded app in Huawei Android OS to monitor the energy consumption of TrajCompressor app an hour. Figure 2.8 plots two curves of energy consumption of TrajCompressor app on the mobile phone with a working time duration of 12 h. One curve corresponds to the case when the visualizer is enabled while the other one corresponds to the case when the visualizer is disabled. It is easy to understand that more energy will be killed if enabling the front-end visualizer. Thus, the visualizer is disabled default to save energy. For both cases, from the figure, we can see that the energy consumption climbs almost linearly as the working time gets longer. As expected, the slop is much bigger when disabling the visualizer. The battery capacity of a HUAWEI P10 smartphone is 3000 mAh, so TrajCompressor can continuously work for around 140 h. In addition, it can work as long as about 280 h if the visualizer is disabled. In summary, the energy consumption of *TrajCompressor* app is acceptable in real-application scenarios.

As a comparison, we also show the energy consumption of the other two apps in Table 2.4. As can be predicted, TrajCompressor app is the most energy-efficient while PRESS app consumes the largest amount of energy, since a more computation time usually implies a higher consumption of electrical power.

Table 2.3 Time cost of different trajectory compression systems

System	VTracer	PRESS	CCF
Average time cost (ms)	146	1483	359
Maximum time cost (ms)	400	9348	671

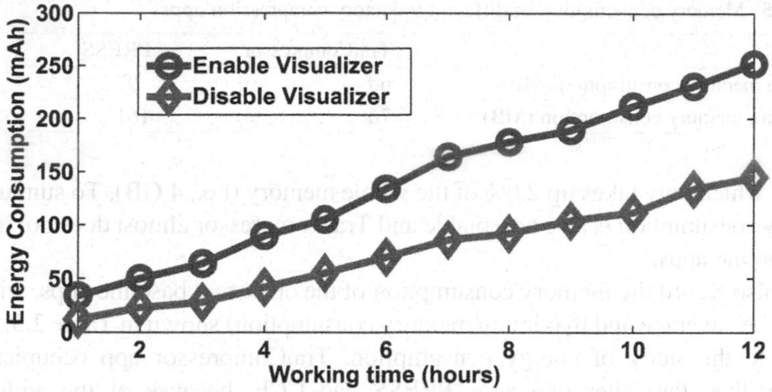
Enable Visualizer
Disable Visualizer
Energy Consumption (mAh)
Working time (hours)
0
50
100
150
200
250
300
2
4
6
8
10
12

Fig. 2.8 Energy consumption results of TrajCompressor app with respect to the working time when the visualizer is disabled and enabled

Table 2.4 Energy consumption of different trajectory compression apps

App	TrajCompressor	PRESS	CCF
Energy consumption (mAh)	210	920	245

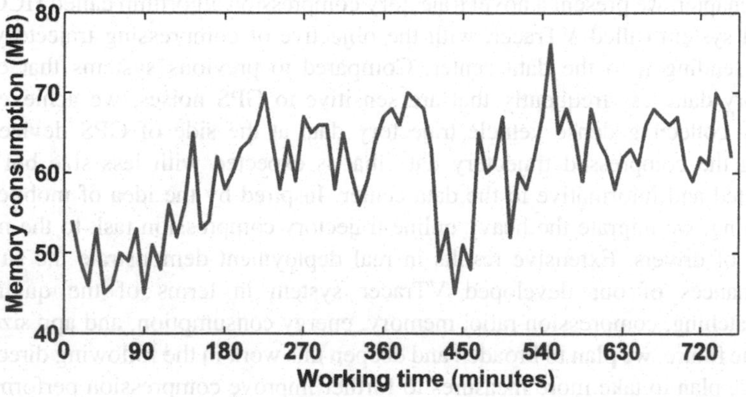
Memory consumption (MB)
Working time (minutes)
40
50
60
70
80
0
90
180
270
360
450
540
630
720

Fig. 2.9 Memory consumption of *TrajCompressor* app with respect to the working time

2.6.3.3 Memory Consumption

Similar to the energy consumption study, we also rely on another embedded app in Huawei Android OS to monitor the memory consumption of *TrajCompressor* (in the unit of MB). Compared to the energy consumption, the memory consumption is more sensitive to the working time, we thus monitor this parameter every 10 min. Figure 2.9 shows the result of memory consumption of the mobile phone with a working time duration of 12 h. As can be observed, the memory consumption fluctuates with the working time. The maximum memory consumption is about

Table 2.5 Memory consumption of different trajectory compression apps

App	TrajCompressor	PRESS	CCF
Average memory consumption (MB)	60	97	86
Maximum memory consumption (MB)	76	161	121

76 MB, which only takes up 2.0% of the whole memory (i.e., 4 GB). To sum up, the memory consumption is also acceptable and TrajCompressor almost does not impact other mobile apps.

We also record the memory consumption of the other two baseline apps, with the results (i.e., average and maximum memory consumption) shown in Table 2.5. Quite similar to the study of energy consumption, TrajCompressor app occupies less memory than the other two apps PRESS and CCF, because of the additional operation of looking up out-edge code (CCF) and computation of shortest-path (PRESS) will consume more computing resource of the mobile phone.

2.7 Conclusions and Future Work

In this chapter, we present a novel trajectory compression algorithm called HCC, and a novel system called VTracer, with the objective of compressing trajectory data before sending it to the data center. Compared to previous systems that collect trajectory data less frequently that are sensitive to GPS noises, we achieve such goal by collecting dense vehicle trajectory data at the side of GPS devices and sending the compressed trajectory data that is expected with less size but more completed and informative to the data center. Inspired by the idea of mobile edge computing, we migrate the heavy online trajectory compression task to the mobile phones of drivers. Extensive results in real deployment demonstrate the superior performances of our developed VTracer system in terms of the quality of map-matching, compression ratio, memory, energy consumption, and app size.

In the future, we plan to broaden and deepen this work in the following directions. First, we plan to take more measures to further improve compression performance, including compression enhancements and more advanced edge coding methods. Second, we also intend to investigate the trajectory compression in the temporal dimension. For instance, the accelerating (decelerating) behaviors of drivers may be retained, which reflect drivers' driving styles and traffic congestion at that time. To be more specific, we can easily calculate the mean accelerated speed (a) of adjacent two consecutive GPS points. Then a series of two-dimensional data (t_i, a_i) can be obtained, where a_i refers to the accelerated speed of adjacent two points p_i and p_{i+1}, and t_i is the starting time of a_i. We could utilize some common methods (e.g. Huffman coding) to compress the two-dimensional data. It should be noted that trajectory data may be collected with equal time interval. Third, to enable more and smarter urban services, we plan to explore the potentials of the combination research of trajectory compression and trajectory summarization [20], due to the capacity making the trajectory data more intuitively and more understandable of trajectory summarization. Last but not least, according to [21–23], we know that deep learning can translate trajectory into embedding representation with lower dimension, which is conductive to further services, e.g. traffic prediction, map matching. The result also inspires us to take it into consideration that combing trajectory compression with deep learning. For instance, we can design a trajectory embedding network, which is capable of representing trajectory compactly and recovering trajectory without spatial loss.

References

1. Ji Y, Zang Y, Luo W, Zhou X, Ding Y, Ni LM. Clockwise compression for trajectory data under road network constraints. In: 2016 IEEE international conference on big data (big data), 2016. p. 472–81.
2. Muckell J, Olsen PW, Hwang JH, Lawson CT, Ravi SS. Compression of trajectory data: a comprehensive evaluation and new approach. GeoInformatica. 2014;18(3):435–60.
3. Song R, Sun W, Zheng B, Zheng Y. PRESS: a novel framework of trajectory compression in road networks. In: Proceedings of the Vldb endowment, 2014. p. 661–72.
4. Chen Y, Jiang K, Zheng Y, Li C, Yu N. Trajectory simplification method for location-based social networking services. In: Proceedings of the 2009 international workshop on location based social networks, 2009. p. 33–40.
5. Lou Y, Zhang C, Zheng Y, Xie X, Wang W, Huang Y. Map-matching for low-sampling-rate GPS trajectories. In: Proceedings of the 17th ACM SIGSPATIAL international conference on advances in geographic information systems, 2009. p. 352–61.
6. Sun P, Xia S, Yuan G, Li D. An overview of moving object trajectory compression algorithms. Math Probl Eng. 2016;2016:6587309, 1–13.
7. Shi W, Cao J, Zhang Q, Li Y, Xu L. Edge computing: vision and challenges. IEEE Internet Things J. 2016;3(5):637–46.
8. Douglas DH, Peucker TK. Algorithms for the reduction of the number of points required to represent a digitized line or its caricature. Cartographica. 1973;10(2):112–22.
9. Han Y, Sun W, Zheng B. COMPRESS: a comprehensive framework of trajectory compression in road networks. ACM Trans Database Syst. 2017;42(2):1–49.
10. Ji Y, Liu H, Liu X, Ding Y, Luo W. A comparison of road-network-constrained trajectory compression methods. In: 2016 IEEE 22nd international conference on parallel and distributed systems (ICPADS), 2016, p. 256–63.
11. Silva A, Raghavendra R, Srivatsa M, Singh AK. Prediction-based online trajectory compression, 2016. p. 1–13.
12. Feng Z, Zhu Y. A survey on trajectory data mining: techniques and applications. IEEE Access. 2016;4:2056–67.
13. Kellaris G, Pelekis N, Theodoridis Y. Map-matched trajectory compression. J Syst Softw. 2013;86(6):1566–79.
14. Chen C, Ding Y, Xie X, Zhang S. A three-stage online map-matching algorithm by fully using vehicle heading direction. J Ambient Intell Humaniz Comput. 2018;9(5):1623–33.
15. Richter KF, Schmid F, Laube P. Semantic trajectory compression: representing urban movement in a nutshell. J Spatial Inform Sci. 2012;4:3–30.
16. Castro PS, Zhang D, Li S. Urban traffic modelling and prediction using large scale taxi GPS traces. In: 2012 International conference on pervasive computing, 2012. p. 57–72.

17. Chen C, Chen X, Wang L, Ma X, Wang Z, Liu K, Guo B, Zhou Z. MA-SSR: a memetic algorithm for skyline scenic routes planning leveraging heterogeneous user-generated digital footprints. IEEE Trans Veh Technol. 2017;66(7):5723–36.
18. Li B, Zhang D, Sun L, Chen C, Li S, Qi G, Yang Q. Hunting or waiting? Discovering passenger-finding strategies from a large-scale real-world taxi dataset. In: 2011 IEEE international conference on pervasive computing and communications workshops (PERCOM workshops), 2011. p. 63–8.
19. Castro PS, Zhang D, Chen C, Li S, Pan G. From taxi GPS traces to social and community dynamics: a survey. ACM Comput Surv (CSUR). 2013;46(2):17, 1–34.
20. Andrae S, Winter S, Strobl S, Blaschke T, Griesebner G. Summarizing GPS trajectories by salient patterns, 2005.
21. Shen Z, Du W, Zhao X, Zou J. DMM: fast map matching for cellular data. In: Proceedings of the 26th annual international conference on mobile computing and networking, 2020. p. 1–14.
22. Zhao K, Feng J, Xu Z, Xia T, Chen L, Sun F, Guo D, Jin D, Li Y. DeepMM: deep learning based map matching with data augmentation. In: Proceedings of the 27th ACM SIGSPATIAL international conference on advances in geographic information systems, 2019. p. 452–5.
23. Cao H, Xu F, Sankaranarayanan J, Li Y, Samet H. Habit2vec: trajectory semantic embedding for living pattern recognition in population. IEEE Trans Mobile Comput. 2020;19(5):1096–108.

Chapter 3
Trajectory Data Protection

3.1 Introduction

Nowadays, numbers of mobile smart devices have penetrated our everyday lives. These smart devices allow users to enjoy location-based services (LBS), or even enable users to carry some sensing tasks to perceive different aspects of the city. Usually, the sensing data can support various applications of the smart city such as monitor urban environment [1], estimate traffic condition [2], and even schedule paths for taxis [3].

As for a mobile device user, no matter he enjoys services or perceives the city, it is inevitable that some kind of geographic information will be sent to the application server, which brings risk his location privacy. At present, there have been lots of research work on how to protect users' location privacy. However, when more and more locations are exposed, they can form a semantic trajectory so that it is more difficult to protect location privacy. Actually, there are already several commonly used models to accurately 'guess' users' locations such as Bayesian inference model, collaborative filtering-based inference model, and hidden Markov-based inference model. Especially, the hidden Markov-based inference model is a powerful method to infer the users' true locations which has been validated by Xiao et al. [4].

In this chapter, we will investigate two trajectory privacy protection methods. Both methods come from the mobile crowd-sensing scenario where participants submit the sensing data to the application server. Since the sensing data contains geographic information, it may cause trajectory privacy leakage of the users. The difference between these two methods lies in how users submit their data. In the first method, we assume that the sensing task is delay tolerant which allows the user to

Part of this chapter is based on the previous work: H. Huang, X. Niu, C. Chen and C. Hu, "A Differential Private Mechanism to Protect Trajectory Privacy in Mobile Crowd-Sensing," 2019 IEEE Wireless Communications and Networking Conference (WCNC), Marrakesh, Morocco, 2019, pp. 1–6.

C. Chen et al., *Enabling Smart Urban Services with GPS Trajectory Data*,
https://doi.org/10.1007/978-981-16-0178-1_3

submit a whole trajectory in a bundle. We call this an offline trajectory publishing. The second method removes the assumption of delay tolerance. When a user records a location, he will upload it before he records the next one. We call this an online, or real-time, trajectory publishing. Note that in the rest of this section, we use 'user' and 'participant' interchangeably.

3.2 Related Work

When we discuss the privacy protection, it is important to determine who will protect the users' privacy. In the first case, all users submit their true trajectories to a server, who usually perturbs the data set before he releases it. This case is called a centralized protection. In another case, we assume that the server is a potential adversary so that each user perturbs his trajectories before submitting them. This case is called a localized protection. As is mentioned above, protecting the trajectory privacy is more challenging than simply protecting individual locations. Also, simply decomposing the correlated locations in the trajectory is not always effective. However, if we regard the whole trajectory as a whole entity, we can design new methods to solve this problem.

In this part, we give a few related works of both centralized and localized trajectory privacy protection.

In the first category, Xiao et al. [4] noticed that there exists temporal and spatial correlations between any successive locations in a trajectory. They assumed that an adversary could use the Hidden Markov Model (HMM) to reconstruct users' trajectories by using the perturbed data set. In order to resist this threat, they designed a differential private mechanism for the server to build a location set based on the HMM and randomly release locations to the adversary. However, the correlation not only exists among the locations in a single trajectory, different trajectories also correlate with each other. In order to tackle correlated trajectories, Ou et al. [5] proposed their method by regarding each trajectory as a vector. When the server releases trajectories, it first finds out all correlated trajectories, then constructs a private candidate set, and last perturbs each location of the trajectory within the candidate set. They also use differential privacy to measure the quantity of privacy that their method can provide.

A representative work of localized trajectory privacy protection comes from Jiang et al. [6]. They mapped a trajectory to a certain point in a high dimension space and use differential privacy to determine the extent of perturbation. When allocating the noise to each location, they added constraint to each location so that the utility of the perturbed trajectory is preserved. Even if each noisy location only costs a small amount of utility, however, the total amount of utility loss may exceed the bound. Recently, Han et al. [7] proposed another method to protect the trajectory from the local side. A user first computes correlation of trajectory points to classify his trajectories containing sensitive points. Then, according to the relevance of location points and the randomized response mechanism, a reasonable candidate set is selected to replace the sensitive points in the trajectory to satisfy the locally differential privacy. They evaluated their method on a real dataset and results validated the effectiveness of their method.

3.3 Preliminary

We discretize the geographic area A into $m \times n$ cells. Each user's location is denoted as the cell which he locates in. All cells are indexed by two ways: state coordinate and map coordinate, which are shown in Fig. 3.1. First, we mark each cell with a serial number i and use c_i to represent its state. For example, c_5 denotes the *5-th* cell in the area. Using this coordinate, a trajectory t can be represented as an ordered tuple $t = <c_1, c_2, \ldots, c_n>$ where c_i denotes the *i-th* location in t. For example, in Fig. 3.1, the illustrating trajectory is $t = <2; 4; 8; 12>$. Second, in order to compute the correlation between two trajectories, we also denote each cell with its index on x axis and y axis respectively. For example, the 3rd cell in Fig. 3.1 is (1, 3). Hence, the trajectory t can also be represented as $t = <x_1, y_1, x_2, y_2, \ldots, x_n, y_n>$, e.g., $t = <2, 1, 4, 1, 4, 2, 4, 3>$ in Fig. 3.1. We call the number of locations in a trajectory as its length.

Now we define the distance between two locations. As is shown in Fig. 3.2a, given two locations l_1 and l_2, we use α to denote the azimuth between them. The horizontal and vertical distances between l_1 and l_2 are $d_\perp(l_1, l_2) = d_E(l_1, l_2) \times cos(\alpha)$ and $d_\parallel(l_1, l_2) = d_E(l_1, l_2) \times sin(\alpha)$ respectively, where $d_E(l_1, l_2)$ is the Euclidean distance between l_1 and l_2. Therefore, the Manhattan distance between l_1 and l_2 is $d_M(l_1, l_2) = |d_\perp(l_1, l_2)| + |d_\parallel(l_1, l_2)|$. Besides, for each trajectory, there usually exist few locations which can depict the skeleton of the trajectory, called key locations. An intuitive way to identify a key location is to check if the user makes a sharp turn

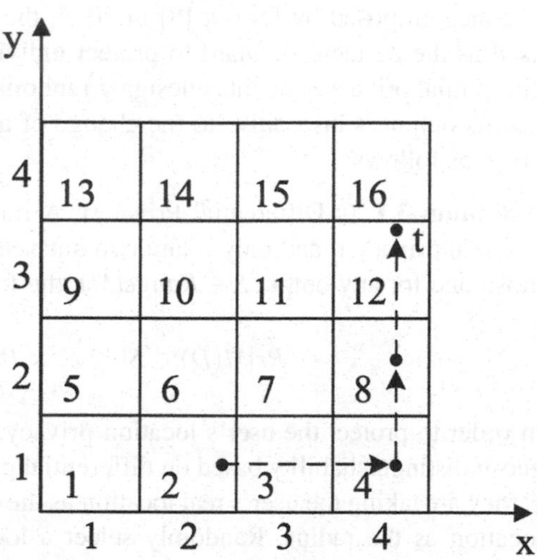
y
4
3
2
1
13
14
15
16
9
10
11
12
5
6
7
8
1
2
3
4
t
1
2
3
4
x

Fig. 3.1 Two coordinate system representations of the users' trajectories

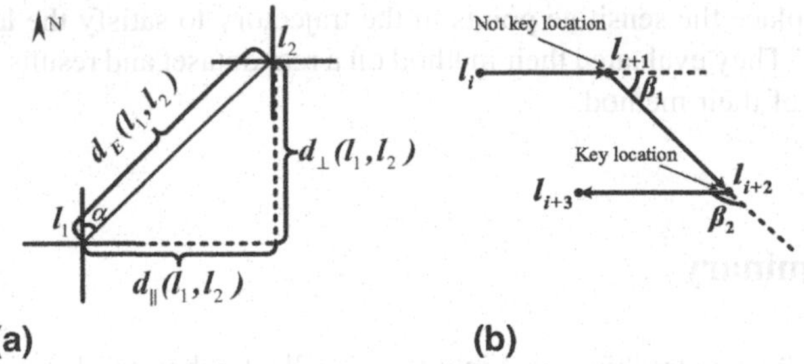
N
l_2
$d_E(l_1,l_2)$
$d_\perp(l_1,l_2)$
l_1
α
$d_\parallel(l_1,l_2)$
(a)
Not key location
l_i
l_{i+1}
β_1
Key location
l_{i+3}
l_{i+2}
β_2
(b)

Fig. 3.2 Illustration of distance between locations and how to determine a key location. (**a**) Distance between c_1 and c_2 and (**b**) Key or not key location

at that location. Formally, for any three consecutive locations l_i, l_{i+1}, and l_{i+2}, we use β to denote the angle between vectors l_il_{i+1} or $l_{i+1}l_{i+2}$. If $\pi/2 \leq \beta < \pi$, then l_{i+1} is a key location. Otherwise, l_{i+1} is not a key location. The examples are shown in Fig. 3.2b.

Besides the distance, we use correlation between trajectories to evaluate the utility of a perturbed trajectory. Before we introduce the definition of correlation, we depict the trend of a trajectory using a vector:

$$N_t = < a_1, a_2, \ldots, a_{n-1} >$$

where each a_i is the angle between the vector a_ia_{i+1} and the due north. Please note that if $a_i < \pi$, we keep its value, and set $a_i = 2\pi - a_i$ if $a_i < \pi$. Return to Fig. 3.1, we can see that $N_t = <90, 0, 0>$. Then the correlation of two trajectories is defined as the cosine of their trend vectors:

$$r_t(t, t') = N_t \bullet N_{t'} / |N_t| \bullet |N_{t'}|$$

Since proposed by Dwork [8] in 2006, the differential privacy has been widely used as the de facto standard to protect individual privacy in datasets. To achieve differential privacy, one must design a randomized mechanism M which guarantees that its output is insensitive to the change of a single datum. A formal definition is given as follows:

Definition 3.1 (ε-Differential Privacy). A randomized mechanism M gives ε-differential privacy if and only if any two datasets D and D' differ in a single datum at most, and for any output $S \in Range(M)$, the following inequality holds:

$$Pr[M(D) \in S] \leq e^{\varepsilon} \times Pr[M(D') \in S].$$

In order to protect the user's location privacy, Andrés [9] proposed the concept of geo-indistinguishability based on differential privacy. For a given privacy parameter ε, they are taking the user's real location as the center and the distance r from the real location as the radius. Randomly select a location as the perturbed location and report it to the server. The level of privacy protection provided by their mechanisms to users is εr. A formal definition is given as follows:

Definition 3.2 (Geo-indistinguishability). A mechanism M satisfies ε-geo-indistinguishability if and only if for all x and x', the following inequality holds:

$$DP(M(x), M(x')) \leq e^{\varepsilon} \times d(x, x').$$

Equivalently, the definition can be formulated as $M(x)(Z) \leq e^{\varepsilon d(x,x'')} \times M(x')(Z)$ for all $x, x' \in X, Z \in \mathbb{Z}$. Note that for all points x' within a radius r from x, the definition forces the corresponding distributions to be at most εr distant.

We find that, in the mobile crowd-sensing scenario, the concept of differential privacy cannot be used directly to protect the privacy of participants from their local side. One of the most important reason is that each user only has his own trajectory, i.e., the datasets D in the above definition, instead of a set of trajectories which are available at the server. It is meaningless to analyze D' which differs from D with only one location. To solve this problem, inspired from the geo-indistinguishability, we define *trajectory-DP* to denote the privacy level that a mechanism can guarantee for users to protect their trajectories. The idea is that, given a predetermined finite set of trajectories T, each time a participant submits his trajectory $t \in T$ to the server, he uses a random mechanism to generate a perturbed trajectory $t' \in T$. We say that the mechanism satisfies *trajectory-DP* is if and only if t and t' differs within a range of a multiplicative factor $e^{-\varepsilon d(t;t')}$. We give the formal definition of *trajectory-DP* as follows:

Definition 3.3 (Trajectory-DP). Given a privacy parameter ε and a distance metric $d(t, t')$ between any two trajectories t and t', a randomized mechanism M: $T \rightarrow T$ satisfies ε differential privacy iff for all trajectories t and t':

$$M(t) \leq e^{\varepsilon dt(t,t')} M(t'),$$

or equivalently,

$$DP(M(t), M(t')) \leq d(t, t'),$$

where $DP(\bullet)$ is the multiplicative distance between two distributions.

Note that the parameter ε specifies the privacy level of a unit distance. For example, if there are $k - 1$ trajectories around the true trajectory t within a unit distance, and if the probability that M maps t to another trajectory t' is p, then the probability that M maps any of the other trajectories to t' is within $[|p - e^{\varepsilon}|, |p + e^{\varepsilon}|]$. From this aspect, we can derive that a smaller ε means a higher privacy level.

The concept of *Trajectory-DP* can be used to evaluate the privacy level of the whole trajectory. However, it does not guarantee that each segment of the trajectory can obtain the same privacy level. In order to solve this problem, we need to decompose the trajectory into pieces of any size. Based on the ε geo-indistinguishability and w-event privacy [9], we propose w-trajectory differential privacy that guarantees w differential privacy for each of sub-trajectory which consists of w successive locations.

Definition 3.4 (w-Trajectory Differential Privacy). A randomized mechanism M provides w-trajectory ε-differential privacy if and only if all sub-trajectories $T \in$ *Range(M)* consisting of successive w locations and all neighboring sub-trajectories T_i and T'_i whose locations pair wisely satisfy ε geo-indistinguishability, and it holds that,

$$\Pr[M(T_i) \in T] \le e^{\varepsilon} \times \Pr[M(T_i')] \in T].$$

Now we introduce the adversary. In this chapter, we consider an honest but curious server who is attempting to retrieve users' exact trajectories based on their submitted data. The server regards each received trajectory as a sequence of states, i.e., locations. His goal is to decode the states to reveal the true locations, i.e., hidden states. Among different inference methods, we assume that the server use Viterbi algorithm [10] to correct each received location and consider it as the truth.

To start with the inference process, we assume that the server maintains three kinds of prior knowledge. The first is a prior distribution P^{-} on A. Note that P^{-} is an $m \times n$ vector, and element, i.e., $p-(c_i = x)$, denotes the probability that a user locates in cell x. Furthermore, the server also maintains a mapping matrix M to represent the mapping from the users' reported locations to the exact locations in which they "really" are. Each entry of M, i.e., $m_{x,y} = P(c_i = y|c_i = x)$, denotes the conditional probability that a user's *i-th* location is "really" in cell y when his reported location is cell x. The third one is a transition matrix $S = [s_{i,\, j}]$ to denote the probability that a node moves from one cell to another. Note that all prior knowledge can be built from a training dataset.

When the server receives a trajectory from a user, he simply regards it as a sequence of observed locations, denoted as $c_1, c_2, \ldots, c_N$. The server's mission is to find the true sequence, denoted as $c'_1, c'_2, \ldots, c'_N$. For the tth location, assume that the server observes it is in cell i, let $\delta_t(i)$ be the maximum probability of the true trajectory $c'_1, c'_2, \ldots, c'_N$:

$$\delta_t(i) = max_{c*1,c*2,\ldots,c*t-1} P(c*_t = i, c*_1, c*_2, \ldots, c*_{t-1}, c_1, c_2, \ldots, c_t)$$
$$= max_{1\le j\le m\times n}\left[\delta_{t-1}(j)\ s_{j,i}\right] m_{ct,i}.$$

We let $c*_t = i$ if it has the maximum trajectory probability, i.e.:

$$c*_t = \arg\ max_{1\le j\le m\times n}\left[\delta_{t-1}(j)\ s_{j,i}\right].$$

Each time the server decodes a trajectory, it updates the prior distribution to be a posterior distribution P^{+}, which will also be P^{-} for the second location. Here we use Bayesian rule for the update:

$$p^{+}(c_i = x) = m_{y,x}p^{-}(c_i = y)/\Sigma_t m_{t,x}p^{-}(c_i = t)$$

Meanwhile, the server also updates the *x-th* row of the matrix *M* and *S*. Using this method, the server corrects each location one by one.

Although locations submitted by users are perturbed, servers or adversaries can still obtain some information about them. To quantify the amount of user's leakage of privacy, we define a metric, i.e., trajectory privacy leakage (*TPL*), as follows:

$$TPL = \sum\nolimits_{i=1}^{n} 1/S(c_i, c'_i)d_M(c_i, c'_i)$$

Here $S(c_i, c'_i)$ is a similarity function which measures the correlation degree of two locations. In our work, we model the similarity under spatial correlation by Gaussian kernel $S(c_i, c'_i) = exp(-d_E^2(c_i, c'_i)/2\sigma^2)$. The parameter σ is a scaling parameter which controls how fast the similarity increases as the distance decreases.

To evaluate the utility of a perturbed trajectory, we compute the expected distance between perturbed trajectory and real trajectory. Given the randomized mechanism *M*, and $P(c_i|c_{i-1})$ denoting the probability that a user moves from l_{i-1} to l_i, we obtain:

$$d(c_i, c'_i) = \Sigma_{ci,c'i\in A}P(c_i|c_{i-1})M(c'_i|c_i)d_M(c_i, c'_i)/\Sigma_{ci,c'i\in A}P(c_i|c_{i-1})M(c'_i|c_i)$$

Then we can calculate the mean expected distance $E(t, t')$ of a certain trajectory as follows:

$$E(t, t') = \sum\nolimits_{i=1}^{n} d(c'_i, c_i)$$

The value of the mean expected distance $E(t, t')$ reflects the utility of the perturbed trajectory. The smaller the $E(t, t')$ is, the more effective information the server can obtain from the trajectory, which indicates more privacy leakage.

3.4 Trajectory Protection Mechanism

In this section, we first introduce the trajectory protection mechanism (TRM), then we present the analysis of its characteristics. In brief, the TRM consists of two phases. In the first phase, TRM computes the total amount of noise and randomly allocate the noise to all coordinates of the locations in the true trajectory. The result of the first phase is a set of candidate trajectories which not only satisfy *Trajectory-DP*, but also preserve utility. In the second phase, TRM uses an exponential method to randomly choose a candidate trajectory as the output.

It is worth noting that the noise affects the distance between the true and perturbed trajectories. Furthermore, the amount of noise is dependent on the parameter ε. A smaller ε leads to a larger noise and vice versa. On the other hand, we use the correlation, i.e., r_t, as the metric to evaluate the utility of the perturbed trajectory because the correlation is a better metric to depict the similarity between two trajectories than the Euclidean distance. As an intuitive example, suppose that t in Fig. 3.1 is a true trajectory and we have other two perturbed trajectories, denoted as $t_1 = <1, 1, 3, 1, 3, 2, 3, 3>$ and $t_2 = <2, 1, 4, 1, 3, 3, 1, 3>$. We can see that $d_t(t, t_1) = d_t(t, t_2) = 1$ which means they have equal privacy level. We can also see that t_1 is more similar to t than t_2 because t and t_1 have the same trend, and $r_t(t, t_1) = 1 > r_t(t, t_2) = 0{:}692$ agrees with this intuition.

We present the TRM using the trajectory perturbation algorithm (TPA). In order to generate a Laplacian random number r as the noise (line 4), we can use the method proposed in [9], which has been proved to satisfy ε-differential privacy. Note that $C_\varepsilon^{-1}(p) = -1/\varepsilon\,(W_{-1}(p - 1/e) + 1)$, where W_{-1} is the Lambert function of -1 branch. From line 6 to 8, we allocate the noise to each coordinate. Since each noise has two directions, i.e., adjust $+1$ or -1 to either x_i or y_i, there are totally 2^p different allocations, where p is the number of locations which needs to add noise. From line 11 to 13, we construct the perturbed trajectory t'. The item $v^T w$ at line 13 determines the amount and direction of noise to each location. If w_i equals to 0, it means that the *i-th* location does not need to be perturbed. If $v_i w_i$ perturbs the true location exceeds the geographic boundary, we simply let $w_i = -1 \times w_i$, which perturbs the location to the opposite direction. For the last step of the first phase, if t' satisfies the utility requirement, we add it to the candidate sets (line 14 and 15).

The second phase starts from line 16. We first update each u_i with its exponential version and then normalize them, which guarantees that the sum of all u_i equals to 1. By randomly choose $r \in (0, 1)$ and check which bin r locates in (line 22–25), we implement the exponential method to choose the candidate trajectory as the output.

Algorithm 3.1. Trajectory Perturbation Algorithm (TPA)

Input:
t - the true trajectory
n - the length (number of locations) of t
ε - the privacy parameter
α - utility parameter
Output:
t' - the perturbed trajectory

1 Initialize a set S as an empty dictionary to store candidate trajectories. Each key is a perturbed trajectory t_i and its value is the utility u_i.
2 Initialize the noise vector v, whose length is $2n$, by zero to store the amount of noise allocated on the according coordinate.
3 Initialize the direction vector w, whose length is also $2n$, by zero to store the direction of the noise.
4 Draw p uniformly in [0; 1) and set $r = C_\varepsilon^{-1}(p)$.
5 Compute the total amount of noise $m = \lceil r * n \rceil$.
6 **while** $(m > 0)$ **do**
7 Randomly choose $i \in \{0, 1; \ldots, 2n - 1\}$ with uniform distribution.

(continued)

$v_i = v_i + 1$, $w_i = 1$, $m = m - 1$.
Let $q = \Sigma_{i=1}^{2n} w_i$.
for $(i = 1; i < 2^q; i++)$ **do**
Transform i into a binary b of q bits. Let b_j be the highest j-th bit of b, which relates to the j-th non-zero element of w from the left.
Let $w_j = 1$ iff $b_j == 0$, and $w_j = -1$ iff $b_j == 1$.
Let $t' = t + v^T w$.
if $u = r_t(t, t') > \alpha$ then
$S = S \cup \{< t', u >\}$
norm = 0, sum = 0
for each $(t_i \in S)$ **do**
$u_i = exp(u_i)$, $norm = norm + u_i$
for each (u_i) **do**
$u_i = u_i / norm$
Randomly choose $r \in (0, 1)$ with uniform distribution.
$i = -1$
do
$i = i + 1$; $sum = sum + u_i$
while $(sum < r)$;
return $t' = t_i$.

After the offline mechanism, we present the overview of the real-time data collection mechanism with trajectory privacy (RDCTP). As is shown in Fig. 3.3, the RDCTP consists of three parts: budget allocation, location perturbation, and optimization. Specifically, budget allocation is designed to allocate privacy budget to each certain location. Since the importance of locations are different, e.g., key locations are more important when considering the trajectory privacy, we need to allocate the privacy budget to the current processing location adaptively. Location

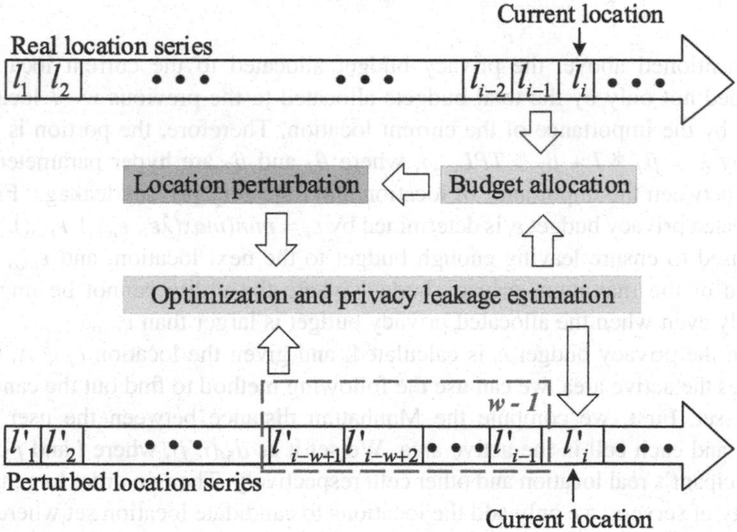
Current location
Real location series
l_1 l_2 • • • • • • l_{i-2} l_{i-1} l_i • • •
Location perturbation
Budget allocation
Optimization and privacy leakage estimation
w - 1
l'_1 l'_2 • • • l'_{i-w+1} l'_{i-w+2} • • • l'_{i-1} l'_i • • •
Perturbed location series
Current location

Fig. 3.3 An overview of RDCTP

perturbation aims at generating candidate location set according to the allocated privacy budget. Instead of choosing a location from the noisy location set randomly, the user formulates optimization models to choose perturbed location to minimize the trajectory privacy leakage of usrs while maximizing the utility of server. Finally, the user uploads the perturbed location to the server. In the following, we describe the main components of RDCTP in detail.

As is shown in Algorithm 3.2, let ε_g denote the total privacy budget within any successive w locations, ε_{min} and ε_{max} denote the lower and upper limits of privacy budget, respectively. Considering a key location, we calculate its importance as $I = |cos(\beta)|$. Otherwise, the importance of current location is $I = 0$. Then, the remaining privacy budget for the location lies in the window $[i-w + 1, i]$, which is $\varepsilon_r = \varepsilon_g - \Sigma_{k=i-w+1}^{i-1} \varepsilon_k$. The reason is that the privacy budget allocated to the current location is affected by the privacy for previous locations; therefore, we need to calculate the sum of location privacy leakage for previous $w-1$ locations to make sure that the whole window satisfies the privacy requirement.

Algorithm 3.2. Budget Location

Input:
The total privacy budget ε_g, window size w, successive w locations $c_{i-w+1}, \ldots, c_i$, the minimum and maximum budget, i.e., ε_{min} and ε_{max} for current location c_i.
Output:
The allocated privacy budget ε_i for c_i.
1: Check whether l_i is a key location. If yes, then $I = |cos(\beta)|$. If not, then $I = 0$;
2: Calculate the remaining budget for perturbation;
3: $\varepsilon_r = \varepsilon_g - \Sigma_{k=i-w+1}^{i-1} \varepsilon_k$;
4: Calculate the trajectory privacy leakage of the successive w - 1 locations;
5: $TPLw\text{-}1=\Sigma i \quad k=i\text{-}w+1 1/S(c_k, c'_k)(d_\perp(c_k, c'_k) + d_{\|}(c_k, c'_k))$;
6: Calculate portion $\lambda = \beta_1 * I + \beta_2 * TPL_{w-1}$;
7: Obtain privacy budget $\varepsilon_i = min(max(\lambda\varepsilon_r ; \varepsilon_{min}); \varepsilon_{max})$.

As mentioned above, the privacy budget allocated to the current location is determined not only by the total budgets allocated to the previous $w-1$ locations, but also by the importance of the current location. Therefore, the portion is determined by $\lambda = \beta_1 * I + \beta_2 * TPL_{w-1}$, where β_1 and β_2 are hyper parameters that balance between the importance of location and trajectory privacy leakage. Finally, the allocated privacy budget ε_i is determined by $\varepsilon_i = min(max(\lambda\varepsilon_r, \varepsilon_{min}), \varepsilon_{max})$, where ε_{min} is used to ensure leaving enough budget to the next location, and ε_{max} is the threshold of the maximum privacy budget where the utility cannot be improved obviously even when the allocated privacy budget is larger than ε_{max}.

When the privacy budget ε_i is calculated, and given the location $c_i \in A$, where A denotes the active area, we can use the following method to find out the candidate location set. First, we compute the Manhattan distance between the user's real location and each cell in the active area. We set it as $d_M(i, j)$, where i and j denote the participant's real location and other cell, respectively. Then, in order to guarantee the utility of servers, we only add the locations to candidate location set where $d_M(i, j) \leq 10$. Next, we use the cumulative distribution function $C_{\varepsilon i}(d_M(i, j)) = 1 -$

$(1 + \varepsilon d_M(i, j))e^{-\varepsilon d_M(i,j)}$ which is proposed by in [11] to get the probability that show how possible a location can be submitted to the server. Finally, we obtain the candidate location set and the probability of each cell to be selected as the perturbed location.

After obtaining the candidate location set from the perturbation section, we try to model the tradeoff between the minimum of the trajectory privacy leakage and the maximum of the server's utility. We notice that the distance between a perturbed location and the real location determines server's utility and participant's trajectory privacy leakage. Therefore, we try to find the optimal distances in active area A, i.e., Euclidean distance, vertical distance, horizontal distance and Manhattan distance, to achieve the best tradeoff between the utility and privacy. We decompose these goals into two sub-problems: utility (UT) sub-problem and privacy leakage (PL) sub-problem.

In the UT sub-problem, its objective is to minimize the expected Manhattan distance between perturbed locations and real locations. We use the Manhattan distance because it is more suitable than the Euclidean distance to represent the actual length of the trajectory. In this sub problem, $P(c_i|c_{i-1})$ denotes the probability that a participant moves to the current location c_i. The differential privacy mechanism $M(c'_k|c_k)$ is used to select the perturbed location. We solve these two sub-problems iteratively and interchangeably, i.e., we fix the PL sub-problem while solving the UT sub problem and vice versa.

UT subproblem:

$$\min_{M,A} \sum_{k=1}^{n} \frac{\sum_{c_k, c'_k \in A} P(c_k|c_{k-1})M(c'_k|c_k)d_M(c_k, c'_k)}{\sum_{c_k, c'_k \in A} P(c_k|c_{k-1})M(c'_k|c_k)}$$

$$s.t.$$

$$\varepsilon_{\min} \leq \varepsilon_i \leq \varepsilon_{\max}$$

$$M(c'|c_1) \leq \exp(\varepsilon_i d(c_1, c_2))M(c'|c_2) \quad c_1, c_2, c' \in A$$

$$d_E(c_1, c_2) = \sqrt{(x_1 - x_2)^2 + (y_1 - y_2)^2}$$

$$d_\perp(c_1, c_2) = d_E(c_1, c_2)\cos(\alpha)$$

$$d_\parallel(c_1, c_2) = d_E(c_1, c_2)\sin(\alpha)$$

$$d_M(c_i, c'_i) = |d_\perp(c_i, c'_i)| + |d_\parallel(c_i, c'_i)|$$

$$P(c_i|c_{i-1}) \geq 0 \qquad c_{i-1}, c_i \in A$$

$$\sum_{c_{i-1}, c_i \in A} P(c_i|c_{i-1}) = 1 \qquad c_{i-1}, c_i \in A$$

$$M(c'_i|c_i) \geq 0 \qquad c_i, c'_i \in A$$
$$\sum_{c'_i \in A} M(c'_i|c_i) = 1 \qquad c_i, c'_i \in A$$

For the PL sub-problem, the objective is to minimize the privacy leakage of the participant. The similarity function $S(l_k, l'_k)$ is used to measure the correlation degree of two locations. We can regard $S(l_k, l'_k)$ as the probability that l'_k is submitted to the server. Note that only using the similarity to choose the perturbation location is not enough, so we aim to find the location that minimizes the expected sum of vertical distance and horizontal distance.

PL sub problem:

$$\min_A \sum_{k=1}^{n} \frac{1}{S(c_k, c'_k)\left(d_\perp(c_k, c'_k) + d_\parallel(c_k, c'_k)\right)}$$
$$s.t.$$
$$S(c_k, c'_k) = \exp\left(-d_E^2(c_k, c'_k)/2\sigma^2\right)$$
$$d_\perp(c_1, c_2) = d_E(c_1, c_2)\cos(\alpha)$$
$$d_\parallel(c_1, c_2) = d_E(c_1, c_2)\sin(\alpha)$$

Through the above modeling, we intuitively choose Benders Decomposition (BD) [12] to solve our problem, and the details are shown in Algorithm 3.3. The user iteratively updates the perturbed location based on the optimization goal of two sub-problems until convergence criterion is satisfied, which indicate that both the change of server's utility and trajectory privacy leakage are less than 10^{-3}.

Algorithm 3.3. Perturbed Location Selection

Input:
Candidate location set.
Output:
Perturbed location.
1: Get the optimal solution l' of the UT sub-problem while fixing the PL sub-problem, obtain the sum O_1 of servers utility and trajectory privacy leakage according to l';
2: Get the optimal solution l'' of the PL sub-problem while fixing the UT sub-problem, obtain the sum O_2 of servers utility and trajectory privacy leakage according to l'';
3: Calculate $\Delta = |O_1 - O_2|$, if $\Delta < 10^{-3}$, randomly choose l' or l'' to submit to the server. Otherwise, repeat steps 1-2.

3.5 Evaluations

Now we present performance evaluations of TPA and RDCTP using the Shanghai taxi trajectory dataset. The whole dataset includes more than 5000 taxis' trajectories in a month. To simplify the scale of experiments, we only extract the trajectories which locate at the central area of the city, which is marked by a red rectangle in Fig. 3.4. This area is 12 km in North-South direction and 16 km in East-West direction. We divide this area into 120 $\times$ 160 cells, where each cell is a square of 100 meters. The true GPS data consist of some important information, which has the format of <. . ., *taxied, longitude, latitude, timestamp,* . . .>. We map the longitude and latitude of each GPS data to the state coordinate, which is defined in Sect. 3.3, and assign each data an index to record its order to build a trajectory as the ground truth. The final format of our data for this work is <*trajectory id, location index, state id, timestamp*>. In order to train the adversary, we randomly take 60% of the data to build the priori distribution and the mapping matrix $\boldsymbol{M}$. The rest 40% of the dataset are used for the performance evaluation.

Before presenting the statistics, we show how the privacy parameter ε can affect the performance of TPA. We randomly choose a trajectory and use TPA to perturb it with different ε. Figure 3.5 shows 4 groups of results. The blue solid line is the true trajectory and perturbed trajectories are shown in red dash lines. We can see that when $\varepsilon = 0.5$, the perturbed trajectories can hardly tell the ground truth. When ε is larger than 1, the trend of true and perturbed trajectories begins to agree. Because of the characteristics of Laplacian distribution, the probability that a random number

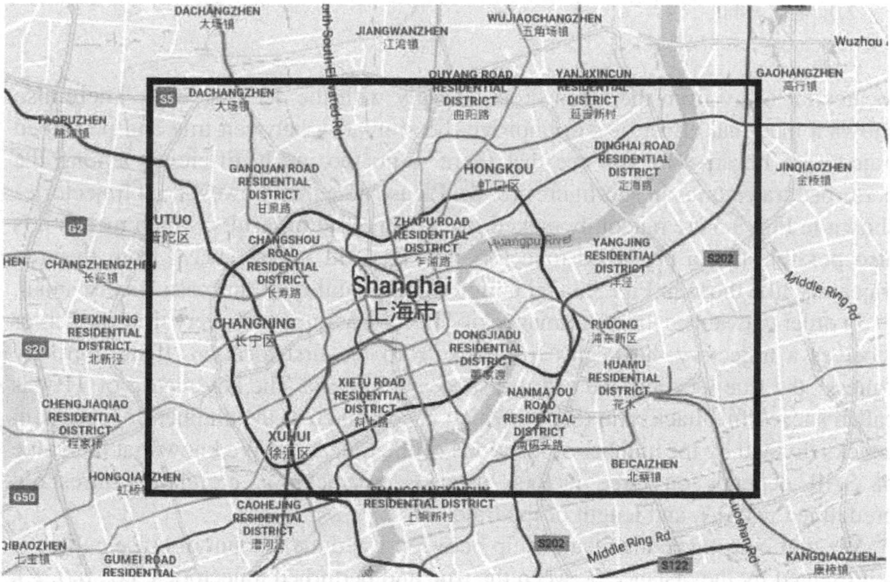
DACHANGZHEN
JIANGWANZHEN
WUJIAOCHANGZHEN
Wuzhou
GAOHANGZHEN
HONGKOU
Shanghai
上海市
CHANGNING
XUHUI
PUDONG
JINQIAOZHEN
BEIXINJING RESIDENTIAL DISTRICT
CHENGJIAQIAO RESIDENTIAL DISTRICT
BEICAIZHEN
KANGQIAOZHEN
Middle Ring Rd

Fig. 3.4 The examined Shanghai metropolitan area

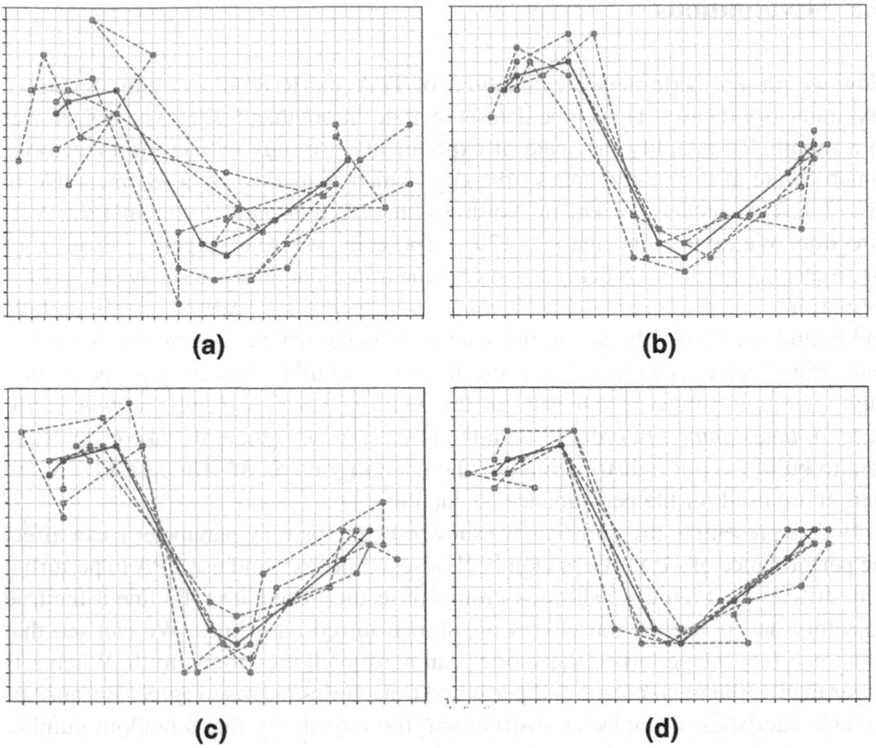
(a)
(b)
(c)
(d)

Fig. 3.5 The illustration of perturbed trajectory with different ε. (**a**) $\varepsilon = 0.5$, (**b**) $\varepsilon = 1.0$, (**c**) $\varepsilon = 2.0$, (**d**) $\varepsilon = 4.0$

locates far away from the mean drops sharply with the decrease of ε. There also shows a marginal effect of ε to improve the similarity between true and perturbed trajectories because we can see that there is no too much difference among the perturbed trajectories in sub-figures (b–d). Please note that, however, all trajectories shown in Fig. 3.5 are randomly picked from a large set of results, which means it is also possible that a perturbed trajectory in Fig. 3.5d is "far away" from the true trajectory like those in Fig. 3.5a, although the probability of this case is very small.

In order to evaluate the performance of TPA, we assume that every time the server receives a trajectory, he begins to attack from the first location. If he correctly "guess" the true location, we call it a successful attack. The key metric of TPA is called successful attack ratio (SAR), which is computed as the number of successful attacks divided by the number of locations of the trajectory. We know that the value of SAR depends on three factors: privacy parameter ε, correlation threshold predefined in TPA, and length of the trajectory.

We first investigate the relationship between ε and the amount of noise, which is represented by the distance between the true and perturbed trajectories, i.e., $d_t(t, t')$. As is shown in Fig. 3.6, due to the nature of Laplace distribution, $d_t(t, t')$ always

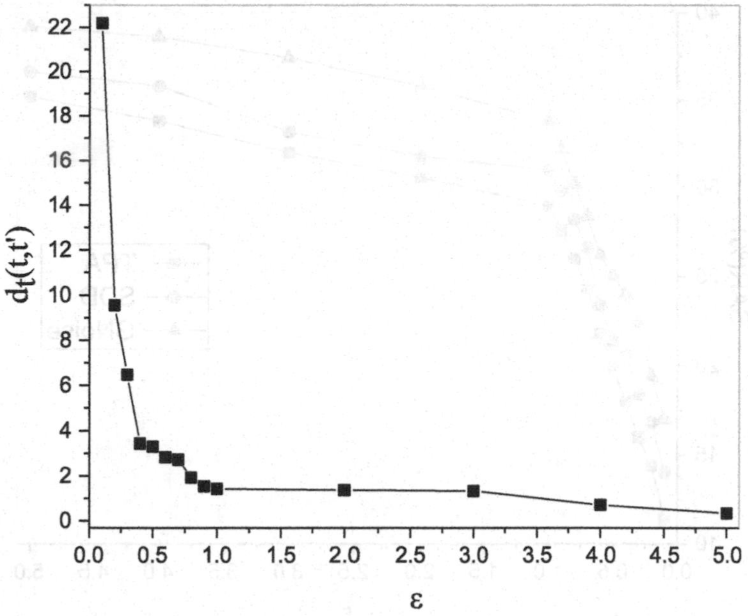
22
20
18
16
14
12
10
8
6
4
2
0
$d_t(t,t')$
0.0 0.5 1.0 1.5 2.0 2.5 3.0 3.5 4.0 4.5 5.0
ε

Fig. 3.6 The amount of noise with different setting of ε

decreases with the increment of ε. But the slope of the curve appears two stages. When ε is less than 0.5, the $d_t(t, t')$ is very large and decrease sharply. The knee point appears near $\varepsilon = 0{:}5$. When ε exceeds 1.0, the $d_t(t, t')$ decreases slowly.

Next, we examine the impact of ε and trajectory length on the SAR. We randomly take 10 trajectories, whose length are all 5, from the testing dataset. For each trajectory, we run 10 times and show the average results in Fig. 3.7. We also compare our method, TPA, with the other two methods, GNoise and SDD, proposed by Jiang [6].

We can see that SAR always increases with ε. But the increment of SAR appears two stages, which divides at $\varepsilon = 1$. The reason has been explained by Fig. 3.7. We can also see that TPA outperforms GNoise and SDD because TPA converts the privacy parameter ε to the distance between the true and perturbed trajectories, which make the perturbed location relatively far away from the true location. So, the server is hard to successfully infer the true data. Meanwhile, we don't worry the appearances of too much outlier locations in the perturbed trajectory because we use correlation to ensure that the trend of the perturbed trajectory is similar to the true one.

Then, we investigate the impact of correlation threshold on SAR. Fixing the $\varepsilon = 0.5$, we tune the correlation from 0.9 to 1.0. The results shown in Fig. 3.8 say that SAR increases with larger correlation threshold. Since the correlation represents the utility of perturbed trajectories, this result also shows the trade-off between utility and privacy. We can also see that when the correlation reaches 1.0, which is its maximum value, the SAR is still below 30%. The reason is that although the

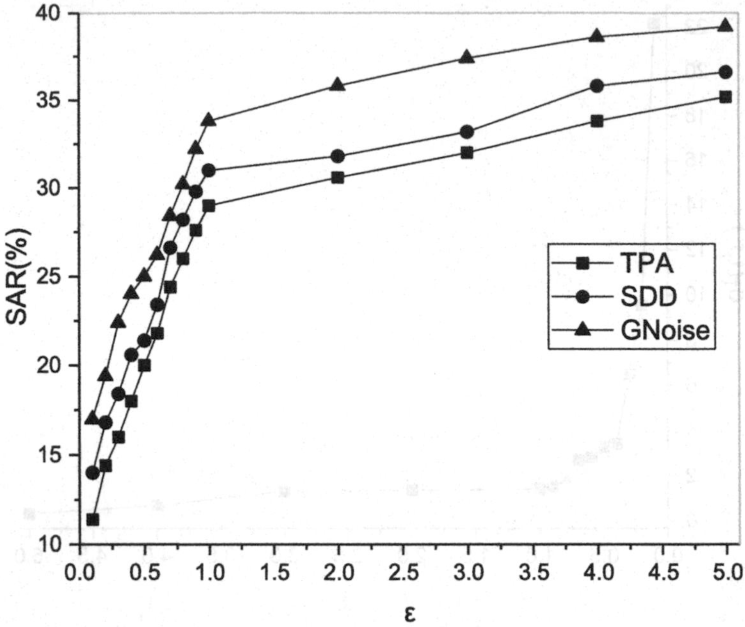
SAR(%)
40
35
30
25
20
15
10
0.0
0.5
1.0
1.5
2.0
2.5
3.0
3.5
4.0
4.5
5.0
ε
TPA
SDD
GNoise

Fig. 3.7 The impact of ε on SAR

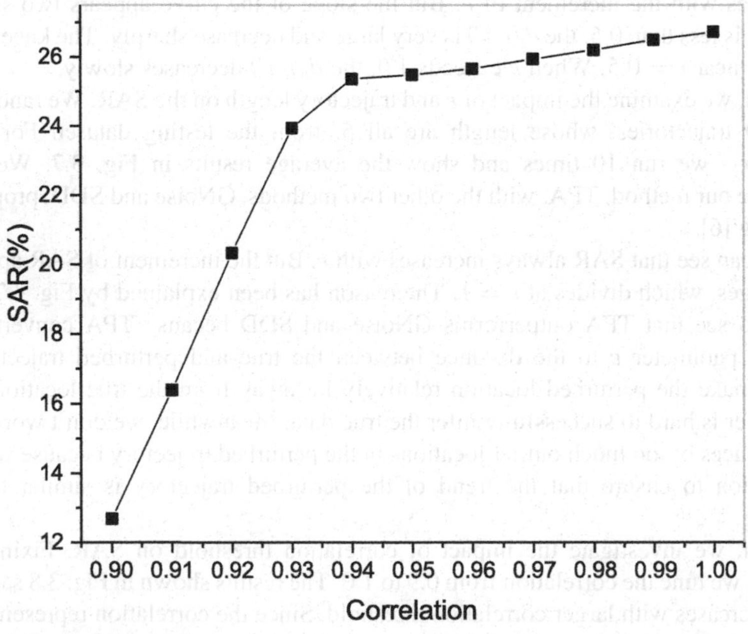
SAR(%)
26
24
22
20
18
16
14
12
0.90
0.91
0.92
0.93
0.94
0.95
0.96
0.97
0.98
0.99
1.00
Correlation

Fig. 3.8 The impact of correlation on SAR

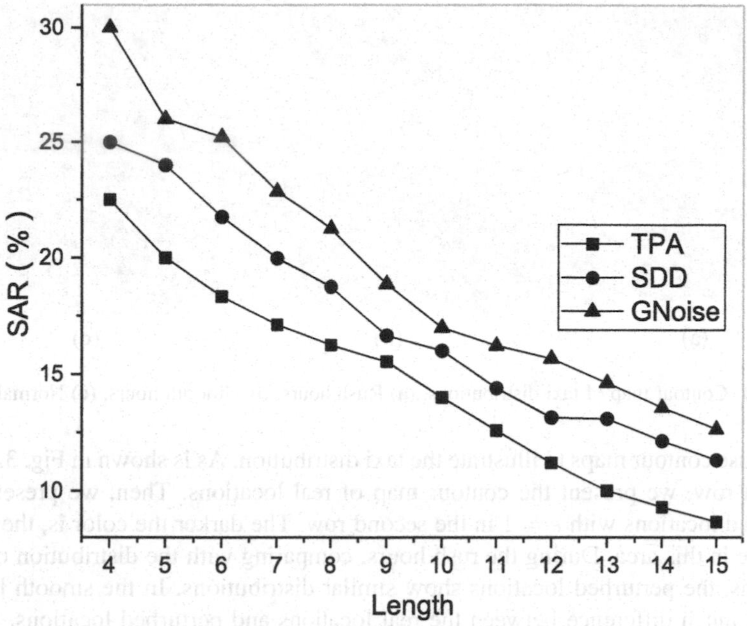
30
25
20
15
10
SAR (%)
4 5 6 7 8 9 10 11 12 13 14 15
Length
TPA
SDD
GNoise

Fig. 3.9 The impact of the length of a trajectory on SAR

perturbed trajectory has the same trend with the true one (correlation = 1.0), there is still large distance between them ($\varepsilon = 0.5$). It is difficult for the adversary to successfully guess a location which is far away from the truth.

The third parameter that could impact the SAR is the length of a trajectory. We fix the $\varepsilon = 0.5$ and correlation to be 0.95, the relationship between SAR and the length of trajectories is shown in Fig. 3.9. The results in this figure show that longer trajectories lead to lower SAR. The reason is that if the server fails to decode the true location, the error will feed back to mapping matrix which in turn affects the future attacks and make them more likely to fail. Figure 3.7 also shows that TPA still outperforms GNoise and SDD.

For the RDCTP algorithm, we use the Mean Absolute Error (MAE) and Mean Relative Error (MRE) as the utility metrics to evaluate the performance of GNoise, SDD and RDCTP. To illustrate how to compute MAE and MRE, let $t = \{c_1, \ldots, c_n\}$ denote the real locations series, and $r = \{r_1, \ldots, r_n\}$ denote the perturbed locations series. The MAE and MRE are defined as follows.

$$MAE(t, r) = 1/n\Sigma_{i=1}^{n} d_M(c_i, t_i)$$

$$MRE(t, r) = 1/n\Sigma_{i=1}^{n} d_M(c_i, t_i)/MAE(t, r)$$

Before validating the utility of our mechanism, we perform a visual illustration. We divide the time of a day into three segments: smooth hours (00:00–7:00), rush hours (7:00–10:00 and 17:00–20:00), normal hours (10:00–12:00 and 13:00–17:00).

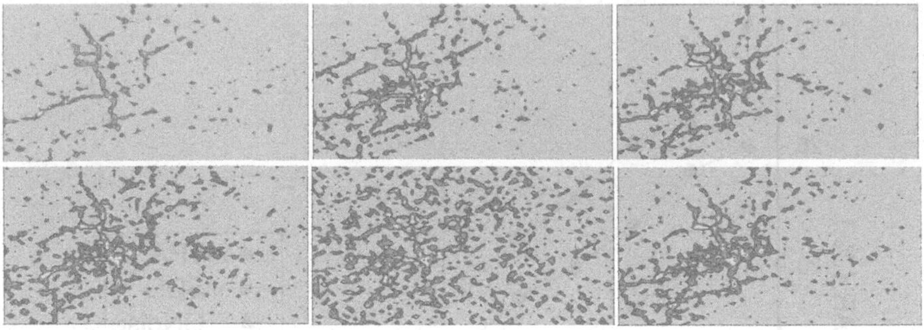
(a)
(b)
(c)

Fig. 3.10 Contour map of taxi distributions. (**a**) Rush hours, (**b**) Smooth hours, (**c**) Normal hours

We use contour maps to illustrate the taxi distribution. As is shown in Fig. 3.10, at the first row, we present the contour map of real locations. Then, we present the perturbed locations with $\varepsilon = 1$ in the second row. The darker the color is, the more taxis are in this area. During the rush hours, comparing with the distribution of real locations, the perturbed locations show similar distributions. In the smooth hours, there is much difference between the real locations and perturbed locations. However, the two figures almost overlap in some places which have the same density. During the normal hours, taxis are more evenly distributed and the two figures are almost identical. We can conclude that the perturbed locations can represent the true locations. Although no one-to-one matching exists between these locations, it is enough to illustrate that the taxi distribution is not changed significantly even when their reallocations are perturbed by RDCTP.

Then, we illustrate the impact of ε and w on the location utility. We randomly take 100 trajectories, with length varying from 5 to 15, from the testing dataset. For each trajectory, we run 100 times and the average results are followed.

First, we investigate the effect of privacy budget ε on the utility. Set $w = 5$ and the length of trajectory is 5. Figure 3.11 shows the utility comparison among GNoise, SDD and RDCTP when ε varies from 0.5 to 5. We can see that both MAE and MRE of GNoise and SDD decrease as ε increases, since the added noise decreases gradually. The utility of RDCTP is better than GNoise and SDD even when ε is small. This is because RDCTP can allocate the privacy budget adaptively.

Then, from technical insights, we use the Manhattan distance between participant's real trajectory and perturbed trajectory to measure the effectiveness of the three algorithms. Assume that the real trajectory of the participant is t, and the length of the trajectory is w. The perturbed trajectories generated by GNoise, SDD and RDCTP algorithms are t_1, t_2, t_3, and the noises added to each location of participant's real trajectory are r_1, r_2, …, r_n. In the GNoise and SDD algorithms, the privacy budget allocated to each location is ε/w, and the noises added to each location of participant's real trajectory are the same. The difference is that the SDD algorithm considers the participant's moving direction when adding noise. The Manhattan distance between participant's real trajectory and perturbed trajectory is $d(t, t_1) \approx d(t,$

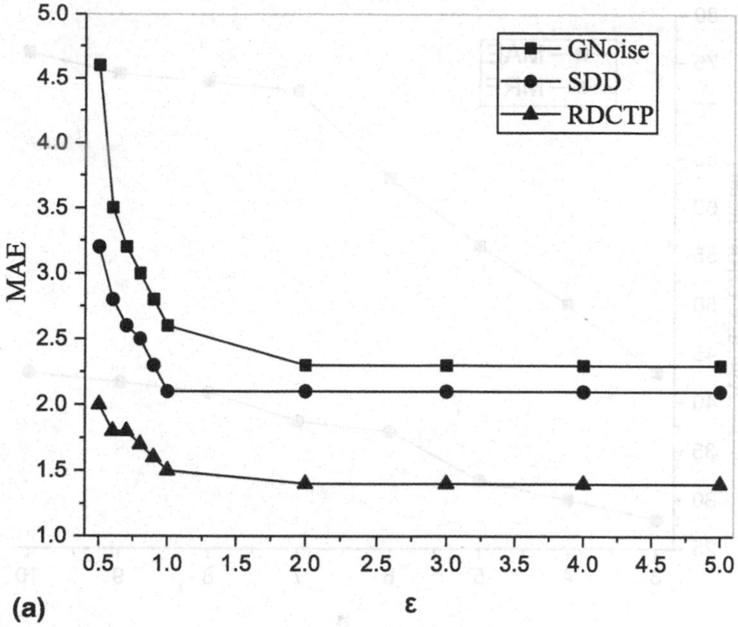
GNoise
SDD
RDCTP
MAE
5.0
4.5
4.0
3.5
3.0
2.5
2.0
1.5
1.0
0.5
1.0
1.5
2.0
2.5
3.0
3.5
4.0
4.5
5.0
(a)
ε

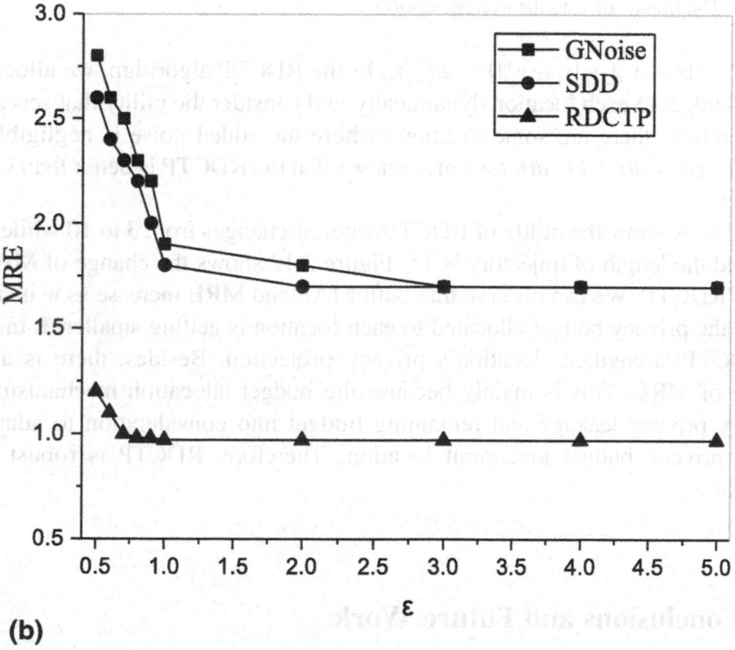
GNoise
SDD
RDCTP
MRE
3.0
2.5
2.0
1.5
1.0
0.5
0.5
1.0
1.5
2.0
2.5
3.0
3.5
4.0
4.5
5.0
ε
(b)

Fig. 3.11 Error comparison when ε changes ($w = 5$). **(a)** MAE and **(b)** MRE

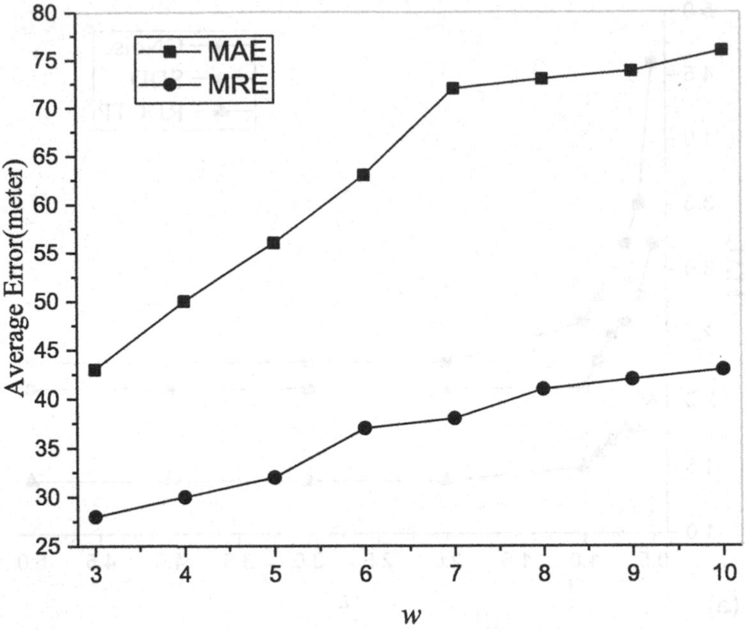
MAE
MRE
Average Error(meter)
80
75
70
65
60
55
50
45
40
35
30
25
3
4
5
6
7
8
9
10
w

Fig. 3.12 The impact of w to the average error

$t_2) = \Sigma_{i=1}^{w} (|x_i - x'_i| + |y_i - y'_i|) = \Sigma_{i=1}^{w} r_i$. In the RDCTP algorithm, we allocate the privacy budget to each location dynamically and consider the utility that servers can get. Therefore, there are some locations where the added noise is negligible. We obtain $d(t, t_3) \leq d(t, t_1) \approx d(t, t_2)$ which shows that the RDCTP is better than GNoise and SDD.

Finally, we show the utility of RDCTP when w changes from 3 to 10 while fixing $\varepsilon = 1$ and the length of trajectory is 15. Figure 3.12 shows the change of MAE and MRE of RDCTP. We can observe that both MAE and MRE increase as w increases, because the privacy budget allocated to each location is getting smaller. It indicates that RDCTP strengthens location's privacy protection. Besides, there is a small increase of MRE. This is mainly because the budget allocation mechanism takes trajectory privacy leakage and remaining budget into consideration to adaptively allocate privacy budget to current location. Therefore, RDCTP is robust to the change of w.

3.6 Conclusions and Future Work

In this chapter, we discuss the trajectory privacy protection in a crowd-sensing scenario which is a typical application of smart city. We have considered both offline and real-time scenarios. In offline scenarios, by assuming that a participant can submit his trajectory in a bundle instead of submitting locations one by one with the evolution of time, we introduce a mechanism, TPA, to protect the trajectory as a whole. The TPS mechanism first computes the total amount of noise to satisfy differential privacy. Then it randomly allocates the noise to each coordinate of the trajectory. In order to improve the utility of the perturbed trajectory, the TPA also considers the correlations between each pair of perturbed and true trajectories. We investigate three factors which may have impact to the successful attack ratio (SAR) and compare TPA with other two methods, i.e., GNoise and SDD. The performance evaluation is based on the Shanghai taxi datasets. Simulation results prove the effectiveness of TPA. The advantages of TPA over GNoise and SDD are also validated through these simulations.

Despite of the above off line scenario, we also introduce RDCTP, a real-time (online) MCS data collection mechanism, which achieves w-event ε-differential privacy for the crowd-sensing participants. Each participant first allocates the privacy budget based on the trajectory privacy leakage in previous $w - 1$ successive locations. Then, he perturbs his location according to the allocated privacy budget and obtains the candidate location set. To guarantee the utility of server, we built an optimization model to select the best location from the candidate set which will be uploaded to server. The experimental results on Shanghai taxi dataset show that RDCTP also achieves high utility.

References

1. Kang X, Liu L, Ma H. Enhance the quality of crowdsensing for fine-grained urban environment monitoring via data correlation. Sensors. 2017;17(1):88, 1–17.
2. Shao L, Wang C, Li Z, Jiang C. Traffic condition estimation using vehicular crowdsensing data. In: IEEE 34th international performance computing and communications conference (IPCCC), 2015. p. 1–8.
3. Chen C, Zhang D, Ma X, Guo B, Wang L, Wang Y, Sha E. Crowddeliver: planning city-wide package delivery paths leveraging the crowds of taxis. IEEE Trans Intell Transp Syst. 2017;18 (6):1478–96.
4. Xiao Y, Xiong L. Protecting locations with differential privacy under temporal correlations. In: ACM Sigsac conference on computer and communications security (CCS), 2015. p. 1298–1309.
5. Ou L, Qin Z, Liu Y, Yin H, Hu Y, Chen H. Multi-user location correlation protection with differential privacy. In: IEEE 22nd international conference on parallel and distributed systems (ICPADS), 2016. p. 422–9.
6. Jiang K, Shao D, Bressan S, Kister T, Tan K-L. Publishing trajectories with differential privacy guarantees. In: International conference on scientific and statistical database management (SSDBM), 2013. p. 12, 1–12.
7. Han Q, Lu D, Zhang K, Du X, Guizani M. Lclean: a plausible approach to individual trajectory data sanitization. IEEE Access. 2018;6:30110–6.
8. Dwork C. Differential privacy. In: ICALP, 2006. p. 112.
9. Andrés ME, Bordenabe NE, Chatzikokolakis K, Palamidessi C. Geo-indistinguishability: differential privacy for location-based systems. In: ACM Sigsac conference on computer and communications security (CCS), 2013. p. 901–14.

10. Press WH, Teukolsky SA, Vetterling WT, Flannery BP. Numerical recipes: the art of scientific computing. 3rd ed. Cambridge: Cambridge University Press; 2007.
11. Kellaris G, Papadopoulos S, Xiao X, Papadias D. Differentially private event sequences over infinite streams. Proc Very Large Data Bases Endowment. 2014;7(12):1155–66.
12. Geoffrion AM. Generalized benders decomposition. J Optim Theory Appl. 1972;10(4):237–60.

Part II
Enabling Smart Urban Services: Drivers

Chapter 4
Hunting or Waiting: Earning More by Understanding Taxi Service Strategies

4.1 Introduction

As one of the typical representatives of digital footprints, the easily available large-scale taxi trajectory data offers a unique window to understand taxi drivers' behaviors in various contexts and exploit the underlying crowd intelligence [1–5]. Taxi drivers' behaviors such as operational and service behaviors are implicitly contained in the taxi GPS trajectory data. For instance, the GPS trajectory data from the drop-off to the next pick-up can tell where they hunt for new passengers after passenger arrivals (i.e., passenger-searching strategies); the GPS trajectory data carrying passengers can tell which routes they choose to deliver passengers (i.e., passenger-delivering strategies). These service behaviors vary from one driver to another, depending on one's service strategy in a given context. For example, after sending passengers to their destinations, some taxi drivers prefer to go back to their familiar and popular areas to search for new passengers while some drivers prefer to wait near the drop-offs for new passengers. Different service strategies adopted by taxi drivers play a direct influence on income. Efficient service strategies can lead to many benefits, such as the increase of income, the efficiency improvement on the taxi service system and market, and the enhancement of the experience of passengers. Moreover, good or efficient service strategies can help reduce carbon emission by either choosing traffic-light routes or increase the occupied/vacant distance ratio. By linking the service strategies and the resulted income of taxi drivers, we can identify efficient and inefficient service strategies, and further improve the service performance of low-income taxi drivers (i.e., earn more).

Part of this chapter is based on a previous work: D. Zhang, L. Sun, L. Li, C. Chen, G. Pan, S. Li, Z. Wu, "Understanding Taxi Service Strategies from Taxi GPS Traces," in IEEE Transactions on Intelligent Transportation Systems, vol. 16, no. 1, pp. 123–135, Feb. 2015, doi: https://doi.org/10.1109/TITS.2014.2328231.

C. Chen et al., *Enabling Smart Urban Services with GPS Trajectory Data*,
https://doi.org/10.1007/978-981-16-0178-1_4

Existing research work on improving the taxi service performance is mainly to build models to explicitly take into account the following factors, including passenger's demands along the search roads, traffic conditions, competition from other drivers [6–10]. There is no analytical model developed to systematically answer which service strategies are efficient/inefficient for both passenger hunting and delivering. In contrast to considering influencing factors separately, we argue that taxi service strategies are the decisions made by taxi drivers combining all the influencing factors. Given that the ultimate goal of taxi drivers is to *earn more* through their operations, we investigate the correlation between each service strategy and the resulting revenue in a specific context and uncover which strategy is beneficial or harmful to the revenue, leveraging the GPS trajectories of a large number of taxis.

Based on the existing work, we group taxi service strategies from three perspectives, detailed as follows.

- **Passenger-searching strategies**. They are about strategies that taxi drivers adopted when finding new passengers. In more detail, a taxi driver may choose to wait at a popular location nearby or go back to a familiar area far away to search for new passengers, according to the experience.
- **Passenger-delivery strategies**. They are about strategies that taxi drivers adopted when sending passengers to their destinations. A taxi driver may choose his own "best" route to deliver passengers, considering traffic conditions, trip fares, and so on.
- **Service area preference**. Given the time context, taxi drivers may prefer to serve in some city regions, considering the traffic, passenger demands, and their familiarity comprehensively.

Before identifying efficient/inefficient taxi service strategies from the large-scale taxi GPS trajectory data, an important task is to link the service strategies to the resulted revenue of taxi drivers, which mainly needs to deal with the following issues:

- **GPS trajectory data separation**. In our dataset, a taxi is commonly shared and operated by two drivers. Taxi drivers take shift handover twice every day (i.e., once in the morning and once in the afternoon), but the exact shift handover time changes from day to day and is not indicated in the GPS records. Thus, it needs to separate the trajectory data to the correct taxi driver by detecting shift handover events.
- **Performance measurement**. The taxi driver may not perform consistently across all time slots. This issue concerns the selection of time granularity. Specifically, if a driver's operation performance is consistent on a daily basis, we simply use the average daily revenue as the performance indicator; otherwise, we use the average revenue at a much finer time granularity as the performance indicator.
- **Taxi service strategies extraction and modeling**. It is usually not easy to describe and represent taxi service strategies, since they consist of a sequence of decision makings and actions at different times and locations. For example, for passenger-searching strategies, a taxi driver may make a sequence of decisions. They just make the new decision once the previous one is ineffective. Here, we

simply investigate the first intention of the taxi driver and use it to represent the passenger-searching service strategy as it corresponds to the very first strategy that the taxi driver takes.

With the above-mentioned challenges, we mainly make the following contributions in this chapter.

- To correctly correspond the trajectory data to a taxi driver, we propose a simple algorithm to detect the shift handover events, i.e., when and where taxi drivers take shift handover. We also propose an algorithm to identify the initial intended location (i.e., the very first service strategy that the taxi driver takes) of taxi drivers after each drop-off event.
- We investigate the taxi service strategies in a complete view, from passenger-searching strategies and passenger-delivery strategies to service-region preference. Particularly, according to the moving speed and distance, we propose three concrete strategies for the passenger searching, including *hunting locally*, *waiting locally*, and *going distant*.
- We discover that most of taxi drivers do not perform consistently on a daily basis (i.e., across all time slots in a day). Thus, the correlation between the service strategies and drivers' performance in terms of revenue should be investigated at the granularity of each time slot.
- We build a feature matrix to represent the taxi service strategies. On top of the feature matrix, we compute the correlation between different service strategies and their corresponding revenue to discover the efficient and inefficient service strategies. To demonstrate the characterizing capacity of our proposed approach, we apply SVR to predict the revenue of taxi drivers by inputting historical strategies and achieve a prediction residual of less than 2.35 RMB/h.

4.2 Related Work

The work on improving taxi drivers' performance is relevant which can be broadly grouped into the following three categories.

The first category of work focus on recommending popular pick-up areas or predicting the taxi demands for taxi drivers [6–9, 11]. The pick-up locations can be easily inferred from the taxi GPS trajectory data according to passenger status from vacant to occupied. Based on the historical trajectory data, with some clustering algorithms, popular pick-up areas along the roads at different time periods can be easily obtained. Such clusters are commonly recommended to taxi drivers to pick potential new passengers at different times. For instance [12], analyzed the pickup patterns of tax drivers in Jeju, Korea, and recommended the popular clusters to vacant taxis to reduce the idling time of taxis. Similarly, in [6], a collection of clustering algorithms including K-means, agglomerative hierarchical clustering, and DBSCAN, together with some additional context such as time, location, and weather, are used to cluster the gathered demand requests into hotspots, and the top-ranked clusters are used to recommend to taxi drivers. Based on the historical records, going a step further [8], predicted the number of pickups in different hotspots which can be used to guide the vacant taxi drivers.

Unlike suggesting hotspots globally, the second category of the research work also considers other practical influencing factors to provides optimized passenger-searching areas and routes. The considered factors include human mobility patterns, passenger-searching possibilities, potential trip lengths, and so on. To name a few [9], checked the surrounding areas only. The knowledge of passenger's mobility patterns and taxi drivers' pickup/drop-off behaviors inferred from taxi GPS trajectory data is used to measure the profitability of each area in terms of fare gains of all occupied trips originated from that area, the number of trips, and the cost from the current location to that area. Based on the historical trajectory data again, in [10], the probability of searching a passenger along a route is computed and used to recommend drivers the profitable locations and routes.

The third category of the research work is about extracting effective taxi service guidelines, instead of suggesting explicit guidance on profitable areas or routes for hunting passengers. For instance [11], concluded that it is normally better to go to adjacent locations in urban areas while it is better to distant locations in suburban areas. In [13], authors uncovered that high-income taxi drivers not only had good skills in choosing the right areas to service around the city but also were good at selecting light-traffic delivery routes when sending passengers, based on the taxi trajectory data collected from the city of Shenzhen, China. Unlike the first two categories of work that aims to provide concrete operations, e.g., go to the recommended areas or routes to hunt for new passengers, the above-mentioned work [11, 13] provided strategic service guidelines, which is closer to our current work in this chapter. Our work still differs, since [11] did not consider the passenger-delivery and area-preference strategies and [13] did not differentiate taxi GPS trajectory data for the paired drivers. Moreover, neither [11] nor [13] built the model to characterize the relationship between taxi service strategies and the corresponding revenue. There was also some work [14, 15] using survey information from taxi drivers to recommend promising waiting/cruising locations.

Our work differs from all previous work as follows. Firstly, we identify a set of taxi service strategies from three perspectives. Secondly, we extract the efficient and inefficient taxi service strategies at different time and location contexts by conducting a correlation analysis study. Finally, to verify the effectiveness of extracted service strategies, we built a model based on SVR to predict the corresponding revenue using the top-ranked service strategies.

4.3 Empirical Study

In this section, we first introduce the taxi GPS trajectory collection method for our study. The large-scale taxi GPS dataset contains about 7600 taxis' one-year GPS trajectories in Hangzhou, a metropolis in China, from April 2009 to March 2010. A

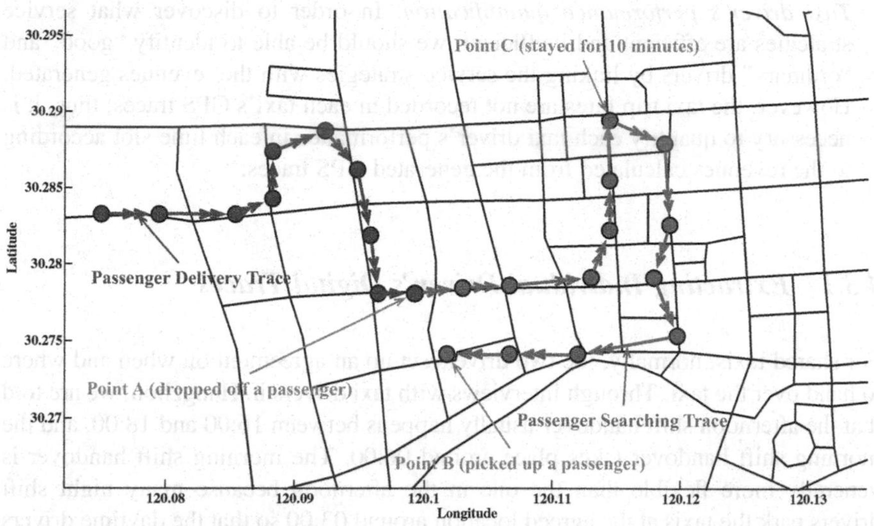
Point C (stayed for 10 minutes)
Passenger Delivery Trace
Point A (dropped off a passenger)
Passenger Searching Trace
Point B (picked up a passenger)
Latitude
Longitude
30.295
30.29
30.285
30.28
30.275
30.27
120.08
120.09
120.1
120.11
120.12
120.13

Fig. 4.1 Visualization of a complete taxi service cycle. The markers denote the sampling points. The segment in blue denotes passenger-delivery process (occupied). The segment in red denotes passenger-searching process (vacant)

GPS device that obtains the real-time taxi information including its longitude/latitude, the time stamp, the passenger status (i.e., "occupied" or "vacant"), the driving speed, and orientation was installed at each taxi. It uploads the data to a cloud server via a telecommunication network at a sampling rate of once per minute. The GPS samples frequency of the device is about one sample per minute. After filtering out suspicious taxi records which caused by device error or network failure, and we obtain 6863 taxis to do the research in the end.

A taxi's moving trajectory over a time interval can be depicted by connecting the GPS points on a 2D plane. For example, Fig. 4.1 shows one taxi's trajectory for a complete passenger-searching and passenger-delivery cycle, where red and blue lines correspond to the taxi's vacant and occupied statuses, respectively.

Generally speaking, when the taxi is at passenger-delivery stage, the status of the taxi is occupied; while the taxi is at passenger-searching stage, the status of the taxi is usually is vacant. The change from red to blue corresponds to a passenger pickup event, and the change from blue to red indicates a drop-off event.

Two data preprocessing tasks need to be done before analyzing the taxi driver's service strategies.

- *Individual taxi driver GPS trace extraction*: Almost all the taxis in Hangzhou are served by two drivers, i.e., one works in daytime and the other in nighttime. As the shift status of a taxi is not recorded in the taxi GPS traces, it is necessary to detect when the shift handover takes place in order to separate the taxi GPS traces of each driver. Only with each taxi driver's GPS traces extracted, individual drivers' service strategies and revenues in different time slots can be known.

- *Taxi driver's performance quantification*: In order to discover what service strategies are efficient and inefficient, we should be able to identify "good" and "ordinary" drivers by linking the service strategies with the revenues generated. However, the taxi trip fares are not recorded in each taxi's GPS traces; thus, it is necessary to quantify each taxi driver's performance in each time slot according to the revenues calculated from the generated GPS traces.

4.3.1 Extracting Individual Driver's Digital Traces

For shared taxis, normally, the two drivers set up an agreement on when and where to hand over the taxi. Through interviews with taxi drivers in Hangzhou, we are told that the afternoon shift handover usually happens between 16:00 and 18:00, and the morning shift handover takes place around 06:00. The morning shift handover is generally more flexible than the one in the afternoon because many night shift drivers park the taxis at the agreed location around 03:00 so that the daytime drivers could take the taxis early if they wish. However, due to issues such as traffic, delay of last passenger deliveries, or personal reasons, it is hard to strictly obey the shift handover agreement every day, particularly for the afternoon shift handover. Thus, the exact shift handover time detection becomes a nontrivial task in order to extract individual taxi driver's GPS traces.

4.3.1.1 Shift Handover Event Detection

The detection of shift handover events is based on two observations: (1) shared taxis go to the same agreed location during the rough shift handover time slot of the day with vacant status; (2) shared taxi park at the agreed shift handover location for some time (with vacant status) to ensure taxi handover.

The detailed shift handover event detection process is elaborated as follows. First, we identify the parking locations in a passenger-searching trajectory using the same method reported in [10]. We divide the city map into grid cells of 50 m × 50 m and split a day into half-overlapping 20-min time frames. In each time frame, we record the days that a taxi visits each grid cell. If the shift handover location falls near the edge of a cell, the shift handover events may scatter in the neighboring cells due to GPS error. To address this problem, in addition to counting the days a taxi visits each cell, we also count the days it visits the neighboring cells. Since we count the visit to neighboring cells, the same shift handover event may correspond to several candidate cells. We can combine these candidate cells into one as they might be produced from the same traces. In order to accommodate early or delayed shift handover events, we expand the time frame for shift handover event detection to 2 h. In real life, there is anomalous shift handover that the two drivers change the shift handover time and location for reasons such as person-al issues or temporarily blocked roads. Therefore, we may be unable to find the shift handover event in all days. In practice, if the number of days that the taxi visits a grid cell during the rough shift handover time slot is greater than 25 in a month, we mark the grid cell as the candidate shift handover location. Furthermore, if it is the only candidate within the 15:00–19:00 time slot, it is said to be the afternoon shift handover location; if it is the only candidate within the 03:00–8:00 time slot, it is the morning shift handover location. If there are more than one candidate cells, we consider it as an uncertain (failed) case.

With the shift handover event detected, the GPS traces of the two shared taxi drivers are naturally separated. For the anomalous shift handover aforementioned, we simply choose the vacant parking location in the digital trace that is closest to the middle of the detected shift handover time slot and separate the digital trace of that day at that point.

One possibility that may cause false shift handover event detection is when some drivers wait for passengers every day at certain popular locations (e.g., railway/bus stations and grand hotels) during a certain period of time. This appears to have a similar pattern as the shift handover event. However, we observe that the majority of these passenger waiting events are followed by an immediate "occupied" status (indicating that a passenger is picked up successfully), whereas the shift handover event seldom takes place at a popular place and the taxi usually has vacant status after the taxi leaves the shift handover location. In such a way, we can filter out most of the false shift handover events.

With the proposed algorithm, we successfully identified the shift handover events (including location and time) for 4773 taxis out of the 6863 taxis. The remaining 2090 taxis do not follow the fixed shift handover pattern, which may be due to the following two reasons: (1) the location and time to hand over the taxi vary from day to day based on two drivers' agreement; (2) there are some taxis served by only one driver. To exclude the error introduced by the shift handover event detection algorithm, we only keep the taxi traces from the taxis whose shift handover events were successfully identified (i.e., the 4773 taxis) for experiments. It is believed that the 9546 individual drivers' traces from 4773 taxis are enough to illustrate our proposed idea and show meaningful statistical results for discovering *efficient* and *inefficient* taxi service strategies.

4.3.1.2 Statistical Study of Taxi Shift Handover Location and Time

The geographical distributions of the morning and afternoon taxi shift handover locations are shown in Fig. 4.2a, b, respectively. It can be observed that most of the taxi shift handover events are not taking place in popular areas such as the downtown and railway stations. Another interesting observation is that many drivers choose to hand over the taxi near bus stops during afternoon shift handover time slot. This may be because the taxi drivers need to take buses for the taxi handover.

The time distributions of morning and afternoon taxi shift handover events are shown in Fig. 4.3a, b, respectively. The morning taxi handover time is almost evenly distributed between 4:00 and 7:20, whereas the afternoon shift handover events

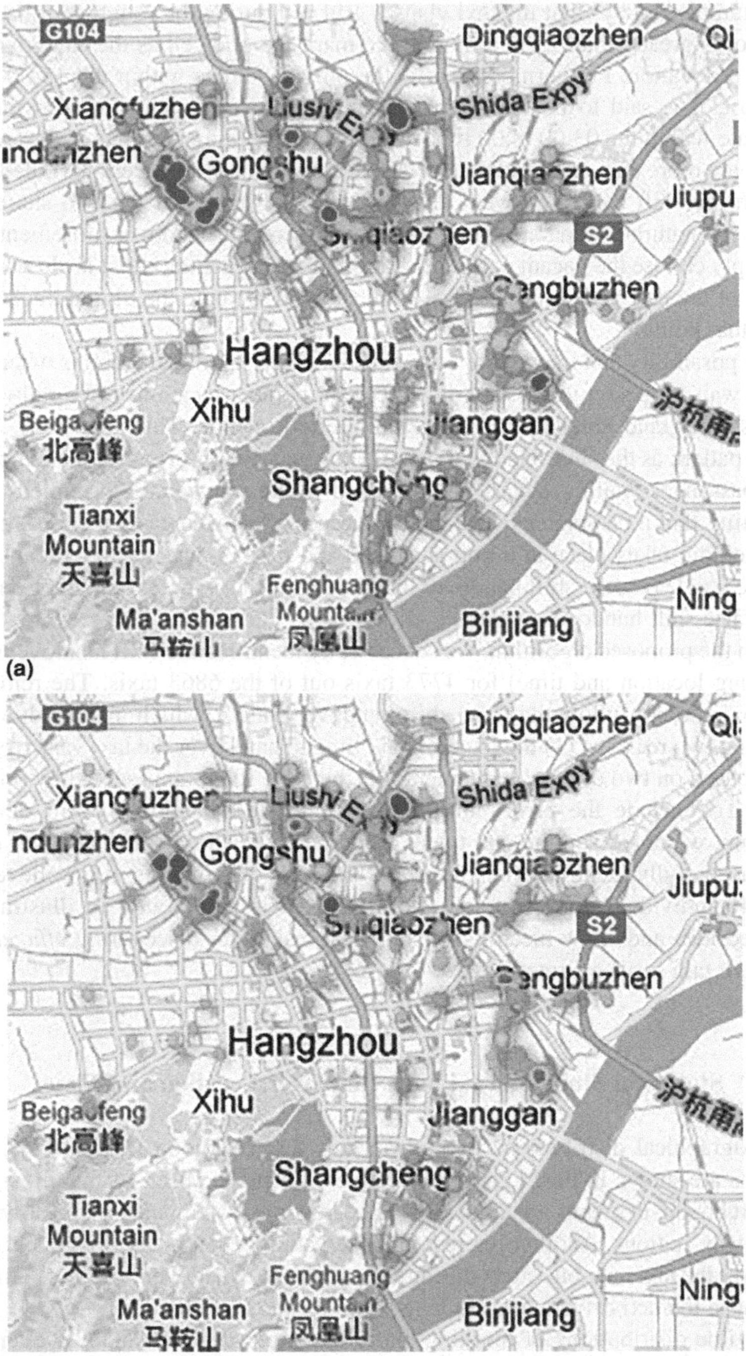
G104
Dingqiaozhen
Shida Expy
Jianqiaozhen
Jiupu
S2
Hangzhou
Xihu
Jianggan
Shangcheng
Tianxi
Mountain
Fenghuang
Mountain
凤凰山
Ma'anshan
马鞍山
Binjiang
北高峰

(a)

(b)

Fig. 4.2 Heat maps of shift handover locations. (**a**) Morning. (**b**) Afternoon

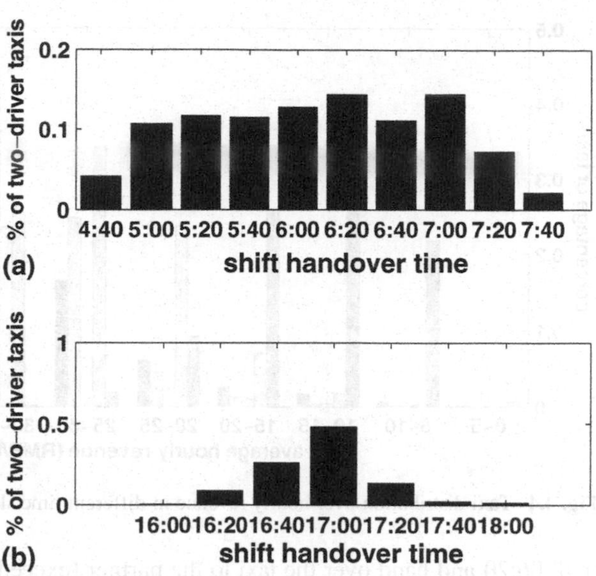
% of two–driver taxis
0.2
0.1
0
4:40 5:00 5:20 5:40 6:00 6:20 6:40 7:00 7:20 7:40
shift handover time
(a)
% of two–driver taxis
1
0.5
0
16:0016:2016:4017:0017:2017:4018:00
shift handover time
(b)

Fig. 4.3 Time distributions of taxi shift handover events. **(a)** Morning. **(b)** Afternoon

mainly take place between 16:40 and 17:20, which explains why people in Hangzhou find it difficult to find taxis around this time period.

4.3.2 *Analyzing Individual Driver's Performance*

Based on the separated digital traces for each individual driver, we can further extract the passenger-delivery traces of each driver according to the passenger status (i.e., "occupied" or "vacant"). We can then estimate the taxi revenue in each time slot by considering the accumulated passenger-delivery distance. By averaging the revenue in each time slot during a long time period (e.g., a month), we are able to measure the performance of a driver. In this section, we will investigate the performance of all the taxi drivers in different time slots and give an empirical study on the factors influencing the performance.

4.3.2.1 Taxi Performance Quantification

First, we split a day into five time slots: *late night* (00:00–05:59), *morning* (06:00–09:59), *noon* (10:00–13:59), *afternoon* (14:00–17:59), and *evening* (18:00–23:59). Since workdays have quite different taxi service patterns from weekends and holidays, in this chapter, we only use the taxi GPS traces in workdays to illustrate our ideas. The performance of a taxi is measured by hourly revenue within each time slot. Since taxi drivers may only work part of the time within a time slot, for example, one night shift driver might work until 4:00, another driver might work

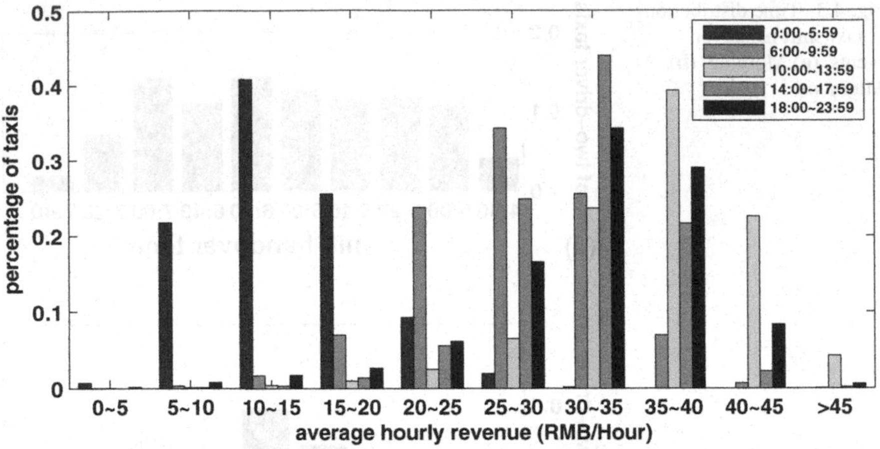
0:00~5:59
6:00~9:59
10:00~13:59
14:00~17:59
18:00~23:59
percentage of taxis
0.5
0.4
0.3
0.2
0.1
0
0~5
5~10
10~15
15~20
20~25
25~30
30~35
35~40
40~45
>45
average hourly revenue (RMB/Hour)

Fig. 4.4 Taxi distribution over hourly revenue in different time slots

until 06:20 and hand over the taxi to the partner (exceeding 20 min in the morning time slot), we only count the taxis that serve more than half of the time slot and discard those taxi GPS traces that serve less than half of the time slot.

The distribution of taxi drivers' performance in different time slots is shown in Fig. 4.4. We can observe that generally speaking, taxi drivers perform best in the *noon* time slot (around 30–45 RMB/h) and worst in *late-night* time slot (around 5–20 RMB/h); they also have better performance in the *evening* and *afternoon* time slots than the *morning* time slot. The distribution of taxi driver performance in each time slot roughly follows a normal distribution, which is compliant with the observation in [13].

Since we aim to investigate taxi service strategies in terms of drivers' performance, we first investigate the consistency of individual driver's performance over different time slots. People might think that good taxi drivers perform well in all the time slots and vice versa. Through our study, it is found that only 110 taxis performing best in the *late night* time slot are among the top 500 taxis performing well during the *nighttime* (containing both the *evening* and *late night* time slots), and only 67 taxis performing best in the daytime are among the top 500 taxis if we count the performance in the three time slots containing *morning*, *noon*, and *evening*. Our study shows that less than 22% of the taxi drivers consistently perform well in the nighttime and only 13.4% in the daytime. Thus, unlike the method proposed in [16], which chooses good/ordinary taxis according to the daily revenue, we investigate taxi drivers' performance in each time slot and expect to obtain more reasonable and accurate results.

For each time slot, we select the 500 top-performing taxis as good samples and 500 taxis in the middle range (around the mean of the normal distributions in Fig. 4.4 as ordinary taxis. The reason why we do not select the taxis in the bottom range is that many factors might cause taxis' extra low performance and these factors are irrelevant to taxi service strategies. By comparing good taxis with ordinary ones, we can discover some simple taxi service patterns that are closely related to revenue. In the following section, we will report a brief empirical study on two influencing factors.

4.3.2.2 Number of Passenger Delivery Trips

Intuitively, a larger number of passenger-delivery trips should give a higher overall revenue. We use the correlation coefficient between the number of passenger-delivery trips P and revenue R

$$\mathrm{corr}(P,R) = \frac{\sum_{i=1}^{n}(p_i - \bar{p})(r_i - \bar{r})}{\sqrt{\sum_{i=1}^{n}(p_i - \bar{p})^2}\sqrt{\sum_{i=1}^{n}(r_i - \bar{r})^2}} \tag{4.1}$$

to validate this intuition, where p_i and r_i are the total number of passenger-delivery trips and the revenue of the taxi driver i, respectively, and n denotes the number of taxi drivers. The results for all the time slots are shown in Table 4.1. The correlation coefficient in most time slots is close to 1, suggesting that the revenue is highly correlated to the number of passenger-delivery trips.

We also compare the average number of passenger-delivery trips for good and ordinary taxis in Table 4.2 in all time slots. The number of passenger-delivery trips per hour is displayed as "Mean ± Std." We can see that good taxi drivers complete more trips than ordinary ones, where the difference is at least 21% in the afternoon time slot and at most 87% in the nighttime slot. These results imply that good drivers are always more efficient in finding next passengers than ordinary ones.

Table 4.1 Correlation between the number of delivery trips and revenue

	Night	Morning	Noon	Afternoon	Evening
Corr	0.96	0.84	0.72	0.61	0.86

Table 4.2 Number of delivery trips per hour in each time slot

	Night	Morning	Noon	Afternoon	Evening
Good	1.3 ± 0.2	2.0 ± 0.2	2.5 ± 0.3	2.1 ± 0.2	2.6 ± 0.2
Ordinary	0.7 ± 0.1	1.4±0.2	2.0±0.2	1.8±0.2	2.0±0.2
Good/Ordinary	1.87	1.4	1.24	1.21	1.34

4.3.2.3 Popular Passenger Pickup and Drop-Off Areas

Another straightforward but important influencing factor on taxi drivers' performance is the taxi operation area. This factor has been considered in a number of studies [6, 9, 12] to guide taxi drivers to find passengers. In Fig. 4.5, we plot the top 99 pickup and drop-off hotspots at different time slots. We can see that the railway station and Zhejiang University are among the popular pickup and drop-off areas across all the time slots. Residential areas become popular at night as many people are returning home. The pickup number decreases greatly from the downtown to

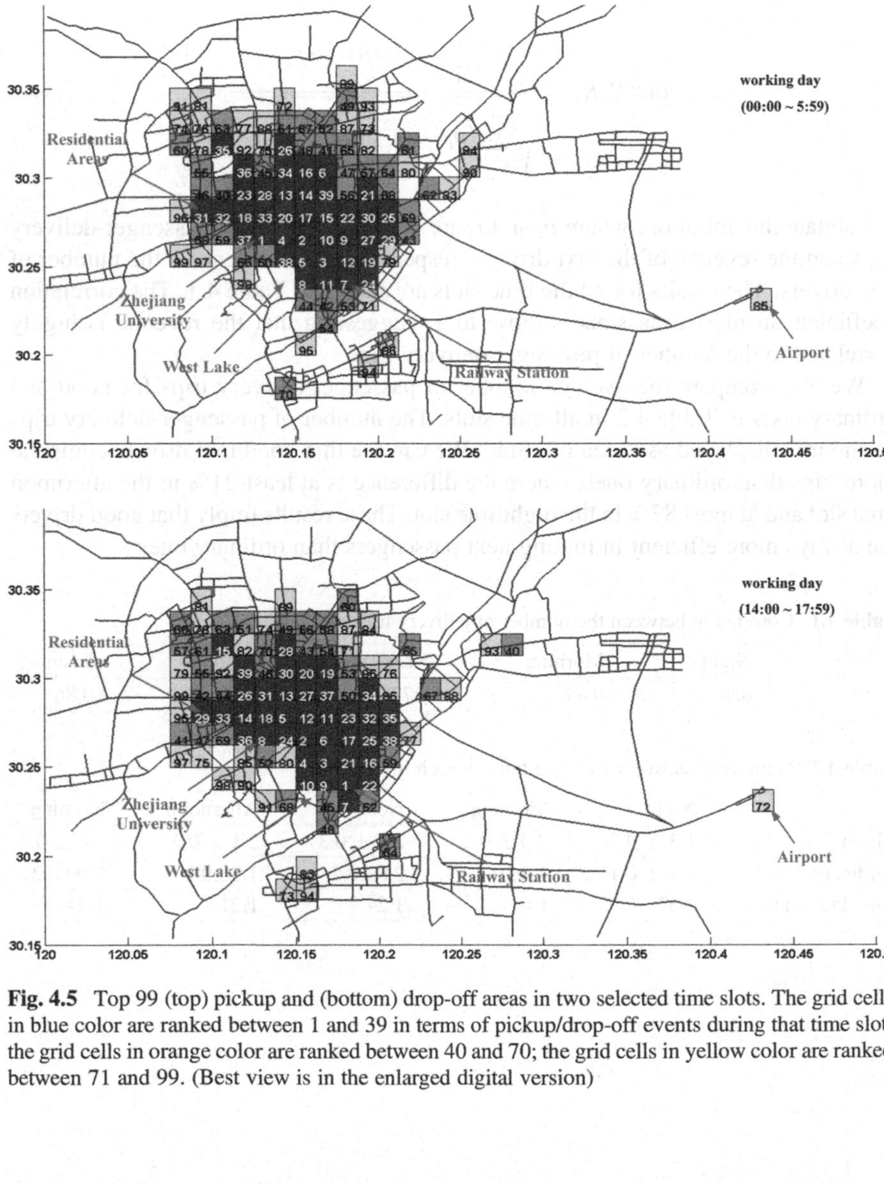

working day
(00:00 ~ 5:59)
Residential Areas
Zhejiang University
West Lake
Railway Station
Airport
working day
(14:00 ~ 17:59)
Residential Areas
Zhejiang University
West Lake
Railway Station
Airport
30.35
30.3
30.25
30.2
30.15
120
120.05
120.1
120.15
120.2
120.25
120.3
120.35
120.4
120.45
120.5

Fig. 4.5 Top 99 (top) pickup and (bottom) drop-off areas in two selected time slots. The grid cells in blue color are ranked between 1 and 39 in terms of pickup/drop-off events during that time slot; the grid cells in orange color are ranked between 40 and 70; the grid cells in yellow color are ranked between 71 and 99. (Best view is in the enlarged digital version) suburb areas. It is noted that the numbers in the grid cells in Fig. 4.5 refer to the ranking of the area in terms of the number of pickup or drop-off events. Apparently, the same grid cell may have different numbers in different time slots (i.e., night, morning, noon, afternoon, and evening) due to the variation of the taxi and passenger demands.

4.4 Taxi Strategy Formulation

We investigate taxi drivers' service strategies from three perspectives: (1) passenger-searching strategies; (2) passenger-delivery strategies; and (3) service-area preference. For passenger-searching strategies, we group drivers' preferences into three possible behaviors, namely, *hunting locally*, *waiting locally*, and *going distant* (i.e., *traveling to a distant location*). For passenger-delivery strategies, we study their average passenger-delivery speed, which potentially reflects the drivers' ability to choose clearer routes when delivering passengers. For service-area preference, we study their preference of service areas in a city. The overview of the taxi service strategies is illustrated in Fig. 4.6.

4.4.1 *Taxi Service Strategy Extraction*

In the following, we introduce how to extract the taxi service strategies from the three perspectives outlined in the beginning of this section.

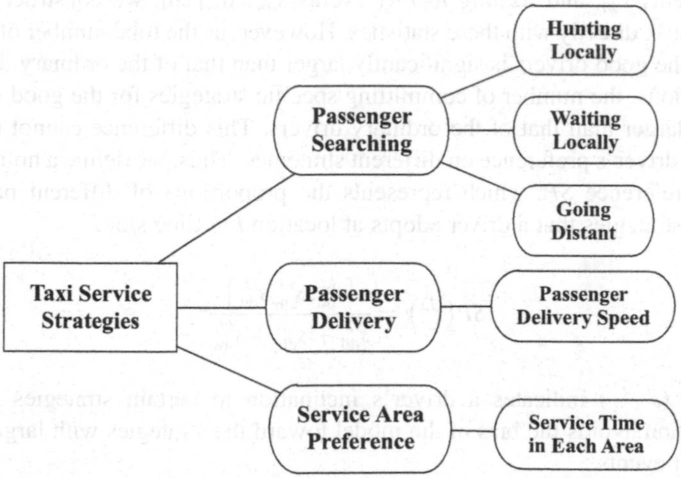
Hunting Locally
Passenger Searching
Waiting Locally
Going Distant
Taxi Service Strategies
Passenger Delivery
Passenger Delivery Speed
Service Area Preference
Service Time in Each Area

Fig. 4.6 Overview of taxi service strategies

4.4.1.1 Passenger-Searching Strategies

After detecting the initial intended paths, we proceed to extract the passenger-searching strategies of individual drivers after dropping off passengers. We use the same partition of the city as that in Fig. 4.5. For each time slot, we select the top 99 busiest cells and treat the rest as an entire nonpopular area and finally obtain 100 location labels. For each location, we are interested in understanding drivers' preferences on three types of passenger-searching strategies after dropping off passengers, i.e., *hunting locally*, *waiting locally*, and *going to distant locations*. The *local* and *distant* properties are determined by comparing the distance d_{drop} between the first and the last cells of the initial intended path to a threshold τ_d. The *hunting* and *waiting* properties depend on whether the trajectory is ended with a waiting event whose time duration t_{wait} is longer than ω_d. In this chapter, we empirically set $\tau_d = 1.5$ km and $\omega_d = 5$ min. The criteria for determining these properties are:

$$d_{\text{drop}} \begin{cases} > \tau_d \text{going distant} \\ \leq \tau_d \begin{cases} t_{\text{wait}} > \omega_d, & \text{waiting locally} \\ t_{\text{wait}} \leq \omega_d, & \text{hunting locally} \end{cases} \end{cases} \tag{4.2}$$

For a specific location, we count the number of *going distant* events s_{dd}, *hunting locally* events s_{dh}, and *waiting locally* events s_{dw}. In [16], we construct the taxi-pattern matrix directly with these statistics. However, as the total number of drop-off events of the good drivers is significantly larger than that of the ordinary drivers, in most locations, the number of committing specific strategies for the good drivers is generally larger than that of the ordinary drivers. This difference cannot reveal an individual driver's preference on different strategies. Thus, we define a notion called strategy preference *SP*, which represents the proportions of different passenger-searching strategies that a driver adopts at location *l* in time slot *t*

$$SP\left(s^{l,t}\right) = \frac{\left[s_{\text{dd}}^{l,t}, s_{\text{dh}}^{l,t}, s_{\text{dw}}^{l,t}\right]}{s_{\text{dd}}^{l,t} + s_{\text{dh}}^{l,t} + s_{\text{dw}}^{l,t}} \tag{4.3}$$

where $SP\ (s^{l,\ t})$ indicates a driver's inclination to certain strategies, and this normalization avoids the bias of the model toward the strategies with large number of drop-off events.

As a driver may have three strategies at each location during different time slots after dropping off passengers, we build a 100-location × 3-strategy × 5-time slot = 1500-dimensional feature vector for each driver. Each dimension of the feature vector corresponds to a specific (location, time, strategy) combination.

4.4.1.2 Passenger-Delivery Strategies

We consider passenger-delivery speed, which is an average speed over all the passenger-delivery trajectories of a driver, as a passenger-delivery strategy. This is based on the observation that a higher value implies that the driver has good skill to choose efficient passenger-delivery routes. We calculate the average passenger-delivery speed in each hour as

$$\text{Speed}_t = \frac{\sum d^t_{\text{delivery}}}{\sum t^t_{\text{delivery}}} \tag{4.4}$$

in which $d^t_{delivery}$ and $t^t_{delivery}$ are the delivery distance and time of a trip in hours t, respectively. Then, for each driver, we build a feature vector with each dimension corresponding to 1 h.

4.4.1.3 Service-Area Preference

As shown in [13], high-revenue drivers are capable of choosing to serve at certain city areas to make high profit and meanwhile avoid heavy traffic in Shenzhen, China. In this chapter, we also investigate the preferences of passenger service areas of different taxi drivers. Unlike passenger-searching strategies, we divide the city into 10 × 5 areas, each of which is about 5 km × 5 km. For each driver, we count the number of passenger-searching trips pft_i in area i in each time slot. The preference of area i is defined as

$$P_i = \frac{pft_i}{\sum_i pft_i} \tag{4.5}$$

which measures a driver's preference of service in each area. For each time slot, we build a 50-dimensional feature vector, with each dimension corresponding to the preference to a particular area.

4.4.2 *Driver–Strategy Matrix Construction*

For each driver, we combine the three types of feature vectors into one-row vector. Then, we construct a driver–strategy matrix *X* by stacking individual drivers' feature vectors in rows (as shown in Fig. 4.7). In the obtained matrix *X*, each row corresponds to one taxi, and each column corresponds to the service strategy in a specific time slot and/or for a specific grid cell. In the next section, the analysis of taxi service strategies is based on this feature matrix.

4.5 Understanding Taxi Service Strategies

In this section, we will leverage the driver–strategy matrix to investigate the taxi drivers' service strategies. Specifically, two studies are conducted: (1) We attempt to uncover effective taxi service strategies and validate their effectiveness in classifying good and ordinary taxi drivers. The discovered service strategies can be used as guidance for taxi drivers. (2) We estimate individual taxi drivers' revenues based on their taxi service strategies. This result can be used to predict a driver's future performance based on the historical driving behaviors.

4.5.1 *Discovering Good Taxi Service Strategies*

For service strategy vector shown in each column of the driver–strategy matrix, we evaluate its impact on taxi drivers' performance by computing its correlation with the revenue of the corresponding taxi drivers. A positive correlation value indicates that this strategy generally brings benefits to drivers' performance, whereas a negative value indicates that the strategy will not help. In the following, we study the impact of the three types of strategies.

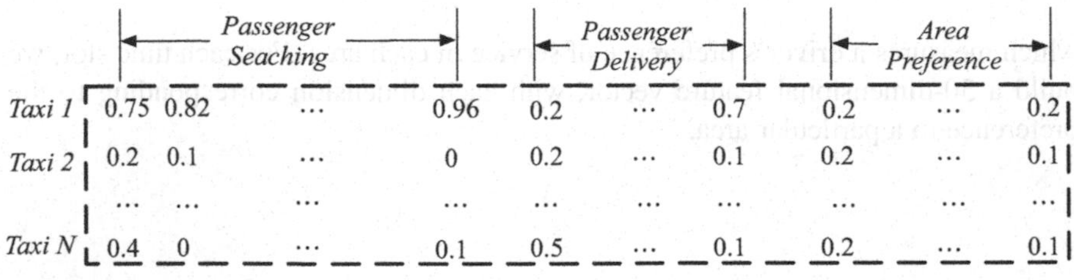
Passenger Seaching
Passenger Delivery
Area Preference
Taxi 1 0.75 0.82 ... 0.96 0.2 ... 0.7 0.2 ... 0.2
Taxi 2 0.2 0.1 ... 0 0.2 ... 0.1 0.2 ... 0.1
... ...
Taxi N 0.4 0 ... 0.1 0.5 ... 0.1 0.2 ... 0.1

Fig. 4.7 Driver–strategy matrix *X*

4.5.1.1 Passenger-Searching Strategies

To understand how different passenger-searching strategies would affect the drivers' revenues generated, we fix the grid cell ID, as shown in Fig. 4.8 (same as the ID numbers in the bottom figure of Fig. 4.5), then compute the correlation between each passenger-searching strategy vector at a specific grid cell and time slot after dropping off passengers and the generated revenues. For the sake of simplicity and clarity, we choose three groups of locations (grid cells) for detailed study, according to the accumulated drop-off density. The first group (Group 1) is composed of grid cells that are ranked top 1–10 through all time slots (marked in blue in Fig. 4.8). The second group (Group 2) is composed of grid cells that are ranked between 40 and 70 through all time slots, marked in orange color. The last group (Group 3) is composed of grid cells that are ranked between 71 and 100 through all time slots, marked in yellow color in Fig. 4.8.

Figure 4.9 shows the correlation values of three different service strategies for specific locations in different time slots. As shown in Fig. 4.9, the correlation value for the local hunting strategy is always positive, independent of the time slots and cell locations. Furthermore, for most time slots and cell locations, the correlation value of the local hunting strategy is the biggest, as compared with that of the other two strategies. The results show that the local hunting strategy always contributes positively to the revenue; it is usually more effective than the local waiting and going to distant strategies. Interestingly, with the drop-off density decreasing, the correlation value of the local hunting strategy is also decreasing. For instance, the correlation value for Group 3 cells is generally smaller than that for Group 2 and Group 1.

For the local waiting strategy, the correlation value can be negative or positive, depending on the cell locations and time slots. Generally speaking, the local waiting strategy has a lower correlation value compared with the other two service strategies,

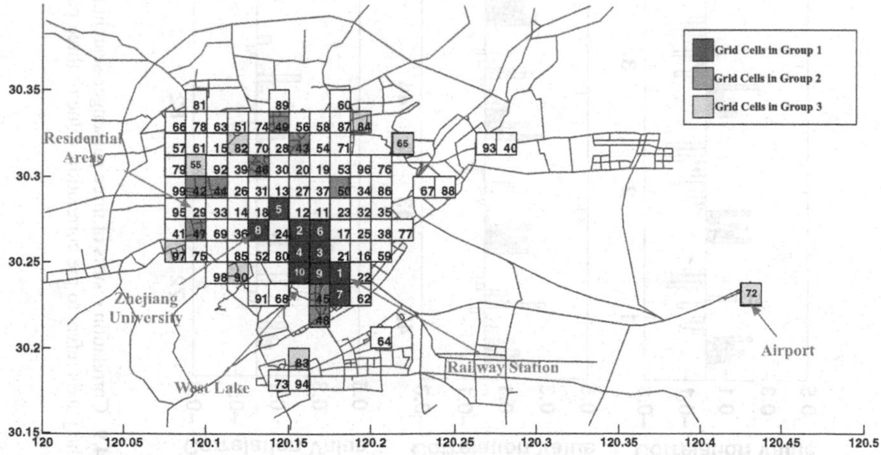
Grid Cells in Group 1
Grid Cells in Group 2
Grid Cells in Group 3
Residential Areas
Zhejiang University
West Lake
Railway Station
Airport
30.35
30.3
30.25
30.2
30.15
120
120.05
120.1
120.15
120.2
120.25
120.3
120.35
120.4
120.45
120.5

Fig. 4.8 Selected three groups of grid cells for detailed study

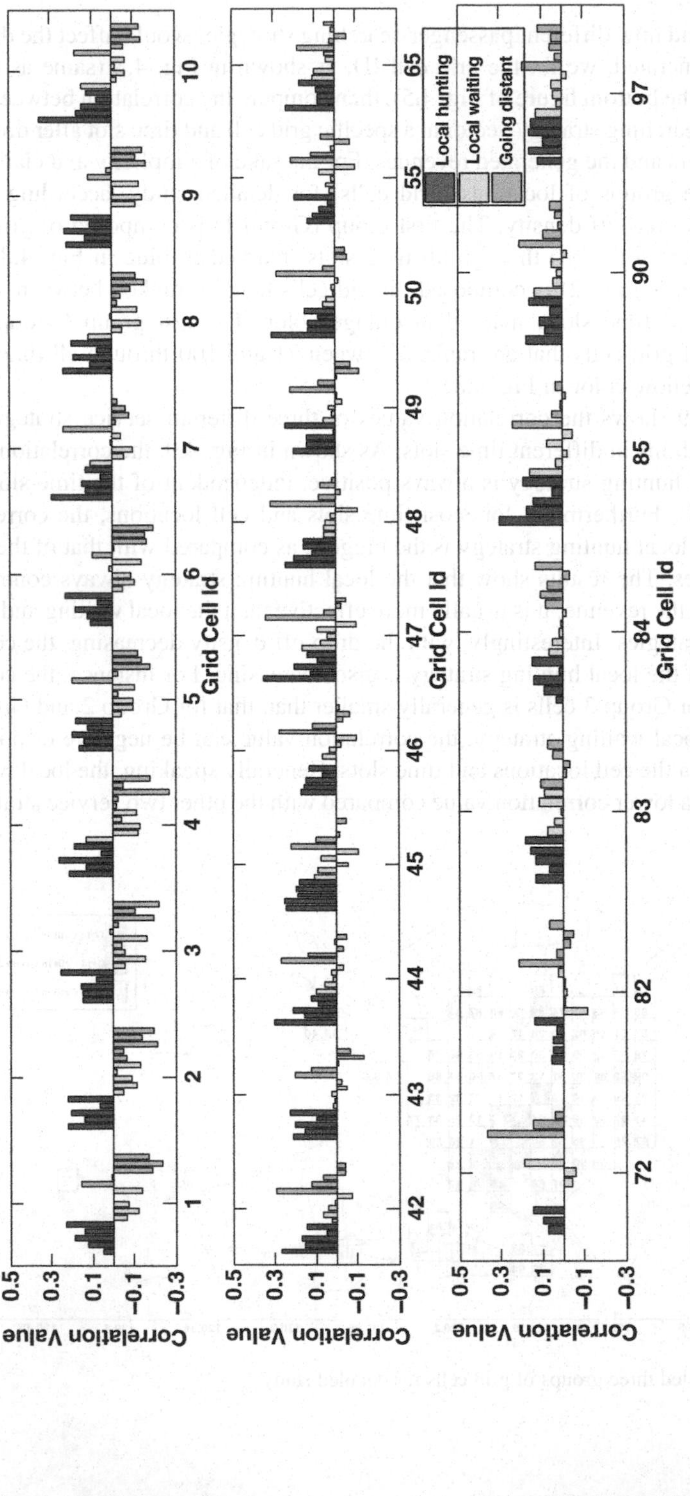
Correlation Value
0.5
0.3
0.1
−0.1
−0.3
1
2
3
4
5
6
7
8
9
10
Grid Cell Id
42
43
44
45
46
47
48
49
50
55
65
72
82
83
84
85
90
97
Local hunting
Local waiting
Going distant

Fig. 4.9 Correlation results of three passenger-searching strategies for three groups of grid cells at five different time slots. The x-axis refers to the grid cell ID and the y-axis refers to the correlation value of three passenger-searching strategies at five time slots except for a few cases such as in the airport (see cell ID 72 in Fig. 4.9) at the afternoon time slot. However, it should be noted that revenue estimation is done based on the passenger-delivery distance without considering the fuel cost; thus, the local waiting strategy should be suggested to drivers although the correlation value for the other two strategies is slightly bigger than local waiting for those time slots and locations. For the case of the airport (i.e., cell 72 in Fig. 4.9), we can see that the correlation value of the going distant strategy is significantly bigger than that of the other two strategies in the morning time slot (06:00–09:59), and the correlation value of the local waiting strategy is biggest among the three in the afternoon time slot (14:00–17:59). This may be due to the following facts. (1) There are few flights reaching the airport in the morning time slot. (2) The airport is far from hot areas with potential passengers. Thus, it is preferable to go back downtown after dropping off passengers at the airport. (3) More flights arrive at the airport in the afternoon time slot; it is thus more effective to wait for new passengers in the airport than to hunt for passengers. Note that there are no correlation values in the airport at late-night and evening time slots; this is probably because there are few flights leaving or reaching the airport during these two periods. Thus, the correlation values for all three strategies are zero.

For the going distant strategy, the absolute correlation value is often bigger than that of the local waiting strategy at most of the grid cells and time slots, showing that this strategy affects the revenue more significantly than the local waiting strategy. For the grid cells of Group 1, it is shown in Fig. 4.9 that the correlation value of the going distant strategy is often negative except for the late night time slot. We can also see that it is positive at the late-night time slot for all the selected grid cells and is as big as that of the local hunting strategy, indicating that taxi drivers could either choose to travel to the nearby grid cells or go to a distant popular grid cell to search new passengers, particularly for the grid cells in Group 2. However, the local hunting strategy is more preferable since it consumes less fuel than going to distant.

4.5.1.2 Passenger-Delivery Strategies

For passenger-delivery strategies, we aim to investigate the influence of the ability to choose clear routes on driver performance. It is easy to understand that both good and ordinary taxi drivers can choose clear routes when the traffic in all the road networks is smooth, for example, at night. However, good drivers usually choose better routes when the road network is congested, which makes their generated revenue higher than ordinary drivers. In this chapter, we use the average passenger-delivery speed of the taxis to represent the capability of taxi drivers in choosing clear routes. We plot the correlation values of the average passenger-delivery speeds in each time slot with drivers' hourly revenue in Fig. 4.10, with respect to the average passenger-delivery speeds. We can easily see that, during late night, the traffic is clear and the correlation values are around zero (i.e., no much influence). In the daytime, when the traffic is more congested, the correlation value is more than 0.1. More specifically, when the difference between the passenger-delivery speed of

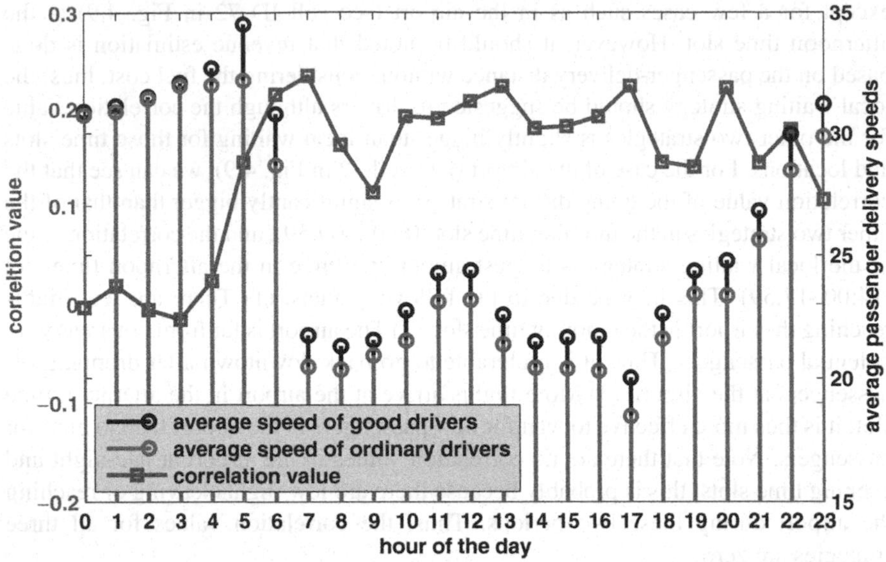

0.3
0.2
0.1
0
−0.1
−0.2
35
30
25
20
15
correltion value
average passenger delivery speeds
average speed of good drivers
average speed of ordinary drivers
correlation value
0 1 2 3 4 5 6 7 8 9 10 11 12 13 14 15 16 17 18 19 20 21 22 23
hour of the day

Fig. 4.10 Correlation between the average passenger-delivery speed and the hourly revenue in each hour of the day (the average passenger-delivery speeds of all good and ordinary drivers are also shown)

good and ordinary taxi drivers is larger, the correlation value becomes bigger. This result confirms that good drivers do have better skills in choosing routes with light traffic.

4.5.1.3 Service-Area Preference

The correlation value between the preference of service area and the hourly revenue rate generally reflects whether an area is worth serving for a taxi. The results are shown in Fig. 4.11. We can see that, in late-night, taxi drivers had better served in the entertainment area (see the red zone in Fig. 4.11a) because there are more passengers who need to be delivered to their homes after night life. Meanwhile, the blue areas are those areas where the taxi demands are few (most of them are suburban areas); hence, it is better for taxis to go back to the top hotspots directly instead of staying in those areas.

Finally, we validate the effectiveness (and ineffectiveness) of the discovered good (and bad) strategies by using them to classify good and ordinary taxi drivers with support vector machine (libsvm [17]). We use the features with the top 1/3 correlation values and those with the bottom 1/3 correlation values to classify the good and ordinary taxi drivers. We use fivefold cross validation to evaluate the classification accuracy. Specifically, we randomly split the good and ordinary taxi drivers into five random groups and choose four groups as the training data set and the remaining one

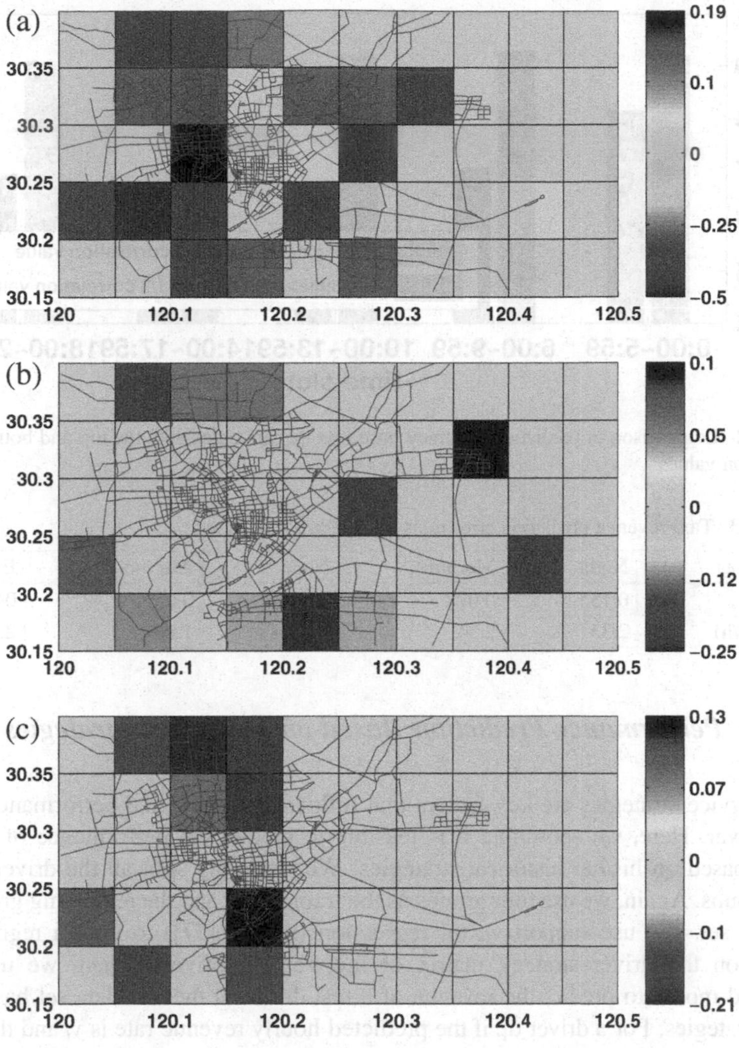
(a)
30.35
30.3
30.25
30.2
30.15
120
120.1
120.2
120.3
120.4
120.5
0.19
0.1
0
−0.25
−0.5
(b)
30.35
30.3
30.25
30.2
30.15
120
120.1
120.2
120.3
120.4
120.5
0.1
0.05
0
−0.12
−0.25
(c)
30.35
30.3
30.25
30.2
30.15
120
120.1
120.2
120.3
120.4
120.5
0.13
0.07
0
−0.1
−0.21

Fig. 4.11 Correlation between taxi service-area preference and the revenue in three selected time slots. (Best view is in the enlarged digital version.) **(a)** 00:00–5:59. **(b)** 10:00–13:59. **(c)** 14:00–17:59

as the test data set. The evaluation results are shown in Fig. 4.12. We can see that the strategies with the top 1/3 correlation values have much better classification accuracy than those with the bottom 1/3 correlation values.

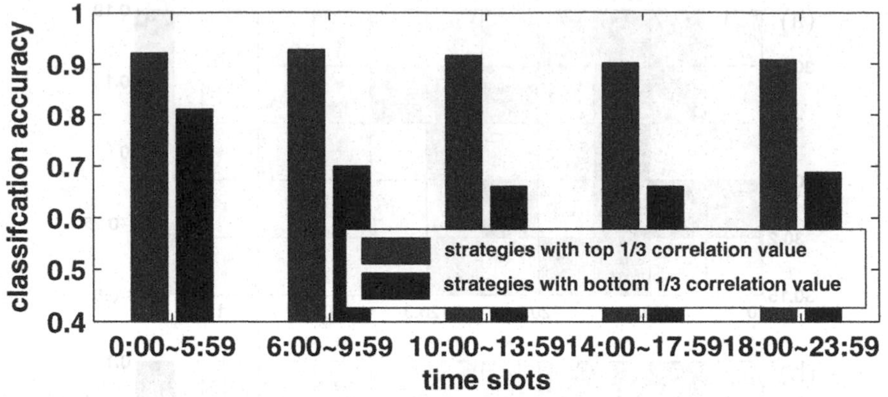
classifcation accuracy
1
0.9
0.8
0.7
0.6
0.5
0.4
strategies with top 1/3 correlation value
strategies with bottom 1/3 correlation value
0:00~5:59
6:00~9:59
10:00~13:59
14:00~17:59
18:00~23:59
time slots

Fig. 4.12 Comparison of prediction accuracy based on the strategies with the top and bottom 1/3 correlation values

Table 4.3 Taxi revenue prediction error rates in different time slots

Time slots	Night	Morning	Noon	Afternoon	Evening
ϵ_r	0.155	0.078	0.056	0.057	0.063
r (RMB/h)	2.35	2.34	2.21	1.96	2.25

4.5.2 Performance Prediction Based on Historical Strategies

Taxi service strategies are key factors that influence the revenue performance of a taxi driver. Here, we show that it is feasible to predict the performance of a taxi driver based on his/her historical strategies. We randomly split all the drivers into five groups. Again, we use four groups as the training set and the remaining group as the test set. We use support vector regression (libsvm [17]) to train a regression model on the driver–strategy matrix of the training drivers. Then, we use the obtained model to predict the revenue of a test driver in the test data set based on their strategies. For a driver d_i, if the predicted hourly revenue rate is y_i and the true value is y_i, then the residual is $r_i = \mid y_i - y_i|$. We use two metrics to evaluate the prediction error: The mean residual $r = mean(r_i)$, which reflects the average accuracy of the prediction, and the relative residual of taxi drivers, which is computed as $r = mean(r_i/y_i)$, indicating the percentage of the misprediction against the true value. A smaller value corresponds to a more accurate prediction result.

The results in all the time slots are shown in Table 4.3. By observing all the time slots, we can obtain the residuals that are less than 2.35 RMB/h, which suggests a quite accurate prediction result. The relative residuals in all the time slots except nighttime are quite small, which fully validate the feasibility of using the extracted taxi service strategies to predict taxi performance. The reason for the relatively low prediction accuracy in nighttime is probably because taxi drivers may choose to sleep for some time late-night, and this behavior influences the revenue performance of the drivers.

4.6 Conclusions and Future Work

It is well worth using the taxi GPS trajectories to model taxi drivers' service behaviors and strategies. In this chapter, by analyzing a real-world large-scale taxi GPS trajectory dataset that contains 7600 taxi driver's activities in Hangzhou, we have researched the taxi drivers' service strategies to offer helpful advice to taxi drivers. First, we have observed two phenomena: 1) Most of the time, the shift handover location and time are fixed; 2) the taxi usually stops temporarily in the handover location and shift handover time slot for handover, while the taxi is vacant status during the whole shift handover process. Given these phenomena, we put forward an algorithm to apart the taxi GPS trajectories of each pair of shared taxi drivers. Then by converting each taxi distance into relevant taxi revenue in each time slot, we have found that a majority of taxi drivers can't always have good performance across different time slots. So, we have put forward an idea to research taxi driver's behaviors in each time slot instead of on a daily basis.

In this chapter, based on three perspectives which include passenger-searching strategies, passenger-delivery strategies, and service-area preference, we come up with a method to model the taxi drivers' service strategies. We first selected the feature matrix to express service strategies. Then, we propose an algorithm to study the original willingness of taxi drivers right after each drop-off event to recognize the passenger strategies. By measuring the correlation between each service strategy and the relevant revenue, we demonstrate both the efficient and inefficient strategies in each time slot and location. In short, we discover that waiting is usually less efficient than hunting in order to find passengers locally, with a few exceptions, such as in the railway station where people go up to fixed places to take a taxi. Taking notice that taxis often take less time on average looking for new passengers by moving from unpopular areas to popular ones. When the traffic is heavy at busy time slots, selecting the routes which traffic is light gives the taxi driver more returns according to the correlation of the taxi average passenger-delivery speed and the revenue. We also discover that some service areas can give taxi drivers higher returns if the taxi drivers go there in different time slots of the day.

The key factor of taxi driver's income is taxi service strategies. We use an SVR algorithm to predict the performance of a taxi driver by analyzing the extracted historical service strategies. The error of our method is less than 2.35 RMB/h, indicating that our proposed method is able to well describe the driving taxi drivers' behavior and performance with extracted service strategies.

In the future, we are prepared to further develop our work from two perspectives. For one aspect, we are willing to do more research about characterizing fine-grained features of taxi behaviors and strategies. For another aspect, we want to search for a more effective method to give the taxi drivers more useful advice based on taxi service strategies.

References

1. Chen C, Zhang D, Li N, Zhou Z. B-Planner: planning bidirectional night bus routes using large-scale taxi GPS traces. IEEE Trans Intell Transp Syst. 2014;15(4):1451–65.
2. Chen C, Jiao S, Zhang S, Liu W, Feng L, Wang Y. TripImputor: real-time imputing taxi trip purpose leveraging multi-sourced urban data. IEEE Trans Intell Transp Syst. 2018;19 (10):3292–304.
3. Wang J, Wang Y, Zhang D, Lv Q, Chen C. Crowd-powered sensing and actuation in smart cities: current issues and future directions. IEEE Wirel Commun. 2019;26(2):86–92.
4. Chen C, Wang Z, Zhang D. Sending more with less: crowdsourcing integrated transportation as a new form of citywide passenger–package delivery system. IT Prof. 2020;22(1):56–62.
5. Guo B, Chen H, Liu Y, Chen C, Han Q, Yu Z. From crowdsourcing to crowdmining: using implicit human intelligence for better understanding of crowdsourced data. World Wide Web. 2020;23:1101–25.
6. Chang H, Tai Y, Hsu JY. Context-aware taxi demand hotspots prediction. Int J Business Intell Data Mining. 2010;5(1):3–18.
7. Lee J, Shin I, Park GL. Analysis of the passenger pick-up pattern for taxi location recommendation, 2008.
8. Xiaolong LI, et al. Prediction of urban human mobility using large-scale taxi traces and its applications. Front Comput Sci. 2012;6(1):111–21.
9. Powell JW, Huang Y, Bastani F, Ji M. Towards reducing taxicab cruising time using spatio-temporal profitability maps. In: Advances in spatial and temporal databases. Heidelberg: Berlin; 2011. p. 242–60.
10. Yuan J, Zheng Y, Zhang L, Xie X, Sun G. Where to find my next passenger. In: Proceedings of the 13th international conference on Ubiquitous computing, New York, NY, USA, Sept 2011.
11. Veloso M, Phithakkitnukoon S, Bento C. Urban mobility study using taxi traces. In: Proceedings of the 2011 international workshop on trajectory data mining and analysis, New York, NY, USA, Sept 2011.
12. Lee J, Shin I, Park G-L. Analysis of the passenger pick-up pattern for taxi location recommendation. In: 2008 Fourth international conference on networked computing and advanced information management, Sept 2008, vol 1. p. 199–204.
13. Liu L, Andris C, Ratti C. Uncovering cabdrivers' behavior patterns from their digital traces. Comput Environ Urban Syst. 2010;34(6):541–8.
14. Takayama T, Matsumoto K, Kumagai A, Sato N, Murata Y. Waiting/cruising location recommendation based on mining on occupied taxi data. In: Proceedings of the 12th WSEAS international conference on mathematical and computational methods in science and engineering, Stevens Point, Wisconsin, USA, Nov 2010. p. 225–9. Accessed 22 Aug 2020.
15. Yamamoto K, Uesugi K, Watanabe T. Adaptive routing of multiple taxis by mutual exchange of pathways. Int J Knowl Eng Soft Data Paradigms. 2009;2(1):57–69.
16. Li B, et al. "Hunting or waiting? Discovering passenger-finding strategies from a large-scale real-world taxi dataset. In: 2011 IEEE international conference on pervasive computing and communications workshops (PERCOM workshops), 2011. p. 63–8.
17. Chang C-C, Lin C-J. LIBSVM: a library for support vector machines. ACM Trans Intell Syst Technol. 2011;2:27.

Chapter 5
GreenPlanner: Planning Fuel-Efficient Driving Routes

5.1 Introduction

Taxi GPS trajectory data can not only reflect the moving paths of taxis, but also implicitly record the vehicle operation behaviours of taxi drivers. For instance, the average moving speed, acceleration or deceleration between two reported locations can be easily derived from the leaving trajectory data. It is well recognized that the driving behaviours and the vehicle's parameters are two factors strongly correlating to the fuel economy. The impact of vehicle's parameters on the fuel consumption can be simply excluded since most taxis in the same city are with the same model or with the similar parameters. Frequent decelerations or accelerations (i.e., abrupt jerk) in a trip generally correspond to the high fuel consumption [1]. For taxi drivers, *saving more* through taking the most fuel-efficient driving path is almost equally important to increase their total income, compared to learning efficient service strategies. To achieve the objective, an essential step is to estimate the potential fuel cost along a driving path for a given driver, i.e., the personalized fuel consumption modelling.

However, taxi GPS trajectory data only cannot tell the amount of fuel consumption along their driving paths for taxi drivers. Fortunately, the ubiquitous deployment of OBD-II (On-Board-Diagnostics) sensors in vehicles, providing the data source to investigate the cause and effect between GPS trajectory data and the fuel consumption [2–4]. The fuel cost goes high when changing speed, i.e., increase or decrease. It is difficult or even impossible to drive at a constant speed, due to the complex road traffic conditions. There are two main groups of driving behaviours (i.e., reactive and proactive driving behaviours), depending on the speed choices *en*

Part of this paper is based on a previous work: Y. Ding, C. Chen, S. Zhang, B. Guo, Z. Yu and Y. Wang, "GreenPlanner: Planning personalized fuel-efficient driving routes using multi-sourced urban data," 2017 IEEE International Conference on Pervasive Computing and Communications (PerCom), Kona, HI, 2017, pp. 207–216, doi: https://doi.org/10.1109/PERCOM.2017.7917867.

C. Chen et al., *Enabling Smart Urban Services with GPS Trajectory Data*,
https://doi.org/10.1007/978-981-16-0178-1_5

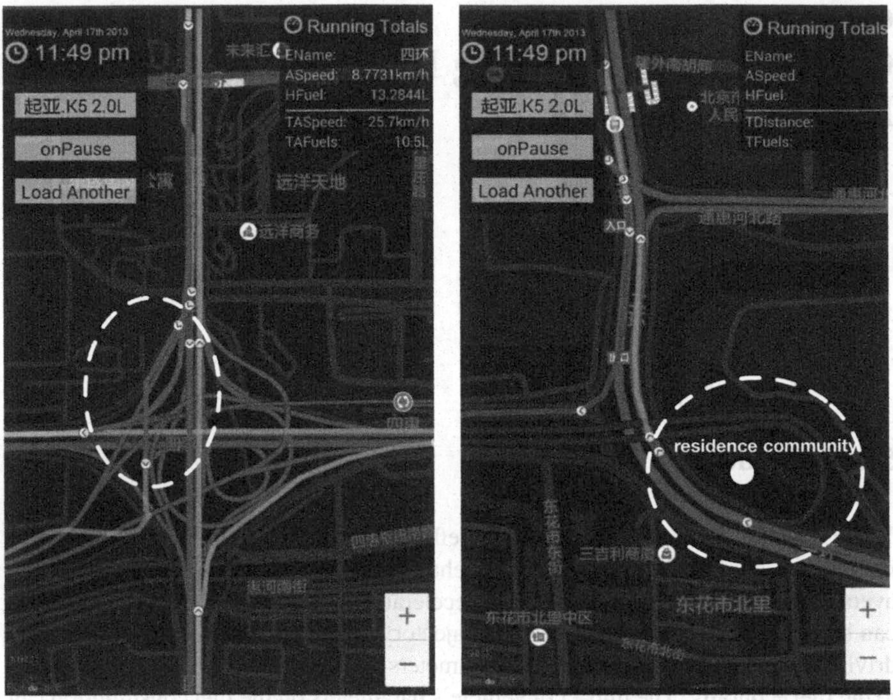
Wednesday, April 17th 2013
11:49 pm
Running Totals
EName:
四环
ASpeed: 8.7731km/h
HFuel: 13.2844L
TASpeed: 25.7km/h
TAFuels: 10.5L
起亚.K5 2.0L
onPause
Load Another
远洋天地
Wednesday, April 17th 2013
11:49 pm
Running Totals
EName:
ASpeed:
HFuel:
TDistance:
TFuels:
起亚.K5 2.0L
onPause
Load Another
residence community
东花市北里

Fig. 5.1 Visualization of the average fuel consumption on the edge. Red color refers to a high value of average fuel consumption, blue color refers to a low value, and yellow color refers to a value in-between. (Best viewed in the enlarged digital version)

route of drivers. *Reactive behaviours* refer to that drivers are forced to slow down or stop when encountering traffic lights and stop signs. For example, drivers decrease the speed when they are approaching to traffic lights/stop signs while increase the speed after passing by them. *Proactive behaviours* refer to that drivers change speed or take actions defensively to prevent any potential risks. For instance, some drivers may slow down proactively in advance when passing by some specific places (e.g., schools, residences, exits and entrances of garages), watch nearby drivers cautiously when merging lanes, leave adequately safe room when passing or driving behind others.

Previous studies have concluded that some common reactive driving behaviours result in the extra fuel cost [5, 6]. For instance, the fuel consumption is normally higher when cars are approaching and leaving traffic lights than other situations [7, 8]. The state-of-art models also considered and integrated reactive behaviours, however, the impact of proactive behaviours has been overlooked. As a matter of fact, experienced drivers take actions proactively to make safe and incident-free trips, at the expense of fuel economy. When driving courteously, the drivers may need to control speed more carefully, probably resulting in more speed variations and thereby an increase of fuel consumption. Figure 5.1 provides readers with an example of this concept. As shown in this figure, the most fuel-cost edges correspond to a merge into a cloverleaf interchange (as marked by the circle on the left), or a POI (as marked by the circle on the right). On edges with some specific features, the drivers are taking proactive actions more frequently, incurring more fuel consumption as a result.

In this chapter, we extend the fuel consumption model to characterize both reactive and proactive driving behaviours. As drivers taking both driving behaviours continuously while driving, and different type of behaviours may cost a varied amount of fuel. In addition, the fuel consumption varies when different drivers taking the same paths since their personalized driving styles are different. Moreover, the fuel consumption also varies even for the same driver taking the same paths at different time since real-time traffic conditions are different. Thus, we model the personalized fuel consumption caused by taking different behaviors separately, which also counts the real-time traffic conditions. More specifically, we mainly make the following main contributions in this chapter.

- We propose a two-phase framework called **GreenPlanner** to model personalized fuel consumption and recommend fuel-efficient routing for drivers.

  In the first phase, we take consideration of both drivers' reactive and proactive driving behaviours when developing the personalized fuel consumption model (PFCM). We also investigate the relationships among PFCM coefficients, car models, fuel consumption performance using data mining techniques. We further derive a general PFCM for a group of similar drivers, instead of deriving the PFCM parameters for each driver separately.

  In the second phase, we collect the real-time traffic information in urban area via MCS with two methods and further compare the performance of different real-time collection methods under varied settings (e.g., different days, different time duration lengths).
- In the evaluation part, all experiments are done by leveraging multi-sourced urban data including the historical GPS trajectory, OBD II data, and the open data extracted from OpenStreetMap to evaluate the framework extensively. Experimental results demonstrate the effectiveness of the proposed two-phase framework in terms of modeling accuracy and pro-viding high-quality driving routes.

5.2 Related Work

5.2.1 Fuel Consumption Modelling

It is not unique to create a fuel consumption model since different types and sources of data are used. The fuel consumption models based on data-driven are a popular branch [9–12]. To name a few, some studies rely on the vehicle speed data to train the model [9, 10], while some studies also include vehicle parameters such as vehicle type, car mark, loading, engine size when building models [9]. In addition to the vehicles, the driving behaviours also influence the fuel consumption largely. With the deployment of GPS and OBD devices being standard in vehicles, it is possible to link the driving behaviours to the resulted fuel consumption, making the building of more accurate and analytical models become a reality [13–15]. The most well-known physical model is GrennGPS [7, 8]. With input parameters regarding streets (e.g., traffic light and stop sign numbers, the average speed and the average congestion level), vehicles (e.g., frontal area and weight), GreenGPS is able to predict the fuel consumption of a given car when driving along an arbitrary path [7]. The model was exanimated based on real-life datasets, and a nice performance had been achieved [8]. Inspired by the observation that some physical features of the routes strongly correlate with a high fuel consumption, we build a comprehensively personalized fuel consumption model. We believe our model is similar to GreenGPS but goes one step further.

5.2.2 Solutions to Saving Fuel Consumption

In general, there are two categories of fuel consumption saving solutions, i.e., non-driver-oriented and driver-oriented respectively. Non-driver-oriented solutions aim to save the fuel consumption for automobile or car manufacturers and city management departments. For instance, recent years have witnessed several types of green vehicles including electric vehicles on markets. On the other hand, an increasing number of car manufacturing technologies are applied to make fuel-efficient cars [16]. To save fuel efficiently, urban traffic management departments improve the deployment and management of urban infrastructure and optimize the traffic flow to reduce travel times and idling times of vehicles when travelling [5, 13]. There are two main subcategories of driver-oriented fuel-saving solutions, including eco-driving and eco-routing.

Eco-driving aims to save fuel consumption by suggesting fuel-efficient driving behaviours, such as less change of speed, tenderly starting, proper gear change moment, and early accelerator off. Both online and offline ways are adopted to train drivers to learn fuel-efficient driving behaviors. For instance [17], investigate the fuel consumption changes before and after attending offline education courses. Unfortunately, they find that some drivers restore to their normal driving behaviours over time due to that their driving habits are actually difficult to change [13, 16, 17]. Based on vehicular social networks, green transportation systems are able to provides user-friendly mobile application to aware drivers' attention on their fuel-consumption driving behaviours, at the cost of disrupting and distracting drivers [18]. Different from the prior studies that provide driving instructions to become fuel-efficient drivers, we target at saving fuel consumption by suggesting fuel-saving driving paths, without changing drivers' behaviors.

Eco-driving aims to save fuel consumption by suggesting the most fuel economical routes. As the fuel consumption is tightly related to the road types, path distance, a general suggestion could be recommending the paths along freeways, or paths with short distance. To the best of our knowledge, there are quite few studies focusing on this research direction [1, 11]. The most similar work to ours is the GreenGPS [7, 8]. Instead of offering the shortest or the fastest driving paths between two given points, GreenGPS system and ours deliver the most fuel-efficient paths to drivers. We both rely on OBD to collect fuel consumption data. Although with the high similarity, our work still differs from the existing approaches. First of all, both proactive and reactive driving behaviours are explicitly and comprehensively modelled in PFCM. Secondly, we collect the real-time traffic information in urban area via MCS with two methods, with which estimating the potential fuel consumption on each edge. Thirdly, we further compare the performance of different real-time collection methods under varied settings (e.g., different days, different time duration lengths). Finally, we propose a personalized fuel-efficient routing solution and also conduct case studies to validate the effectiveness of the proposed route-planning algorithm.

5.3 Basic Concepts and Problem Formulation

5.3.1 Basic Concepts

Definition 5.1 (OBD-II Data Reading). An OBD-II Data Reading is a data log (the black-filled squares shown in Fig. 5.2) recording the information about the distance that has been covered (in km), the instantaneous speed (in km/h), and the fuel usage

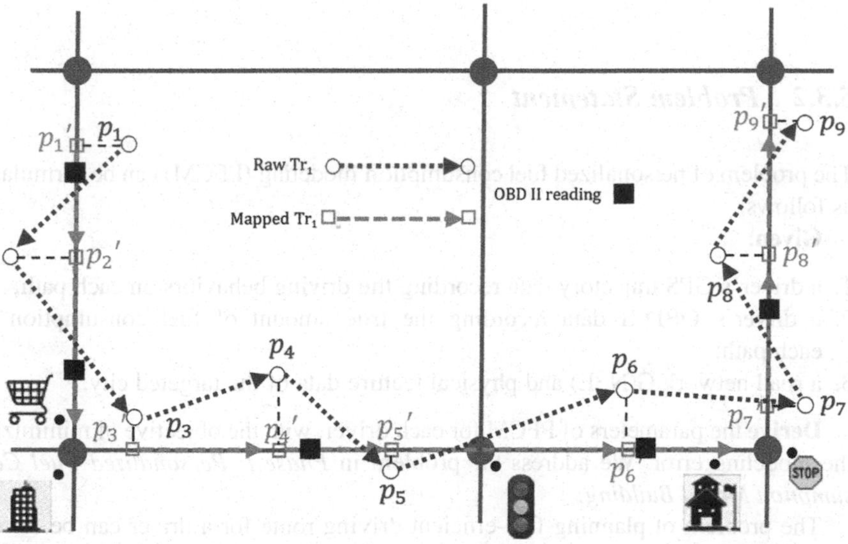
Raw Tr_1
OBD II reading
Mapped Tr_1
p_1'
p_1
p_2
p_2'
p_3
p_3'
p_4
p_4'
p_5
p_5'
p_6
p_6'
p_7
p_7'
p_8
p_8'
p_9
p_9'
STOP

Fig. 5.2 Illustration of main concepts used in the chapter. Note that the exact location of the vehicle at the OBD-II sampling time is not informed (in liter) of the vehicle at the sampling time, which can be denoted by $obd_i = (t_i, v_i, distCov_i, fu_i)$.

Thus, the average fuel consumption (AFC, l/100 km) between any two given OBD-II data readings (obd_j, obd_i, $t_j > t_i$) can be computed as:

$$AFC(obd_i, obd_j) = \frac{fu_j - fu_i}{distCov_j - disCov_i} \times 100 \tag{5.1}$$

Definition 5.2 (Average Fuel Consumption Between Two Given GPS Points). The average fuel consumption between two given GPS points is not straightforward because we are not aware of the exact amount of fuel usage of the vehicle of a given GPS data point, which is due to the fact that the sampling time and rate of OBD-II data and GPS trajectory data are not consistent (as illustrated in Fig. 5.2). To obtain the average fuel consumption (AFC) between any two given GPS data points, we first use the fuel usage value of the time-nearest OBD-II data reading to approximate that of each GPS point, then compute it according to Eq. (5.1).

Definition 5.3 (Physical Features). The physical features of a given path contain a number of attributes (the black dots shown in Fig. 5.2):

1. the number of traffic lights/stop signs along the path;
2. the number of all neighboring edges that connect the edges within the path;
3. the number of some specific POIs within the range of the path. At the current stage, we only consider schools, residential zones, shopping malls and big companies. However, more POIs can be extracted and easily integrated.

5.3.2 Problem Statement

The problem of personalized fuel consumption modeling (PFCM) can be formulated as follows:

Given:

1. a driver's GPS trajectory data recording the driving behaviors on each path;
2. a driver's OBD-II data recording the true amount of fuel consumption on each path;
3. a road network G(N, E) and physical feature data of the targeted city.

Derive the parameters of PFCM for each driver, with the objective of minimizing the modeling error. We address the problem in *Phase I: Personalized Fuel Consumption Model Building*.

The problem of planning fuel-efficient driving route for a driver can be formulated as follows:

Given:

1. a user's driving route query (t_0, S, D), in which t_0 refers to the departure time, S and D refer to the source node and destination node, respectively;
2. real-time traffic information collected from other drivers via Mobile Crowdsensing (MCS) on each edge;
3. the driver's PFCM.

Find the most fuel-efficient driving route (the sequence of edges) for the given driver from S to D and estimate the potential fuel consumption on it. We address the problem in *Phase II: Fuel-Efficient Driving Route Planning.*

5.4 Personal Fuel Consumption Model Building

The objective of building personalized fuel consumption model (PFCM) is to estimate the potential amount of fuel cost when a driver has traveled a certain path while performing the driving behaviors recorded by the GPS trajectory data. Fuel is burned to generate the driving force to overcome the necessary force (F_N) and the extra force (F_E). Specifically, the necessary force (F_N) is the driving force generated to overcome the total force including the friction force caused by the road, the gravitational force caused by the vehicle, and the air resistance, which is *unavoidable*. Ideally, vehicle only needs to overcome the necessary force when the driver moves on a straight road at a constant speed. However, because the driver may encounter traffic congestions, traffic lights/stop signs while driving, the extra force (F_E) is induced. In addition, due to uncertainty, drivers would slow down in advance when traversing some specific POIs as discussed, leading to an *increase* of the extra force.

The fuel consumption rate of a vehicle at time t is proportional to the power generated by its engine at that time. Let fr and P denote the fuel consumption rate and the instantaneous power, respectively. Thus the fuel consumption rate at time t can be computed by Eq. (5.2).

$$fr(t) = \beta P(t) \tag{5.2}$$

The instantaneous power (P) can be calculated by Eq. (5.3).

$$P(t) = (F_N + F_E)v(t) \tag{5.3}$$

where $v(t)$ is the instantaneous speed at time t.

By substituting Eq. (5.3) to Eq. (5.2), we can derive the fuel consumption rate as follows.

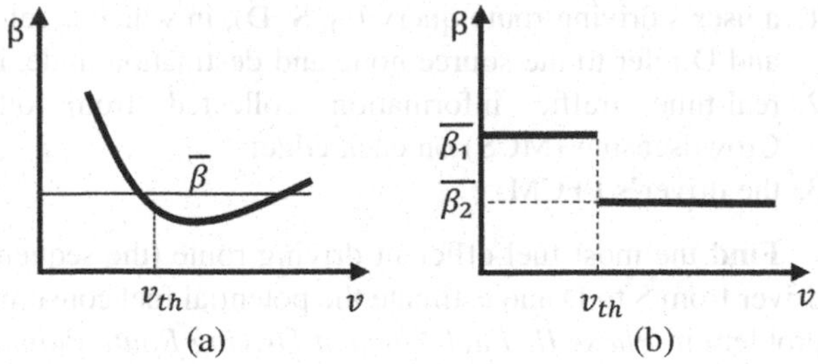
β
$\bar{\beta}$
v_{th}
v
(a)
β
$\overline{\beta_1}$
$\overline{\beta_2}$
v_{th}
v
(b)

Fig. 5.3 The relation between β and the vehicle speed v at a given engine speed. (**a**) Representation of the relation in reality. (**b**) Representation of the relation used in our model

$$fr(t) = \beta(F_N + F_E)v(t) = \underbrace{\beta F_N v(t)}_{\text{Necessary}} + \underbrace{\beta F_E v(t)}_{\text{Extra}} \tag{5.4}$$

where the first part is defined as the *necessary fuel consumption rate* ($fr_n(t)$), the second part is defined as the *extra fuel consumption rate* ($fr_e(t)$); β is the specific fuel consumption (SFC). SFC refers to the *quantity of fuel consumed to produce one unit of power in one unit of time,* and varies under different travel and engine speeds. Figure 5.3a shows the details on how β differs with the travel speed at a given engine speed. β is much larger when the vehicle travels at a low speed, indicating that fuel is consumed more quickly since the fuel cannot be fully burned at that speed. Previous work simply uses a single average value $\bar{\beta}$ to approximate the SFC, which does not reflect the reality when the speed is slow. Going a step further, we use two coefficients to distinct the cases of slow and normal speeds. In more detail, $\bar{\beta}_1$ is used to approximate the SFC when the travel speed is less than vth and $\bar{\beta}_2$ is used to approximate when the travel speed is above v_{th}, as demonstrated in Fig. 5.3b. Here, we set $v_{th} = 10$ km/h.

5.4.1 Necessary Fuel Consumption Rate

The necessary fuel consumption rate is used to estimate the necessary driving force (F_N) that overcomes the road frictional force (F_{Nf}), the gravitational force (F_{Ng}), and the air resistance (F_{Nr}). The relationship of the above forces can be represented as follows.

$$F_N = F_{Nf} + F_{Ng} + F_{Nr} \tag{5.5}$$

The road frictional force F_{Nf} can be characterized by the gravitational force acting on the vehicle and is estimated according to Eq. (5.6).

$$F_{Nf} = \mu mg \cos(\theta) \tag{5.6}$$

where μ is the coefficient of friction, m is the mass of the vehicle, g is the gravitational acceleration, and θ is the ground slope of the road.

The gravitational force F_{Ng} can be approximated using the following equation:

$$F_{Ng} = mg\sin(\theta) \tag{5.7}$$

The force caused by the air resistance F_{Nr} can be estimated by the following equation:

$$F_{Nr} = \frac{1}{2}\varphi C_r A v^2(t) \tag{5.8}$$

where ϕ is the coefficient of the air resistance, A is the frontal area of the car, C_r is the air density, and v is the speed of the vehicle at time t.

Finally, we can obtain the necessary fuel consumption model by simple derivations as follows:

$$\begin{aligned} fr_n(t) &= \beta F_N v(t) \\ &= \beta\left[\mu mg\cos(\theta) + mg\sin(\theta) + \frac{1}{2}\varphi AC_r v^2(t)\right]v(t) \end{aligned} \tag{5.9}$$

We note that in most cases, the slope (θ) of the road is small, and fr_n can be simplified as follows:

$$fr_n(t) = \beta\left[\mu mg + \frac{1}{2}\varphi AC_r v^2(t)\right]v(t) \tag{5.10}$$

5.4.2 Extra Fuel Consumption Rate

In addition to the necessary fuel consumption rate, the vehicle has to consume some extra fuel to generate extra force, the accurate value of which is difficult to get. Here, we simply estimate the extra fuel consumption using Eq. (5.11).

$$\begin{aligned} fr_e(t) &= \beta F_E v(t) \\ &= \beta ma(t)\Big(\underbrace{\gamma + k_1' \mid path \cdot ts \mid}_{\text{Reactive}} + \underbrace{k_2' \mid \text{path} \cdot n \mid + k_3' \mid \text{path} \cdot \text{poi} \mid}_{\text{Proactive}}\Big)v(t) \end{aligned} \tag{5.11}$$

where a(t) and v(t) are the acceleration and the speed of the vehicle at the time t. respectively; γ is a constant coefficient which is used to model the case of traffic congestions; |path.ts|, |path.n|, |path.poi| correspond to the number of traffic lights/stop signs, the number of neighboring edges, and the number of specific POIs of the given path respectively. The first two parts of the equation are used to estimate the extra fuel consumption while drivers have to change driving behaviors *reactively*. The last two parts are used to approximate the extra fuel consumption when drivers change driving behaviors *proactively*.

5.4.3 Personalized Fuel Consumption Model (PFCM)

The total fuel consumption rate is simply the sum of the necessary and extra fuel consumption rates, as shown in Eq. (5.12).

$$fr(t) = fr_n(t) + fr_e(t) \tag{5.12}$$

Therefore, the total fuel consumption fc of a given path for a driver under his/her driving behaviors recorded by the GPS trajectory data (i.e., **PFCM**) can be modeled by integrating the total fuel rate fr(t) during the time period $[t_1, t_n]$, as follows,

$$\begin{aligned} fc = \int_{t_1}^{t_n} fr(t)dt = \int_{t_1}^{t_n} \beta v(t)\{\left[\mu mg + \frac{1}{2}\varphi AC_r v^2(t)\right] \\ + ma(t)\left(\gamma + k_1' \mid \text{path.ts}| + k_2'|\text{path.n}| + k_3'|\text{path.poi}\ |\right)\}dt \end{aligned} \tag{5.13}$$

where t_1 and t_n refer to the time when the driver enters and leaves the given path, respectively.

In fact, since the speed and acceleration information cannot be continuously measured, we are not able to estimate the continuous fuel consumption rate. Only the discrete fuel consumption rates at the sampling times (same to the GPS trajectory data) can be estimated. Thus, it is impossible to calculate the total fuel consumption (i.e., the integration in Eq. (5.13) directly).

We mainly take two steps to approximate the total fuel consumption of a given path with only a limited and discrete points, detailed as follows.

Step 1: Divide the area under the curve (fr(t)) into several partitions with equal time interval (i.e., Δt), based on the sampling time information, as shown in Fig. 5.4. We note that although the essence of calculating the total fuel consumption is to get the area under the curve, how fr(t) changes during the time intervals is unknown.

Step 2: Approximate the area of each partition ($\hat{fc_i}$) based on the driving information provided by the GPS trajectory data (e.g., time, speed), as shown in Eq. (5.14),

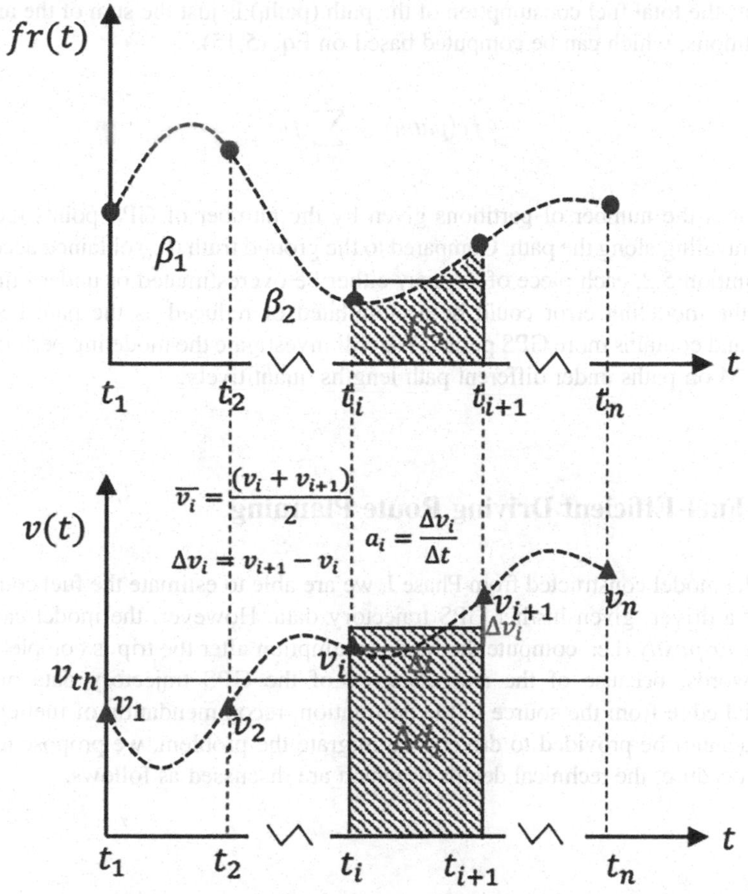

fr(t)
β1
β2
fc6
t
t1
t2
ti
ti+1
tn
v̄i = (vi + vi+1)/2
ai = Δvi/Δt
Δvi = vi+1 − vi
v(t)
vi+1
vn
Δvi
vi
Δt
vth
v1
v2
Δd
t
t1
t2
ti
ti+1
tn

Fig. 5.4 Total fuel consumption rate at different sampling times (top); speed information at different sampling times (bottom)

$$\hat{fc}_i = \beta\Big(k_1 \Delta d_i + k_2 \overline{v}_i^3 \Delta t + k_3 \overline{v}_i \Delta v_i + k_4 \overline{v}_i \Delta v_i \mid \text{path.ts} \mid \\ + k_5 \overline{v}_i \Delta v_i \mid \text{path.n} \mid + k_6 \overline{v}_i \Delta v_i \mid \text{path.poi} \mid \Big) \tag{5.14}$$

where Δt is the sampling time interval of GPS trajectory data. Definitions on Δv_i, Δd_i, $\overline{v}_i$ are shown in the bottom figure in Fig. 5.4. k_1, k_2, k_3, k_4, k_5 and k_6 are constant coefficients but vary among different drivers, which can be easily derived from the driver's historical data (i.e., GPS trajectory data and OBD-II data). The information about the number of traffic lights/stop signs, the neighboring edges, and the specific POIs of the path can be extracted from OpenStreetMap (OSM) platform. Note that we need to derive two βs (i.e., β_1 and β_2) in our model based on $\overline{v}_i$ and v_{th}. For instance, β_1 should be used to model the first partition since $\overline{v}_1 < v_{th}$, as shown in Fig. 5.4.

Thus, the total fuel consumption of the path ($path_i$) is just the sum of the areas of all partitions, which can be computed based on Eq. (5.15),

$$\hat{fc}(path_i) = \sum_{i=1}^{n} \hat{fc}_i \tag{5.15}$$

where n is the number of partitions given by the number of GPS points recorded when traveling along the path. Compared to the ground truth (fc_i) obtained according to Definition 5.2, each piece of $\hat{fc}_i$ may either be overestimated or underestimated. Thus, the modeling error could be accumulated or reduced as the path becomes longer and contains more GPS points. We will investigate the modeling performance of PFCM on paths under different path lengths quantitively.

5.5 Fuel-Efficient Driving Route Planning

With the model constructed from Phase I, we are able to estimate the fuel consumption of a driver, given his/her GPS trajectory data. However, the model can only operate *ex-postly* (i.e., compute the fuel consumption after the trip is completed). In other words, because of the *unavailability* of the GPS trajectory data on each potential edge from the source to the destination, recommendation of fuel-efficient routes cannot be provided to drivers. To migrate the problem, we propose a three-step procedure, the technical details of which are discussed as follows.

5.5.1 Real-Time Traffic Information Collection

The core of the real-time traffic information collection is to predict the information about GPS trajectory (e.g., the number of sampling points, the speed information of each point) on an edge if the driver would travel along it, which is proved to be quite challenging [15]. Here, to simplify the issue, we assume that we just have abundant real-time GPS trajectory data generated by other drivers on most edges during a time period T. We believe that the assumption is reasonable, as in recent years there are an increasing number of mobile crowdsensing apps that share real-time GPS trajectory data and traffic information, such as Waze4. It is obviously that the longer the T, he higher portion of edges having trajectory data left by other drivers. For edges having no trajectory data left during the time period, we simply assume that the driver would traverse them with a constant speed. One potential negative effect if given a longer T is that the traffic information may not represent the true situation. The data is input into the **PFCM** and we get the potential fuel consumption on that edge for the given driver.

More specifically, for an edge, after the driver submits his request at time t_0, we first retrieve all the real-time GPS trajectory data contributed by all passing-by drivers on that edge during the time period $[t_0, t_0 - T]$. We adopt two methods to select *a single trajectory* among all the trajectories on the edge, and use it to represent the case if the given driver travels on it.

- **Method 1**: we just keep the trajectory with the most similar cumulative travel distance ($\sum_{i=1}^{n} \Delta d_i$, where n is the number of GPS points on the edge) to the actual edge length.
- **Method 2**: for all the trajectories on the edge, we first classify the trajectories into three categories according to fuel consumption performance of the contributed drivers (i.e., high, mediate, economic). From historical data, it is trivial to label the driver who submits the request. Then, among trajectories with the same label to the driver, we apply the same principle as in **Method 1** to determine the trajectory.

The travel time on the edge can also be easily obtained by computing the time difference between the last GPS entry leaving the edge and the first GPS entry entering the edge. The selected GPS trajectory data is denoted as a pair of *driverId* and *edgeId*, e.g., <dr_i, e_j>. The effectiveness of collecting real-time information methods via the Mobile Crowdsensing (MCS) manner under varied settings for route-planning will be evaluated.

5.5.2 Fuel Consumption Estimation

With the real-time traffic information collected, based on PFCM for each driver, we are able to estimate the potential fuel consumption on a path ($Path_i$) to be travelled for the given driver (dr_g) by summing the fuel consumed on all edges $\left(e_k|_{k=1}^{N}\right)$ belonging to the path, as shown in Eq. (5.16).

$$\hat{fc}(dr_g, Path_i) = \sum_{k=1}^{N} \hat{fc}(dr_g, \langle dr_o, e_k \rangle) \tag{5.16}$$

5.5.3 Fuel-Efficient Path-Planning

From a source (S) to a destination (D), to obtain the most fuel-efficient path for a specific driver, an intuitive idea is to enumerate all possible paths, then estimate and compare the potential fuel consumption for each path based on the two steps discussed earlier (i.e., *Real-time Traffic Information Collection and Fuel*

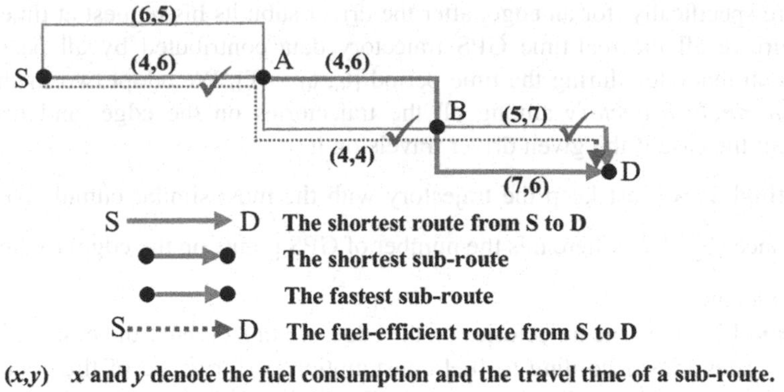
(6,5)
S
(4,6)
A
(4,6)
B
(5,7)
(4,4)
D
(7,6)
S → D The shortest route from S to D
The shortest sub-route
The fastest sub-route
S·······→ D The fuel-efficient route from S to D
(x,y) x and y denote the fuel consumption and the travel time of a sub-route.

Fig. 5.5 An illustrative example of the fuel-efficient route planning algorithm

Consumption Estimation) for each path, and pick up the one with the least fuel as the recommendation to the driver. However, the intuitive solution is not practical because discovering all possible paths for a given source-destination (SD) pair is a well-known NP-hard problem.

To offer a high-quality path in terms of fuel consumption for the driver within an acceptable time, we develop a heuristic algorithm. The basic idea of the heuristic is illustrated in Fig. 5.5, which can be summarized as follows.

- For a given SD pair, find the shortest path via Dijkstra's algorithm.
- Divide the shortest path into a number of *sub-routes* (a subsequence of edges) with similar distances. Measure the sub-routes with two metrics, one is the potential fuel consumption and the other is the travel time. In the example presented in Fig. 5.5, there are three sub-routes for the shortest path from S to D.
- For each sub-route, find the path with the shortest travel time (i.e., *the fastest path*) from the starting node to the ending node. If the potential fuel consumption of the new path is less than the path with the shortest distance, then it will be recommended as the driving route for this partition. Otherwise, the shortest sub-route will be recommended. We demonstrate the process in Fig. 5.5. By connecting the recommended sub-routes, we obtain the final suggested driving route.

Remark The shortest path is initiated as the driving route for the given SD pair because intuitively more fuel is consumed if the driver travels for a longer distance. However, in some cases, driving along the fastest path could cost less fuel, that is why we compare the potential fuel consumption between the shortest path and the fastest one [7]. The motivation of dividing the full path into several sub-routes is that PFCM achieves the best performance (i.e., minimizes the fuel consumption estimation error) when each sub-route approximates a certain value.

5.6 Evaluations

5.6.1 *Experimental Setup*

5.6.1.1 Data Preparation

In our experiments, we use several urban datasets in the city of Beijing, China. The datasets are gathered from multi-sources, i.e., the road network, the physical feature data extracted from OpenStreetMap, the GPS trajectory, and the OBD-II data generated by 595 taxis from September 15th to 21st, 2015. Some basic information about the cars can be found in Table 5.1. The mean travel distance and the mean working time for each taxi were 177.8 km and 14.59 h, respectively. We use 30% of the total data to train the PFCM and derive the parameters, and use the rest data to test the model. Note that some necessary data pre-processing techniques such as *de-noising, data correction/interpolation, map matching* on the GPS trajectory and OBD-II data are applied before the modeling.

Two major metrics, i.e., the error and the mean error, are defined to evaluate the modeling accuracy.

$$\text{error}_i = \frac{\hat{fc}(\text{path}_i) - fc(\text{path}_i)}{fc(\text{path}_i)} \times 100\% \tag{5.17}$$

$$m\_\text{error} = \frac{\sum_{i=1}^{m} \text{error}_i}{m} \tag{5.18}$$

where $\hat{fc}(\text{path}_i)$ is the fuel consumption on the given path estimated via the built model; $\text{fc}(\text{path}_i)$ is the true fuel consumed on that path, which is obtained by computing the fuel usage between the first GPS point entering and the last one leaving the given path; m is the total number of paths evaluated.

5.6.1.2 Baseline Models

To show the effectiveness of our proposed **PFCM**, we com-pare it with the following four baseline models.

Table 5.1 Basic information about the cars in the experiment

Car make	Car model	Car number
Volkswagen	Passat	188
Toyata	Camry	185
BMW	BMW5	22
Buick	LaCROSSE	66
Buick	GL8	85
Hyundai	KIA	31

- **GreenGPS**—This model was first proposed in [7] and further matured in [8]. It mainly differs from our **PFCM** in two aspects. One is that **GreenGPS** does not consider the extra fuel consumption when drivers change their driving behaviors proactively (e.g., slow down in advance) when traverse some specific POIs. The other is that **GreenGPS** does not account for the fact that the specific fuel consumption (i.e., SFC, β) would vary when driving at different speeds.
- **$\textbf{PFCM}_\textbf{v}$**—This model is a variant of our proposed **PFCM**, in which only a single β is derived.
- **AFCM**—The average fuel consumption model (**AFCM**) simply multiply the average fuel consumption (C_i) of the driver in history with the total distance of the path to approximate the total fuel consumed on the path.
- **CFCM**—The constant fuel consumption model (**CFCM**) first derives a constant average fuel consumption (C) which minimizes the following Eq. (5.19). Then, the estimation of the fuel consumption on a path is computed by multiplying C with the path distance, regardless of the detailed path and the individual driving behaviors.

$$\arg min \left[C \times \sum_{i=1}^{n} td_i - \sum_{i=1}^{n} fc(i) \right] \tag{5.19}$$

where td_i is the total travel distance of the ith driver; fc(i) is the true value of fuel consumption of the ith driver during the same day; n is the total number of taxis.

5.6.1.3 Evaluation Environment

All the evaluations in the section are run in MATLAB on an Intel Core i5-4460 PC with 8-GB RAM and Windows 10 operation system.

5.6.2 Evaluation on PFCM

5.6.2.1 Comparison Results to Baselines

We compare the model accuracy of our proposed **PFCM** to four baseline models, with results shown in Fig. 5.6. The x-axis refers to the absolute error value of the model ac-curacy and y-axis refers to the corresponding Cumulative Distribution Functions (CDFs) of the number of paths. As shown in the figure, **PFCM** achieves the best performance consistently. To be more specific, PFCM is able to get a remarkable accuracy on the estimation of fuel consumption. For instance, over 80% of paths are with an absolute error less than 25%, much better than that of **$\textbf{PFCM}_\textbf{v}$**. This demonstrates the effectiveness of differentiating the slow and normal

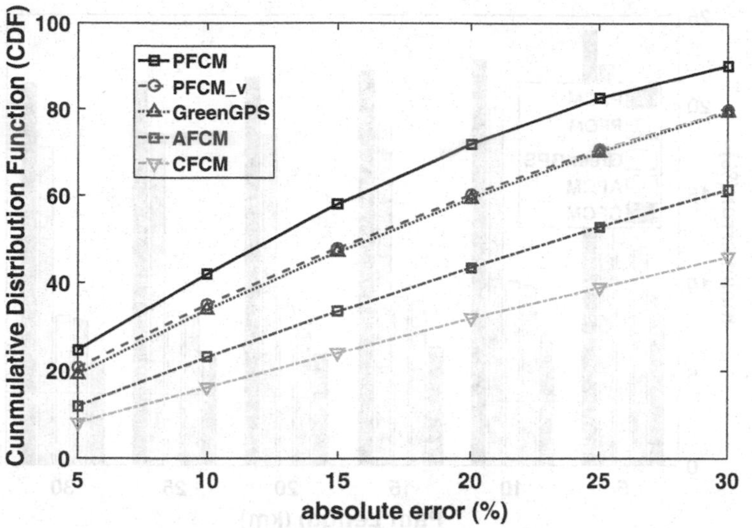
Cunmulative Distribution Function (CDF)
100
80
60
40
20
0
5
10
15
20
25
30
absolute error (%)
PFCM
PFCM_v
GreenGPS
AFCM
CFCM

Fig. 5.6 Comparison results among **PFCM** and four baseline models

driving cases. Moreover, the variant version of **PFCM**, **PFCM$_v$** also achieves a promising accuracy. The **PFCM$_v$** performs slightly better than the state-of-art model (i.e., **GreenGPS**), showing the effectiveness of introducing extra fuel consumption caused by changing driving behaviors proactively.

The two simple models (i.e., **AFCM** and **CFCM**), how-ever, fail to provide a meaningful accuracy. Because these two models only consider the length of a path, and information such as real-time traffic conditions, road network topology, and driving behaviors, which is of crucial importance for the fuel consumption, is not included in the modeling process.

5.6.2.2 Model Performance on Paths Under Varied Lengths

In order to better understand how models perform on paths of different lengths qualitatively, we bin the paths based on their length and compute the mean errors. The corresponding results are shown in Fig. 5.7. For all models, the model accuracy gets higher as path length increases. As can be predicted, **PFCM** again performs consistently better that the others, and **AFCM** and **CFCM** suffer the worst performance under all path lengths. In more details, PFCM achieves a mean fuel consumption error less than 8% when the path length is 10 km, and a quite stable mean error which is close to 5% when the path length is longer than 15 km.

With the above results, drivers are not only able to control their expectation of the modeling accuracy, but also the quality of the fuel-efficient driving routes. The driving routes suggested by Phase II of the system are of higher quality and more

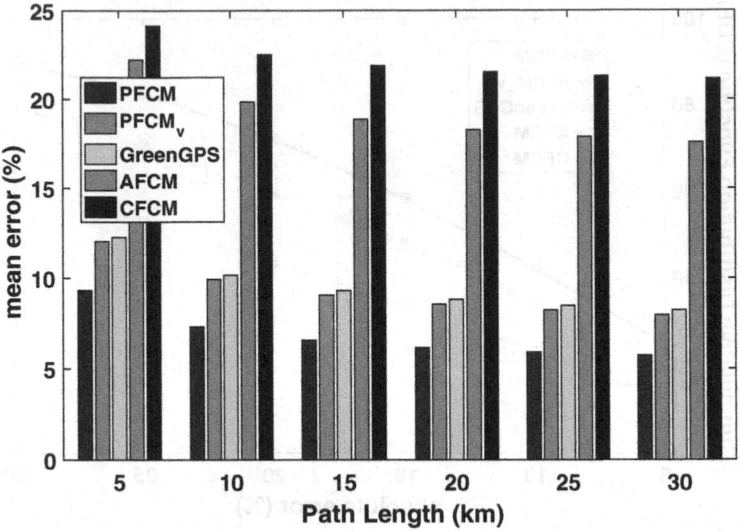
PFCM
PFCM_V
GreenGPS
AFCM
CFCM
mean error (%)
Path Length (km)
0
5
10
15
20
25
30

Fig. 5.7 Model performance of **PFCM** and other four baselines by varying the length of the path

advisable if destinations are distant, while the suggested driving routes are more error-prone when the destinations are close.

5.6.3 A Deep Look at PFCM Coefficients

Based on the historical driving trajectories, with our proposed **PFCM**, one can derive the coefficients (i.e., $k_1 \sim k_6$) for each driver. To get an in depth understanding of the proposed **PFCM**, we investigate the relationships among the fuel consumption performance, car models and **PFCM** coefficients. More specifically, we aim at addressing the concrete question: Do **PFCM** models for drivers who have similar fuel consumption performance present the similar coefficients?

To answer the question, we mainly take a four-step process, detailed as follows. (1) We represent a sample by the PFCM coefficients (i.e., $k_1 \sim k_6$) of each driver. (2) For each sample, we apply the popular principal component analysis (PCA) to get the corresponding latent features in a lower dimension space (i.e., three). (3) We apply k-means to cluster all samples in the latent space with $k = 3$, and get three clusters. (4) For each cluster, we calculate the average fuel consumption by averaging the fuel consumptions of all drivers in the cluster, as well as the standard deviation. As it is known to all, samples within the same cluster have similar **PFCM** coefficients. We then compute the average fuel consumption and standard deviation for each cluster. If the standard deviation within the same cluster is small and the average fuel consumption varies among different clusters, then we can safely draw the conclusion that **PFCM** model's coefficients are highly correlated with the

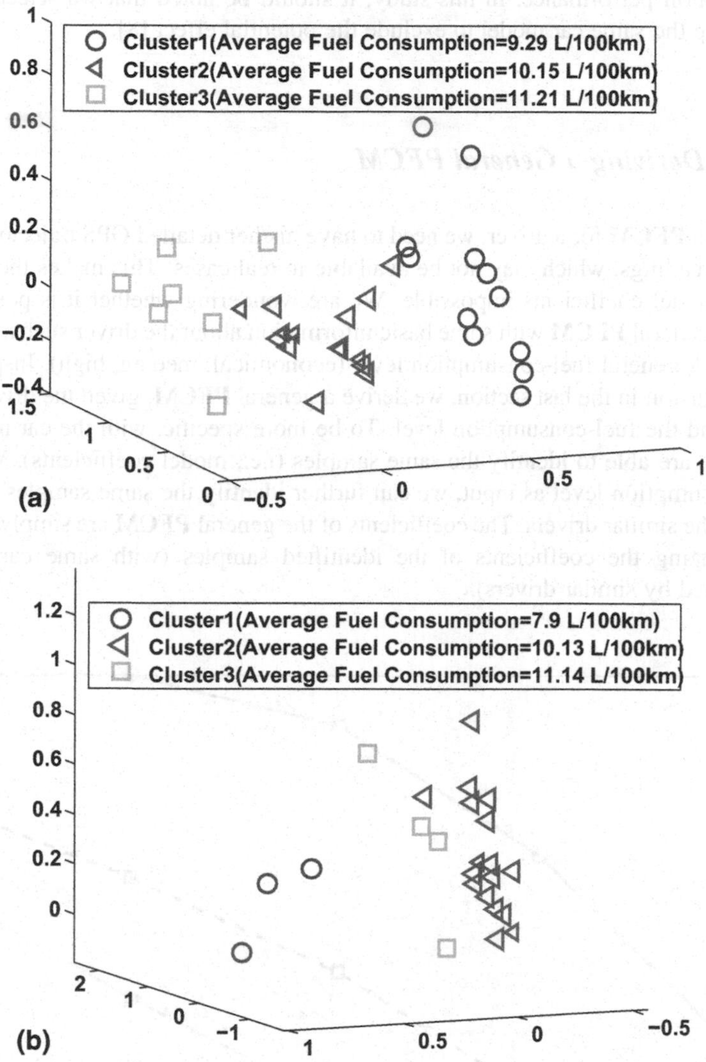
Cluster1(Average Fuel Consumption=9.29 L/100km)
Cluster2(Average Fuel Consumption=10.15 L/100km)
Cluster3(Average Fuel Consumption=11.21 L/100km)
(a)
Cluster1(Average Fuel Consumption=7.9 L/100km)
Cluster2(Average Fuel Consumption=10.13 L/100km)
Cluster3(Average Fuel Consumption=11.14 L/100km)
(b)

Fig. 5.8 Clustering results of PFCM coefficients for two car models, (**a**) Camry; Number: 34 and (**b**) Passat; Number: 25 respectively

fuel consumption performance of drivers, i.e., drivers who have similar fuel consumption performance would present similar coefficients. Results shown in Fig. 5.8 verify our conclusion. For *Camry* cars, the standard deviations for three clusters are 0.7317, 1.495, 0.4698, respectively. For Passat cars, the standard deviations for three clusters are 0.7954, 1.1111, 0.3883, respectively. The three clusters roughly correspond to economical, median, and high fuel consumption drivers. What is more, the conclusion also inspires us to build a general model for drivers with similar fuel consumption performance. In this study, it should be noted that we select drivers who drive the same car model to exclude the potential effect [8].

5.6.4 Deriving a General PFCM

To build a **PFCM** for a driver, we need to have his/her detailed GPS trajectories and OBD-II readings, which may not be available in real cases. This makes the derivation of model coefficients impossible. We are wondering whether it is possible to derive a general **PFCM** with some basic information about the driver such as his/her car model, general fuel-consumption level (economical, median, high). Inspired by the conclusion in the last section, we derive a general **PFCM**, given the driver's car model and the fuel-consumption level. To be more specific, with the car model as input, we are able to identify the same samples (i.e., model coefficients). With the fuel-consumption level as input, we can further identify the same samples contributed by the similar drivers. The coefficients of the general **PFCM** are simply derived by averaging the coefficients of the identified samples (with same car model, contributed by similar drivers).

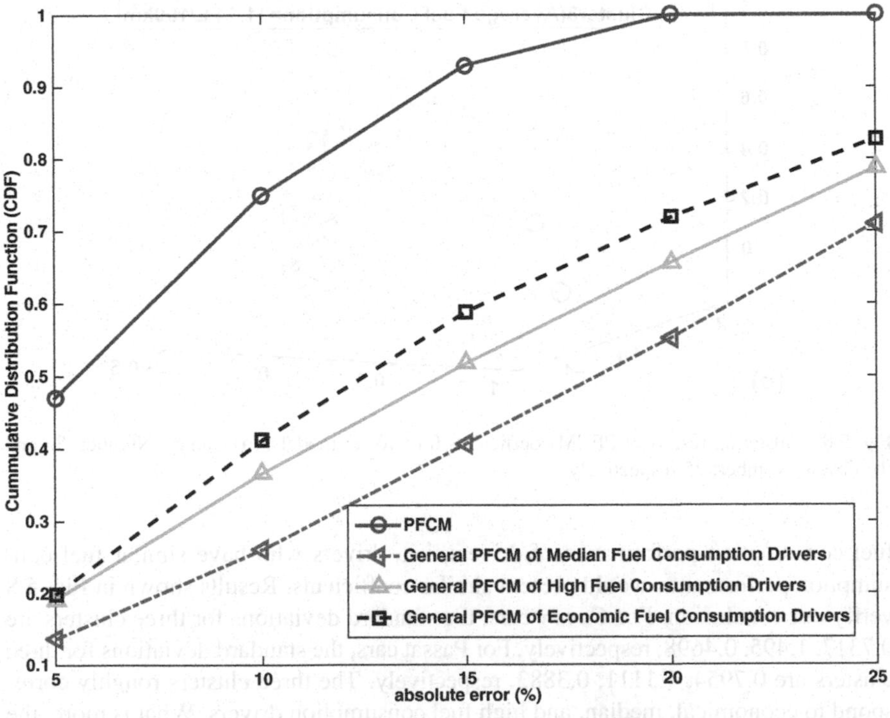

Cummulative Distribution Function (CDF)
absolute error (%)
PFCM
General PFCM of Median Fuel Consumption Drivers
General PFCM of High Fuel Consumption Drivers
General PFCM of Economic Fuel Consumption Drivers

Fig. 5.9 Model accuracy of general **PFCM** for three levels of drivers

We report the model accuracy of the general **PFCM** in Fig. 5.9. For reference, the model accuracy of the **PFCM** is also provided. As expected, the model accuracy of general **PFCM** is lower than that of **PFCM**. Moreover, compared to the other two groups, the accuracy of the general PFCM for median fuel consumption drivers is the worst. We argue that this is probably because that the number of median drivers is the biggest, and also with the biggest standard deviation of the fuel consumption. Simply averaging the coefficients to derive the general **PFCM** may not able to model the median group at a high accuracy.

5.6.5 Evaluation on Path-Planning Algorithm

5.6.5.1 Effectiveness of Real-Time Traffic Information Collection Method

For a given path with recorded GPS trajectory and OBD-II data, we know the actual fuel consumed on it, which can be used as the ground truth to compare with. Meanwhile, at the given starting time of the path, with the proposed collection methods, we are also able to collect the real-time information (the potential GPS trajectory data) on each edge of the path. Then, based on PFCM, the amount of the potential fuel consumption can be computed according to Eq. (5.16). If the result is close to the ground truth, then we can claim that the real-time information collection via MCS is effective. Otherwise, it is ineffective.

We compute the *m_error* for all studied paths with length from 5 to 30 km based on the real-time information collected via MCS of **Method 1**, under different days and different time durations (T). Different days here refer to the number of days considered when counting the trajectories contributed by others on the edge in the real-time information collection. For instance, 1 day indicates that only trajectories generated yesterday on the edge is counted. The results are presented in Fig. 5.10. As a reference, results based on **PFCM** with the actual GPS trajectory data as input are also provided. We can see that with the real-time information collected via MCS, **PFCM** achieves an acceptable accuracy, with a mean absolute error less than 14% when the path length is 5 km, and an increasingly better accuracy when the path length is larger than 10 km, for all considered days. As expected, the estimation accuracy using the real-time information is worse than the corresponding one when the actual GPS trajectory data is used under all path lengths. Moreover, **PFCM** achieves the best accuracy when considering trajectories generated in the last 3 days. In order to gain the rationale behind, we propose the concept of the edge sparseness, which is defined as the ratio of the number of edge having no trajectory left by others in the considered days to the number of all edges in the road network. The edge sparseness should increase as the number of days grows. As discussed earlier, we assume drivers travel on edges at a constant speed if no other drivers left any trajectory on that edge during the considered time, which may be not correct in real cases. Therefore, it can be predicted that the overall accuracy of the model would

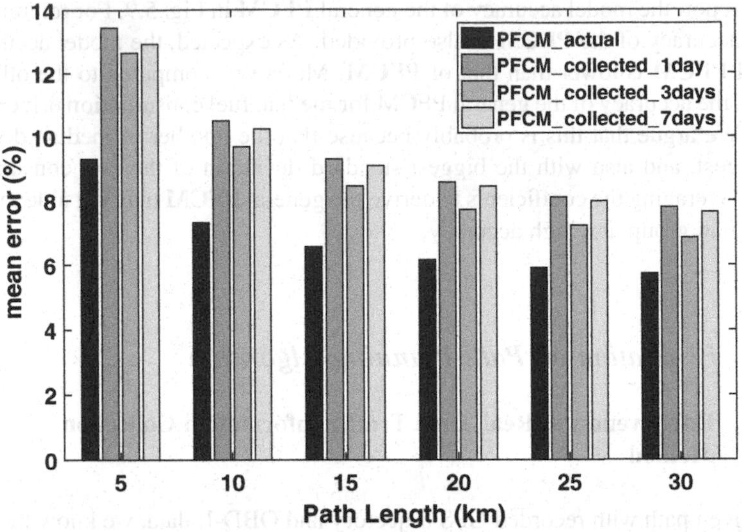
PFCM_actual
PFCM_collected_1day
PFCM_collected_3days
PFCM_collected_7days
mean error (%)
0
2
4
6
8
10
12
14
5
10
15
20
25
30
Path Length (km)

Fig. 5.10 Results of the effectiveness of the proposed **Method 1** of varied path lengths under different days

become worse if containing two many such edges (i.e., the edge sparseness is high). That is the reason why **PFCM** achieves better accuracy if collecting others' trajectories during the most recent 3 days than only 1 day. It is quite strange that PFCM achieves worse accuracy when collecting others' trajectories during the last week (7 days) at first glance. We argue that this is probably because the traffic patterns in work days are significantly distinctive from that in weekend. Note that we set T = 2 h in this study.

We also compare the **PFCM** performance under different lengths of time duration (i.e., T), as shown in Fig. 5.11. As can be seen, **PFCM** achieves the best accuracy when setting T = 0.5 h for all path lengths. This is probably due to the fact that the traffic is more likely to remain unchanged during a much narrower time window.

The performance of **PFCM** when adopting different real-time collection methods (i.e., **Method 1** and **Method** 2) are also compared, with the result shown in Fig. 5.12. As can be observed, the *m_error* is quite close between the two different methods. **Method 1** achieves slightly better accuracy for all path lengths, which demonstrates that the fuel consumption may be more tightly related to the driving distance, when comparing to the driving style.

5.6.5.2 Determination of Length of Sub-route

We investigate the accuracy of **PFCM** by varying the length of the sub-route qualitatively. The results are shown in Fig. 5.13, in which a valley point can be observed. More specifically, we first vary the length of sub-route from 0.5 to 1.6 km,

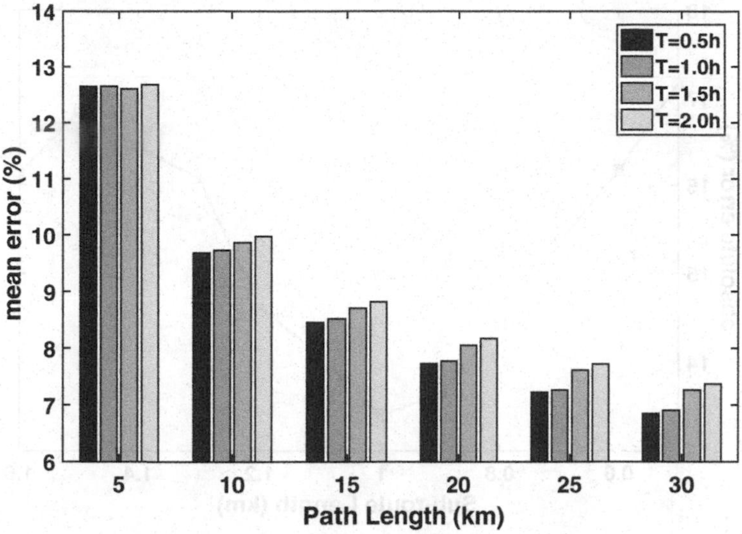
T=0.5h
T=1.0h
T=1.5h
T=2.0h
mean error (%)
Path Length (km)

Fig. 5.11 Results of the effectiveness of the proposed **Method 1** of varied path lengths under different Ts

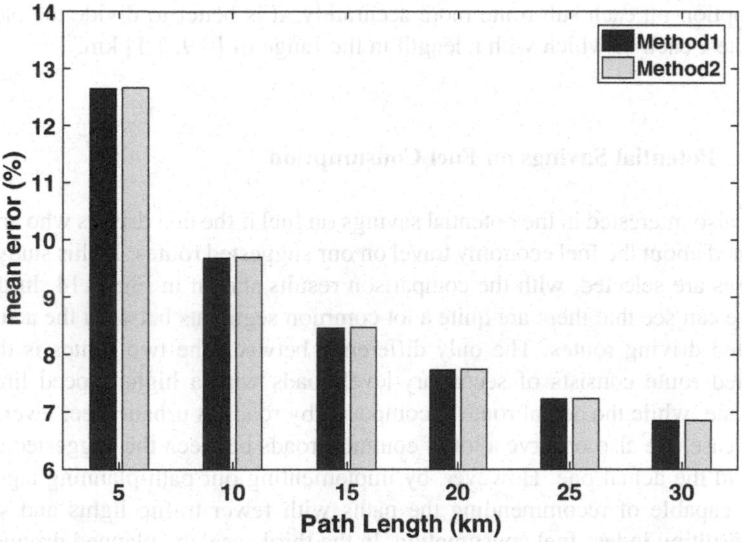
Method1
Method2
mean error (%)
Path Length (km)

Fig. 5.12 Comparison results between two different real-time information collection methods

with an equal interval of 0.1 km, then calculate the metric of *m_error* under each length of all sub-routes. As shown in the result, the model accuracy gets better when the path length goes from 0.5 to 1 km, and becomes worse as the path becomes longer. We can draw the conclusion that the best modeling performance is achieved

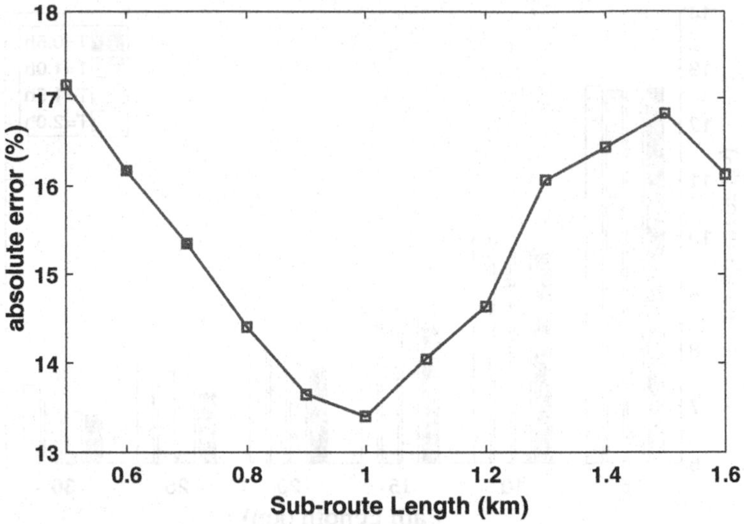
18
17
16
15
14
13
absolute error (%)
0.6
0.8
1
1.2
1.4
1.6
Sub-route Length (km)

Fig. 5.13 Model accuracy of **PFCM** by varying the length of the sub-route

when the length of sub-route is around 1 km. Therefore, to estimate the potential fuel consumption on each sub-route more accurately, it is better to divide the path into sub-routes, each of which with a length in the range of [0.9, 1.1] km.

5.6.5.3 Potential Savings on Fuel Consumption

We are also interested in the potential savings on fuel if the taxi drivers who are more concerned about the fuel economy travel on our suggested routes. In this study, three real cases are selected, with the comparison results shown in Fig. 5.14. In the first case, we can see that there are quite a lot common segments between the actual and suggested driving routes. The only difference between the two routes is that our suggested route consists of secondary-level roads with a higher speed limit and wider lane, while the actual route is composed by roads at urban-street level. In the second case, we also observe a lot of common roads between the suggested driving route and the actual one. However, by implementing our path-planning algorithm, we are capable of recommending the paths with fewer traffic lights and specific POIs, resulting in less fuel consumption. In the third case, the planned driving route is quit different from the actual one. Though the suggested route includes a slightly larger number of traffic lights/stop signs, it distributes much fewer specific POIs along.

We further provide the quantitative information about the suggested driving routes and the actual driving routes for the three cases discussed above. The information includes the total travel distance (in km) and the total amount of fuel consumption (in liter), the results of which are shown in Table 5.2. The savings on

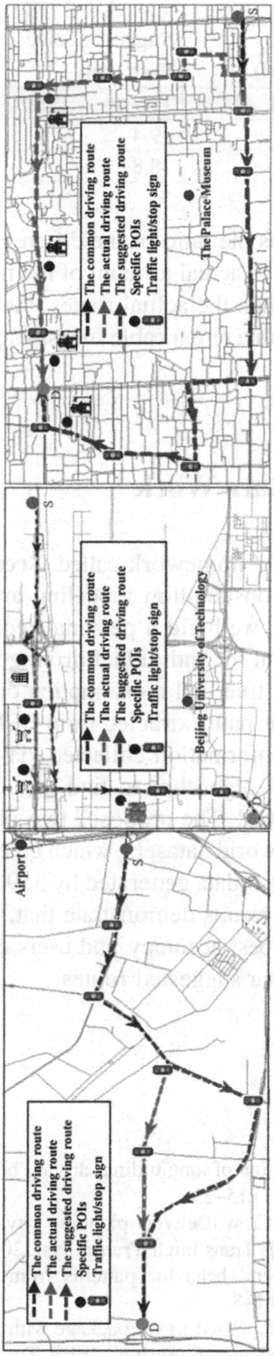
The common driving route
The actual driving route
The suggested driving route
Specific POIs
Traffic light/stop sign
Airport
S
D
Beijing University of Technology
The Palace Museum

Fig. 5.14 Comparison results between the suggested driving routes and the actual driving routes for three cases, respectively Case I (left); Case II (middle); Case III (right) The road network fragment gets denser from left to right, indicating the driver may encounter more complex traffic conditions

Table 5.2 Quantitative information about the suggested driving routes and the actual driving routes

	Actual route		Suggested route		Potential
Case	Dist. (km)	Fuel (l)	Dist. (km)	Fuel (l)	Savings (%)
I	10.0	1.258	11.7	0.992	21.1
II	7.9	1.6	9.4	1.29	19.38
III	10.1	1.37	8.8	1.08	21.17

the fuel consumption is defined as the ratio of the difference of total fuel consumption from two types of routes to the actual amount of the fuel consumption. Though the suggested routes are longer than the actual routes, the fuel consumption on the suggested routes is much less, with a remarkable fuel saving around 20% on average for all three cases.

5.7 Conclusions and Future Work

In this chapter, we present a novel framework called **GreenPlanner** to address the issues of the personalized fuel consumption modeling and the fuel-efficient path-planning. In the modeling phase, we build a personalized fuel consumption model (PFCM) for each driver, based on the individual driving behaviors along the path recorded by GPS trajectory data, true fuel consumption on the corresponding path, and the physical features along the path extracted from OSM platform. In the second phase, with the real-time traffic information collected via the mobile crowdsensing manner, we proposed a heuristic algorithm to find the most fuel-efficient for the given driver based on his/her PFCM. The proposed two-phase framework had been extensively evaluated using real-world datasets, which consist of road network, POI, the GPS trajectory data and OBD-II data generated by 559 taxis in a week in the city of Beijing, China. Experimental results demonstrate that, compared to the baseline models, our model achieves the best accuracy and users could save a considerable amount of fuel if driving along our suggested routes.

References

1. Wang J, Lu M, Li K. Characterization of longitudinal driving behavior by measurable parameters. Transp Res Rec. 2010;2185(1):15–23.
2. Chen C, Zhang D, Ma X, et al. CrowdDeliver: planning city-wide package delivery paths leveraging the crowd of taxis. IEEE Trans Intell Transp Syst. 2016;18(6):1478–96.
3. Liang ACR. Uncovering cabdrivers' behavior patterns from their digital traces. Comput Environ Urban Syst. 2010;34(6):541–8.
4. Yu Z, Xu H, Yang Z, Guo B. Personalized travel package with multi-point-of-interest recommendation based on crowdsourced user footprints. IEEE Trans Hum Mach Syst. 2016;46 (1):1–8.

5. Chang X, Chen BY, Li Q, Cui X, Tang L, Liu C. Estimating real-time traffic carbon dioxide emissions based on intelligent transportation system technologies. IEEE Trans Intell Transp Syst. 2013;14(1):469–79.
6. Jiang L, Chen X, He W. SafeCam: analyzing intersection-related driver behaviors using multi-sensor smartphones. In: IEEE international conference on pervasive computing and communications (PerCom), 2016. p. 1–9.
7. Ganti R, Pham N, Ahmadi H, Nangia S, Abdelzaher T. GreenGPS: a participatory sensing fuel-efficient maps application. In: ACM international conference on mobile systems, applications, and services, 2010. p. 151–164.
8. Saremi F, Fatemieh O, Ahmadi H. Experiences with GreenGPS—fuel-efficient navigation using participatory sensing. IEEE Trans Mob Comput. 2016;15(3):672–89.
9. Ahn K, Rakha H, Trani A, Van Aerde M. Estimating vehicle fuel consumption and emissions based on instantaneous speed and acceleration levels. J Transp Eng. 2002;128(2):182–90.
10. Chan TL, Ning Z. On-road remote sensing of petrol vehicle emissions measurement and emission factors estimation in Hong Kong. Atmos Environ. 2004;38(14):2055–66.
11. Qian J, Eglese R. Fuel emissions optimization in vehicle routing problems with time-varying speeds. Eur J Oper Res. 2016;248(3):840–8.
12. Zhang J, Zhao Y, Xue W, Li J. Vehicle routing problem with fuel consumption and carbon emission. Int J Prod Econ. 2015;170:234–42.
13. Hu X, Leung V, Li KG. Social drive: a crowdsourcing-based vehicular social networking system for green transportation. In: ACM international symposium on design and analysis of intelligent vehicular networks and applications, 2013. p. 85–92.
14. Kang L, Qi B, Dan J, Banerjee S. EcoDrive: a mobile sensing and control system for fuel efficient driving. In: ACM international conference on mobile computing and networking, 2015. p. 358–371.
15. Shang J, Zheng Y, Tong W, Chang E, Yu Y. Inferring gas consumption and pollution emission of vehicles throughout a city. In: ACM SIGKDD international conference on knowledge discovery and data mining, 2014. p. 1027–1036.
16. Wijayasekara D, Manic M, Gertman D. Data driven fuel efficient driving behavior feedback for fleet vehicles. In: IEEE international conference on human system interaction (HSI), 2015. p. 75–81.
17. Beusen B, et al. Using on-board logging devices to study the longer-term impact of an eco-driving course. Transp Res D Transp Environ. 2009;14(7):514–20.
18. Alam S, Mcnabola A. A critical review and assessment of eco-driving policy & technology: benefits & limitations. Transp Policy. 2014;35:42–9.

Part III
Enabling Smart Urban Services: Passengers

Chapter 6
iBOAT: Detecting Anomalous Trajectories On-the-Fly

6.1 Introduction

The proliferation of taxi GPS trajectory data has opened up valuable opportunities in enabling smart urban services from several perspectives. However, few of them are in a passenger-centered fashion, since the taxi trajectory data can only reveal limited information regarding passengers compared to drivers. Specifically, the accumulatively collected taxi GPS trajectory data contains the entire driving routes of taxi drivers, but only the trajectories left within delivering events (i.e., passenger-carrying trajectories) are relevant to passengers.

In recent years, the detection of anomalous trajectories has attracted increasing research efforts. Such work can benefit smart services like the surveillance of adverse events in urban/maritime transportation systems [1–3]. As mentioned before, the passenger-carrying trajectories can show how passengers are delivered explicitly, i.e., the routes chosen by taxi drivers. It is well known that different routes may correspond to different distances and traffic jam duration, resulting in the difference in taxi fares and time cost for taxi passengers. Unfortunately, in reality, many greedy taxi drivers would overcharge passengers by deliberately taking unnecessary detours [4]. Hence, detecting such dishonest driving behaviors is quite important to maintain the reputation of taxi service in cities. In this chapter, we aim to detect anomalous passenger-delivering trajectories to enable the fraud alert service for passengers.

Traditionally, fraudulent behaviors are passively detected through the complaints from passengers. However, such approach is not effective enough for the reason that passengers in an unfamiliar city might even not notice the frauds. Fortunately, those

---

Part of this chapter is based on a previous work: C. Chen, D. Zhang, P. S. Castro, N. Li, L. Sun, S. Li, Z. Wang, "iBOAT: Isolation-Based Online Anomalous Trajectory Detection," in IEEE Transactions on Intelligent Transportation Systems, vol. 14, no. 2, pp. 806–818, June 2013, doi: https://doi.org/10.1109/TITS.2013.2238531.

C. Chen et al., *Enabling Smart Urban Services with GPS Trajectory Data*,
https://doi.org/10.1007/978-981-16-0178-1_6

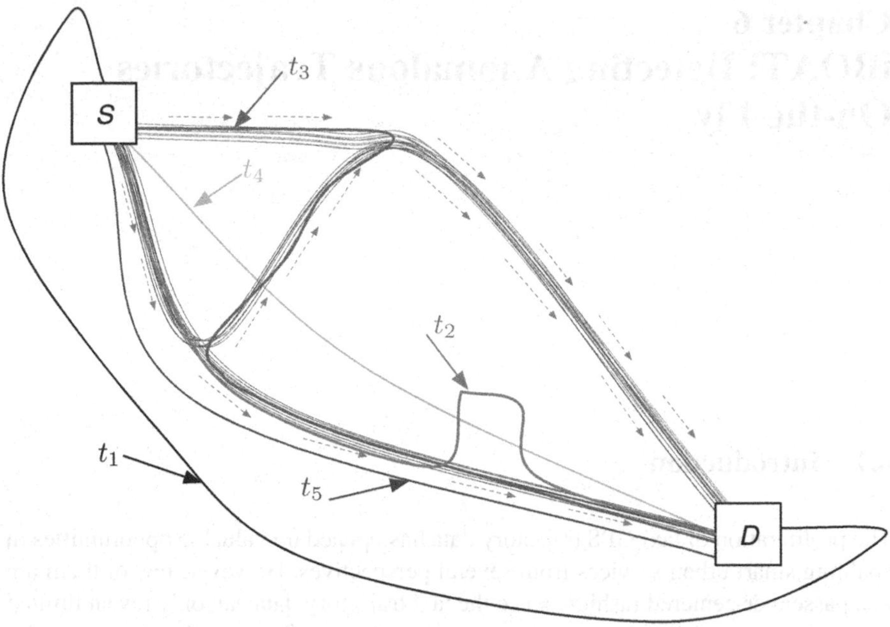
S
D
t_1
t_2
t_3
t_4
t_5

Fig. 6.1 Example taxi trajectories between *S* and *D*

anomalous trajectories are often infrequent and differ from the majority of other "normal" ones. In this sense, a possible solution to identify them is to conduct a comparison between them and large-scale historical traces. Generally, an effective anomalousness detection method should include four characteristics:

1. Accurate classification: It means the algorithm should achieve a great identification accuracy and a small false-alarm probability.
2. Subtrajectory specificity: The method should not only be able to find those anomalous trajectories, but also tell which segments cause the anomalousness.
3. Real-time response: It is expected that the anomalous behavior can be detected in real time, so as to trigger the alarm as soon as possible.
4. Characterizing the anomaly degree: A scoring mechanism is necessary to be exploited to measure the anomalous intensity of trajectories, which could be employed to rank a set of trajectories.

In this chapter, we only consider trajectories traveling from a specified area S to another area D. Figure 6.1 shows three groups of "normal" trajectories between S and D, along with five anomalous trajectories (t_1 through t_5). We can find that some anomalous ones show very distinct routes (t_1, t_4, t_5), and some show unnecessary detours within the trajectory (t_2, t_3). Moreover, some "abnormal" traces can be caused by fraudulent behaviors from greedy drivers (t_1 and t_2 in Fig. 6.1). At meantime, they might be shortcuts or newly founded paths taken by seasoned drivers

(t_4 and t_5 in Fig. 6.1). Generally, it is no trivial task to detect those anomalous trajectories. The underlying challenges are as follows.

1. Several different normal paths might exist between S and D. The clusters are often with different densities and separated from each other. Additionally, normal trajectories would also show different driving distances, so that it is difficult to detect all the anomalous trajectories by simply utilizing driving distance or density.
2. Characterizing anomalous trajectory is not a straightforward task due to the diversity of anomalousness. For example, t_1, t_2, t_3, t_4 and t_5 in Fig. 6.1 are regarded as anomalous for quite different reasons.
3. The concept of anomalous trajectory might drift over time because of the changes of road networks like newly built or blocked roads. Hence, it necessitates the capability in capturing these changes and incorporating them into the model.
4. The uncertainty of GPS trajectories is a serious problem that is usually caused by the low and unstable collection frequency of GPS devices.

With the aforementioned characteristics and challenges in mind, we propose a new anomalous trajectory detection approach in this chapter. The main contributions are listed as follows.

1. An isolation-based online anomalous trajectory (*iBOAT*) detection method is proposed to solve the above four challenges while still showing the aforementioned four characteristics. First, we convert the anomalous trajectory detection task to discovering the anomalous ones from the majority with regard to an S-D pair. Specifically, we evenly partition the urban map into grid cells, then cluster taxi trips between the S-D pair. Next, every trajectory is represented as a sequence of visited cells. Compared with the normal taxi trajectories, those "few" and "different" trajectories are defined as anomalous ones.
2. We perform an empirical evaluation comparing *iBOAT* and other state-of-the-art methods with the real-world GPS traces collected from 7600 taxis for 1 month. Results show the superior performance of our method, and demonstrate its effectiveness in identifying anomalous segments.
3. A systematic analysis is conducted based on the detection results. We find that most anomalous rides are caused by deliberate fraudulent decisions by greedy taxi drivers. Furthermore, we also offer evidence to reject potential excuses of cunning drivers for their fraud.

6.2 Related Work

Existing studies on the anomalous trajectory detection mainly focused on addressing specific forms of anomaly. For example, trajectories with very long travel distances are regarded as anomalous ones in [5]. In [6], the combination of distance, density and the Dempster–Schafer theory is used to identify dishonest taxi trips. Similar to that work, Lee et al. [7] first evenly divide a trajectory into several segments, then employed the distance and density based methods to identify the anomalousness of each segment. However, it is difficult to correctly identify anomalousness by simply utilizing the driving distance and density. Bu et al. [8] proposed one kind of outlier detection algorithm to find outlier from trajectory streams by using the clustering mechanism. Ge et al. [9] formulated an analogous task of the detection of streaming outliers, in which the anomaly degree is derived from the streaming direction and trajectory density. Finally, several learning approaches have been adopted to detect the anomalous trajectory [2, 10, 11]. But such approaches often necessitate large-scale labelled training data which is hard to collect. Li et al. [12] solved a related but different issue. They identify abnormal road sections by detecting sharp changes between current data and historical trends. Their method detected what could be flagged as global anomaly incidents. These incidents affected many taxis, thus their method cannot find abnormal behavior on a single level. Different from these studies, we target to detect those few anomalous taxi trajectories from the whole trajectory set with regard to a specific S–D pair, which show distinct routes from the majority.

Based on the idea of isolating anomalies [13], our previous work [14, 15] proposed a method that identifies trajectories as anomalous when they follow paths that are rare with respect to historical trajectories. Differences between them and the study in this chapter lie in three points.

First, this chapter presents a new anomalousness scoring method which examine the anomalous segments together with the number of involved/supporting trajectories. Those anomalous ones would be scored high in ranking if they show a long distance segment and less support. Second, the effect of the anomaly threshold and the size of the set of historical trajectories on the detection performance are investigated. The trade-off between the performance (i.e., accuracy) and cost (i.e., time and memory) is controlled by developers. At last, we also analyze the underlying motivations behind the fraud behaviors, so that other applications are able to be exploited with the iBOAT algorithm.

6.3 Preliminaries and Problem Definition

Our GPS trajectory data was collected from 7600 taxis in Hangzhou, China, and the sample rate is about 60 s. Figure 6.2 shows the trajectories for one taxi during a month; the red lines indicate when the taxi is occupied, whereas the blue lines indicate when it is vacant. In this study, we will only use occupied trajectories since fraud detection is one of the motivations for this study, and fraud can only be committed with a passenger.

Definition 6.1 A trajectory t consists of a sequence of points $\langle p_1, p_2 \cdots, p_n \rangle$, where $p_i \in \mathbb{R}^2$ is the physical location (i.e., latitude/longitude). We will use t_i to reference position i in t, and for any $1 \leq i \leq j \leq n$, $t_{i \rightarrow j}$ denotes the subtrajectory $\langle p_i, \ldots, p_j \rangle$.

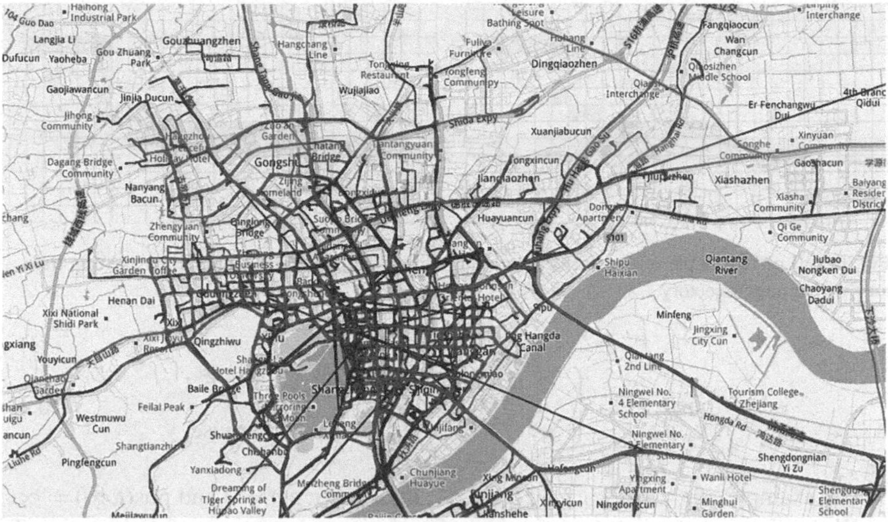
Guo Dao
Langjia Li
Dufucun
Yaoheba
Gou Zhuang Park
Gaojiawancun
Jinjia Ducun
Jihong Community
Dagang Bridge Community
Nanyang Bacun
Zhengyuan Community
Henan Dai
Xixi National Shidi Park
Youyicun
Westmuwu Cun
Feilai Peak
Shangtianzhu
Pingfengcun
Yanxiadong
Qingzhiwu
Baile Bridge
Gongshu
Wujiajiao
Dingqiaozhen
Xuanjiabucun
Huayuancun
Hangda Canal
Chunjiang Huayue
Fangqiaocun
Wan Changcun
Er Fenchangwu Dui
Xinyuan Community
Gaoshacun
Xiashazhen
Xiasha Community
Qi Ge Community
Qiantang River
Jiubao Nongken Dui
Chaoyang Dadui
Minfeng
Jingxing City Cun
Ningwei No. 4 Elementary School
Ningwei No. 3 Elementary School
Tourism College, Zhejiang
Shengdongnian Yi Zu
Wanli Hotel
Minghui Garden
Ningdongcun
Hangzhou Dadao
Interchange

Fig. 6.2 Traces of a taxi in Hangzhou city during a month, where red or blue indicates the taxi is occupied or vacant

Points p_i exist in a continuous domain; therefore, directly dealing with them is difficult. To mitigate this problem, we assume that we have access to a finite decomposition of the area of interest. Specifically, we decompose the city area into a matrix G of grid cells, and we define $\rho : \mathbb{R}^2 \rightarrow G$ as a function that maps locations to grid cells. The criterion for choosing the grid cell size is to ensure the accuracy of the anomalous trajectory detection while maximizing the grid cell size. We experimented with different grid cell sizes and found that 250 m $\times$ 250 m is the biggest grid size with the set detection accuracy.

Definition 6.2 Mapped trajectory $\bar{t}$, which is obtained from trajectory t, consists of a sequence of cells $\langle g_1, g_2, \ldots, g_n \rangle$, where for all $1 \leq i \leq n$, $g_i \in G$, and $\overline{t_i} = \rho(t_i)$. We will write $g \in \bar{t}$ when $\overline{t_i} = g$ for some $1 \leq i \leq n$.

Henceforth, we will only deal with mapped trajectories; therefore, we will drop the mapped qualifier. Because of the rate at which GPS entries are received and the small size of our grid cells, the mapped points (black squares in Fig. 6.3) may not be adjacent, thereby leaving gaps. We augment all the trajectories to ensure that there are no gaps in the trajectories by (roughly) following the line segment (green line in left panel) between the two cells in question and "coloring" the cells underneath (gray cells in figure). Whereas the original trajectory consisted only of the black grids in Fig. 6.3, the augmented trajectory consists of both the black and gray grids.

Let $\mathbb{T}$ denote the set of all mapped and augmented trajectories. Define function $\mathbb{T} \times G \rightarrow \mathbb{N}^+$, given that trajectory t and element g returns the first index in t that is equal to g, as follows:

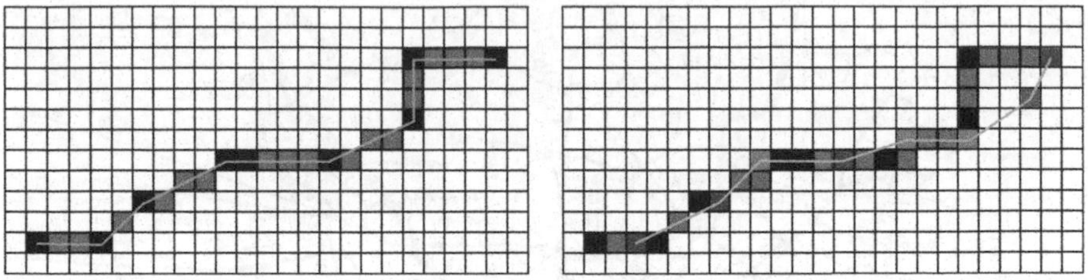

Fig. 6.3 (Left) Example of a trajectory with augmented cells. (Right) Comparing existing trajectory with a new trajectory

$$\text{pos}(t, g) = \begin{cases} \text{argmin}_{i \in \mathbb{N}^+}\{t_i = g\}, & \text{if } g \in t \\ \infty, & \text{otherwise} \end{cases} \tag{6.1}$$

For example, if $t = \langle g_1, g_2, g_3, g_5, g_3, g_8 \rangle$, then $\text{pos}(t, g_3) = 3$ and $\text{pos}(t, g_7) = \infty$.

We will be comparing an ongoing trajectory against a set of trajectories T. Because of the low sampling rates, two taxis following the same path may have points mapping to disjoint cells. In the right panel of Fig. 6.3, we display the augmented trajectory from the left panel, along with a new trajectory (colored squares and green line). Some of the grid cells of the new trajectory fall on the augmented path (blue squares), whereas others fall in "empty" grid cells (orange and red cells). Because of the simplicity of the augmentation method, there is the possibility that the augmented path was not completely accurate; therefore, we must account for this type of error. If a grid cell of the new trajectory is adjacent to one of the augmented cells, we consider it as if it were along the same path (orange cells), whereas if it is not adjacent to any augmented cells, we consider it as following a different path (red cell). For this purpose, we define $N : G \rightarrow \mathcal{P}(G)$ as a function that returns the adjacent neighbors of a grid cell (each nonborder grid cell will thus have nine neighbors, including itself). For a grid cell g and trajectory t, we let $N(g) \in t$ denote the fact that at least one of the neighbors of g is in t, and $\text{pos}(t, N(g))$ return the first index in t that is equal to one of the neighbors of g. For instance, given the grid cells in Fig. 6.4 and sample trajectory $t = \langle g_1, g_2, g_3, g_4, g_{11}, g_{12} \rangle$, we would obtain $\text{pos}(t, N(g_9)) = 2$ (since $g_9 \in N(g_2)$).

Problem statement: We say that subtrajectory t is anomalous with respect to T (and the fixed S-D pair) if the path that it follows rarely occurs in T. Given a fixed S-D pair (S, D) with a set of trajectories T between them and an ongoing trajectory $t = \langle g_1, g_2, \ldots, g_n \rangle$ going from S to D, we would like to verify whether t is anomalous with respect to T. Furthermore, we would like to identify which parts of the trajectory are anomalous.

Definition 6.3 We define function hasPath: $\mathcal{P}(\mathbb{T})$ (where $\mathcal{P}(X)$ is the power set of X) that returns the set of trajectories that contain all of the points in t in the correct order. Note, however, that the points need not be sequential; it suffices that they appear in the same order, as given in the following:

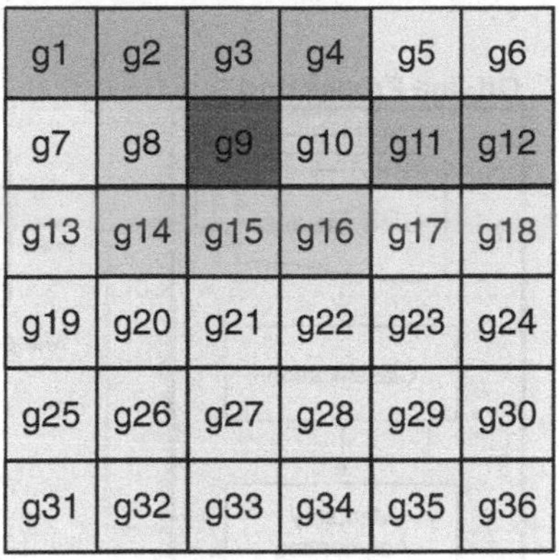
g1 g2 g3 g4 g5 g6
g7 g8 g9 g10 g11 g12
g13 g14 g15 g16 g17 g18
g19 g20 g21 g22 g23 g24
g25 g26 g27 g28 g29 g30
g31 g32 g33 g34 g35 g36

Fig. 6.4 Sample trajectory used to illustrate a cell's neighbors

$$\text{hasPath}(T,t) = \left\{ t' \in T \;\middle|\; \begin{array}{l} (i)\forall 1 \le i \le n.N(g_i) \in t' \\ (ii)\forall 1 \le i < j \le n \\ \text{pos}(t', N(g_i)) < \text{pos}(t', N(g_j)) \end{array} \right\} \tag{6.2}$$

For instance, if $T = \{t1, t2, t3\}$, where $t1 = \langle g_1, g_2, g_3, g_4, g_5, g_8, g_9, g_{10}\rangle$, $t2 = \langle g_1, g_2, g_4, g_5, g_6, g_8, g_{10}\rangle$, and $t3 = \langle g_1, g_3, g_4, g_3, g_6, g_8, g_{10}\rangle$, and an ongoing trajectory $t = \langle g_1, g_2, g_5, g_8\rangle$, then hasPath$(T, t) = \{t1, t2\}$. Given these definitions, we can specify when two trajectories are identical, given our augmentation method.

Definition 6.4 Given threshold $0 \le \theta \le 1$, trajectory t is θ-anomalous with respect to a set of trajectories T if

$$\text{support}(T,t) = \frac{|\,\text{hasPath}(T,t)\,|}{|\,T\,|} < \theta \tag{6.3}$$

6.4 Isolation-Based Online Anomalous Trajectory Detection

Having defined the necessary preliminaries, we are ready to present our method for anomalous trajectory detection. The process is split into an offline preprocessing phase and an online detection phase (see Fig. 6.5). In the offline phase, we receive a set of historical trajectories, which we classify and index using a sophisticated but highly efficient method. This allows us to respond to the online algorithm's queries in real time. In the online phase, we process a series of incoming GPS points from each occupied taxi and provide an indication as to whether each point is anomalous or not. Once this ongoing trajectory is completed, we add it to our historical database.

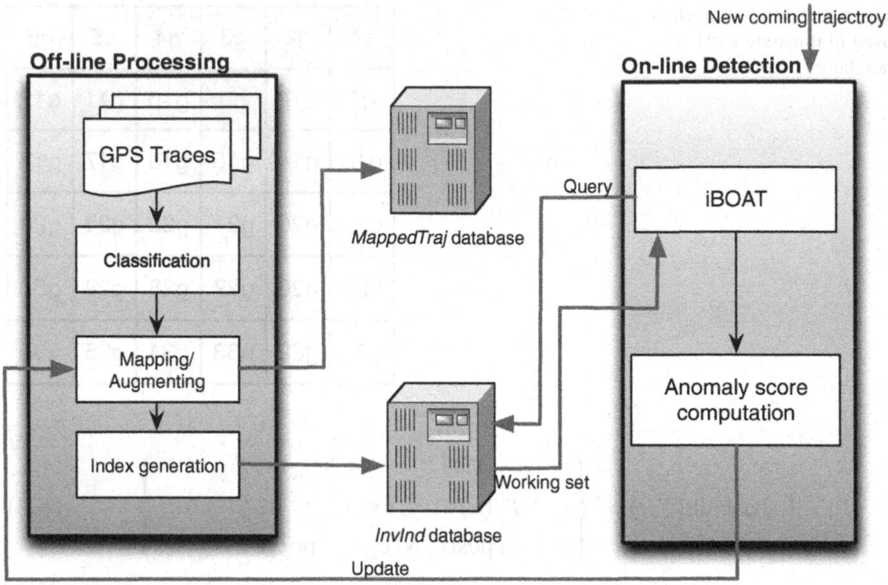
New coming trajectroy
Off-line Processing
On-line Detection
GPS Traces
Classification
Mapping/
Augmenting
Index generation
MappedTraj database
Query
iBOAT
Anomaly score
computation
Working set
InvInd database
Update

Fig. 6.5 Overview of our approach

6.4.1 Offline Preprocessing

In the offline phase, the collection and classification of historical trajectories are employed to obtain the "normal" trajectories between S-D pairs. Note that those historical trajectories must be accessible in an efficient manner to provide a real-time response.

We begin by grouping the trajectories according to S-D pairs and the time of occurrence. It is important to separate trajectories according to the time of occurrence since the "normalcy" of routes may depend on traffic patterns. We index the set of historical trajectories using a triple $\langle \text{sg}, \text{eg}, \text{time} \rangle$, where sg is the starting grid cell, *eg* is the end grid cell, and time is the time at which the trajectory occurred. Note that to avoid the unnecessarily fine granularity of time, we divide time into coarser bins. Each set indexed by a triple may contain both normal and anomalous trajectories. Once the trajectories have been classified, we map and augment them. For each trajectory, we store the resulting mapped grid cells (in the correct order) in a record in the MappedTraj database, and we index the records by their (unique) trajectory number and the time of occurrence.

To determine the anomalousness of a new mapped GPS point, we must be able to access all trajectories that contain this mapped point (or some point in its neighborhood) in the same time bin. Using MappedTraj for this purpose would be terribly inefficient as it would imply searching through all trajectories for each new point. Instead, we make use of the inverted index mechanism [16] for fast retrieval of relevant trajectories. For this mechanism, we maintain a second database where we maintain a record for each possible grid cell; the elements of each record are trajectory–position pairs, indicating the trajectories where the indexing grid cell appears, along with its position in that trajectory. For instance, consider the following trajectories:

$$t_1 : g_1 \rightarrow g_5 \rightarrow g_8 \rightarrow g_{10}$$
$$t_2 : g_1 \rightarrow g_2 \rightarrow g_5 \rightarrow g_8 \rightarrow g_5 \rightarrow g_9$$
$$t_3 : g_2 \rightarrow g_8 \rightarrow g_9$$

In the inverted index database InvInd, the record indexed by grid cell g_1 will be $\text{InvInd}(g_1) = (t_1, 1), (t_2, 1)$, the record indexed by g_5 will be $\text{InvInd}(g_5) = \{(t_1, 2), (t_2, 3), (t_2, 5)\}$, and the record indexed by g_9 will be $\text{InvInd}(g_9)\{(t_2, 6), (t_3, 3)\}$. Thus, if a new GPS point maps to g_9, by accessing $\text{InvInd}(g_9)$, we will immediately know that this grid cells occurs in trajectories t_2 and t_3.

We now have an efficient mechanism for accessing the trajectories that contain a particular grid cell. In Sect. 6.5, we will incrementally use this (as new GPS points arrive) to determine the anomalousness of an ongoing trajectory.

6.4.2 *iBOAT*

Our *iBOAT* detection method is based on the idea of isolating trajectories. Anomalous (sub-)trajectories will be isolated from the majority of routes, whereas normal trajectories will be supported by a large number of trajectories. The less support a trajectory has, the higher its degree of anomalousness would be. In [15], Zhang et al. determine the anomalousness of a trajectory once the trajectory is completed. This is unfortunate, since it prevents one from providing alerts to the passenger while a trajectory is ongoing. On the other hand, using purely density-based methods as described earlier will most likely result in inaccurate classifications. We aim to overcome this problem by using an adaptive working window that provides us with historical contexts to better determine the anomalous-ness of the incoming trajectory. We will use the definition of θ-anomalousness presented in Sect. 6.3 to describe our proposed algorithm.

6.4.2.1 Basic Idea

The basic idea of *iBOAT* is to maintain an adaptive working window of the latest incoming GPS points to compare against the set of historical trajectories. As a new incoming point is added to the adaptive working window, the set of historical trajectories is pruned by removing any trajectories that are inconsistent with the subtrajectory in the adaptive working window. New points continue to be added to the working window as long as the support of the subtrajectory in the adaptive

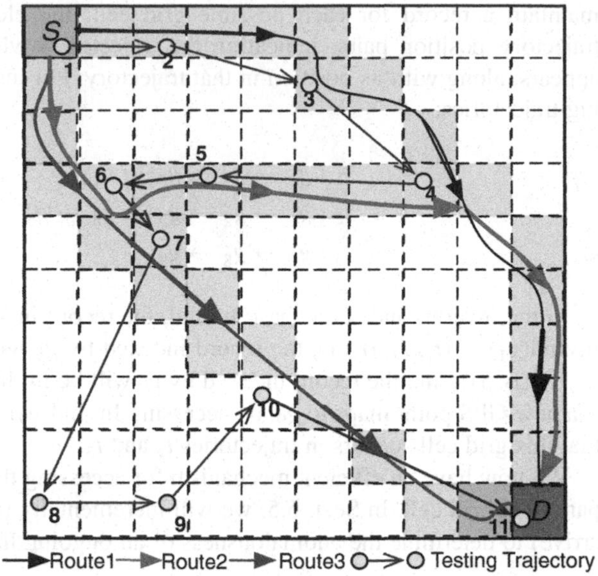
S
D
1
2
3
4
5
6
7
8
9
10
11
Route1 Route2 Route3 Testing Trajectory

Fig. 6.6 Running example for *iBOAT*

working window is above θ. If the support drops below θ, then the adaptive working window is reduced to contain only the latest GPS point. We outline this approach in Algorithm 6.1.

We maintain a working set of trajectories (initially equal to T) and an adaptive working window w. After $i - 1$ entries are received, our partial trajectory t (adaptive working window) consists of $\langle p_1, p_2, \ldots, p_{i-1} \rangle$, and we have a working set T_{i-1}. Upon arrival of point p_i, we map it to grid cell g_i (line 8) and concatenate g_i to the adaptive working window w (line 9). We then compute support(T_{i-1}, w) (line 10). If its value is less than θ, the trajectory points contained in the adaptive working window are said to have anomalies; then, point p_i is considered anomalous. Therefore, it is added to the set of anomalous points χ(line 13), and we reset the working set (line 14) and the adaptive working window (line 15); otherwise, we set T_i = hasPath(T_{i-1}, w) (line 11). This procedure is repeated as long as the trip does not reach the destination. Note that $T_0 = T$ and that, every time an anomalous point is encountered, the working set is reset to the original trajectory set T. This resetting is what enables our adaptive algorithm to accurately detect anomalous subtrajectories in real time with finer granularity than the fixed window approach (with $k > 1$). Additionally, by reducing the working set with each incoming point, the adaptive approach has a computational advantage over the fixed window approach.

To illustrate the process, we will use a running example, as shown in Fig. 6.6. We assume there are three common routes that drivers take when delivering passengers from S to D. There are 100 taxi drivers who have taken Route 1 (in black), 200 taxi drivers have taken Route 2 (in red), and 150 drivers have taken Route 3 (in blue). The test trajectory is depicted using the numbered yellow circles and the purple line (indicating the order of arrival of the points). We can immediately see that, although the test trajectory visits only "common" cells in the initial part, it does so in reverse order between points 4 and 7. In the beginning, the adaptive working window will grow to contain points $\langle g_1, g_2, g_3, g_4\rangle$ since this subtrajectory has enough support. However, when g_5 is added to the adaptive working window, the support of this subtrajectory drops below the threshold; thus, g_5 is considered as an anomalous point, and the new adaptive working window contains only g_5. The size of the adaptive working window would not increase (only containing the single latest GPS point) until receiving g_7; now, the adaptive working window will be $\langle g_6, g_7\rangle$. Again, the working window will shrink to contain only a single point throughout the anomalous section ($\langle g_8, g_9, g_{10}\rangle$). When the trajectory is completed, *iBOAT* will return $\chi = \{g_5, g_6, g_8, g_9, g_{10}\}$ as the set of anomalous points.

We can also consider a simple variant of *iBOAT*: maintaining a fixed-sized window. In this approach, the sliding window consists only of the most recent k points. Specifically, given a set of trajectories T and an ongoing trajectory $t = \langle p_1, p_2, \ldots, p_n\rangle$, we verify whether the last k-sized subtrajectory from t occurs with enough frequency in T to determine if it is anomalous. Note that, when $k =$ 1, we have the density method used for comparison in [17]. Following the example in Fig. 6.6, we have the following results for different values of k:

$$\chi = \begin{cases} \{g_8, g_9\}, & \text{if } k = 1 \\ \{g_5, g_6, g_8, g_9, g_{10}\}, & \text{if } k = 2 \\ \{g_5, g_6, g_7, g_8, g_9, g_{10}, g_{11}\}, & \text{if } k = 3 \end{cases} \tag{6.4}$$

Note that the size of χ depends on the value of k. For the same anomalous trajectory, the larger k is, the larger χ would be. While the size of χ for $k = 2$ and iBOAT is closer to that of the real anomaly segments, it produces larger χ when k increases, leading to excessive counting of the anomaly segments. However, in the case of $k = 1$, the size of χ is much smaller than that of the real anomaly segments. This explains why the anomaly detection algorithm with $k = 2$ and adaptive window outperforms that with $k = 1$ or $k \geq 3$. However, in some specific cases, as shown in Sect. 6.5, the proposed *iBOAT* method with an adaptive window can detect certain anomalous trajectories that the fixed sliding window methods are not able to detect, making *iBOAT* the most effective anomaly detection approach.

6.4.2.2 Anomaly Score

The anomalous score would be applied to rank the trajectories and present alarm once the trip is finished. Intuitively, a trajectory with smaller support and longer anomalous distance should be ranked higher; therefore, we compute this score based on the length of the anomalous subsection, and the density in each anomalous subsection, rather than only summing the length of each anomalous part [14]. We weigh the support according to function σ, which is a logistic function (shown in Fig. 6.7), as follows:

$$\sigma(x) = \frac{1}{1 + e^{\lambda(x-\theta)}} \tag{6.5}$$

Here, λ is a temperature parameter and θ is the aforementioned threshold. For our experiments, we choose $\lambda = 150$. This function will assign a larger weight to very low supports, and the weight will drop to zero for values above θ. The advantage of using this weighting function is that it smoothes the cutoff point imposed by the chosen threshold θ; in a sense, it plays a similar role as sigmoid functions in neural networks. For each incoming point p_i we compute its score as shown in line 17 of Algorithm 6.1. We add the score for the previous point to the distance just traveled multiplied by the weighted support (note that we also do for the fixed window approach).

Given the way the ongoing score is computed, once the trajectory is completed after n steps, we have the final score as given by the following equation, which is a weighted sum of the distance between points:

$$\text{score} = \text{score}(n) = \sum_{i=2}^{n} \frac{1}{1 + e^{\lambda(\text{support}(i)-\theta)}} \text{dist}(p_i, p_{i-1}) \tag{6.6}$$

where dist: $\mathbb{R}^2 \times \mathbb{R}^2 \rightarrow \mathbb{R}$ is the standard sphere distance between two points.

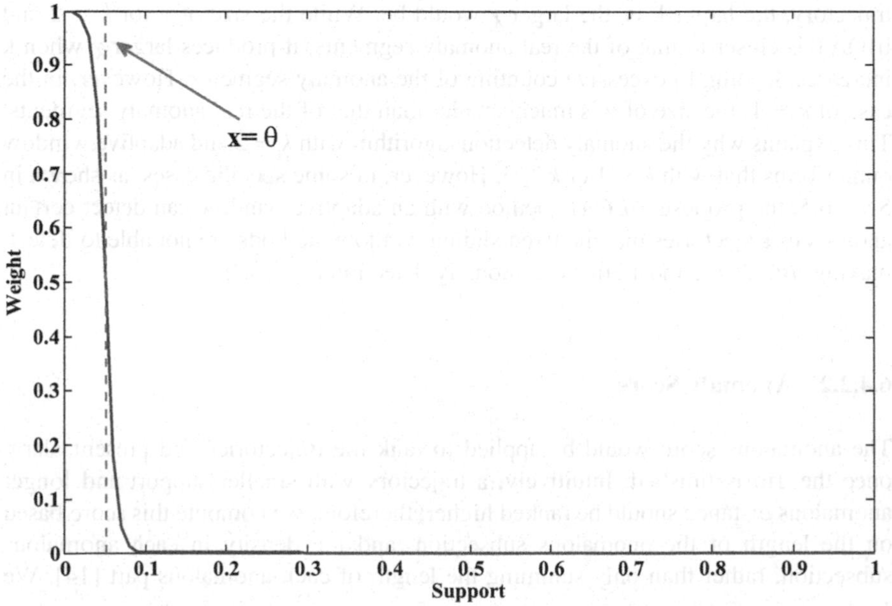
x= θ
Weight
Support
0
0.1
0.2
0.3
0.4
0.5
0.6
0.7
0.8
0.9
1

Fig. 6.7 Weighting function σ

Algorithm 6.1. *iBOAT* with adaptive window

Input: incoming trajectory $-t = \langle p_1, p_2, \ldots \rangle$
T - set of mapped and augmented historical trajectories
θ - anomaly threshold
Output: score; χ - set of anomalous points
$\chi \leftarrow \varnothing$ // initialization
$T_0 \leftarrow T$
$i \leftarrow 0$ // Position in incoming trajectory
$w \leftarrow \varnothing$ // Adaptive window from t
score(0) $\leftarrow$ 0
while the testing trajectory is not completed **do**
    $i \leftarrow i + 1$
    $g_i = \rho(p_i)$
    $w \leftarrow w \cdot g_i$
    support(i) = $|$ hasPath(T_{i-1}, w) $|$ / $|$ $T_{i-1}$$|$
    $T_i \leftarrow$ hasPath(T_{i-1}, w) // working set reduced
    if *support*(i) $< \theta$ **then**
        $\chi \leftarrow \chi \cup p_i$
        $T_i \leftarrow T$
        $w \rightarrow g_i$
    end if
    score(i) = score(i − 1) + σ(support(i)) $*$ dist(p_{i-1}, p_i)
end while

6.5 Empirical Evaluations

6.5.1 Data Sets

The datasets contain 7,350,000 taxi trajectories collected by 7600 taxis in a month. We picked nine S-D pairs[1] (T-1 through T-9) with sufficient trajectories between them (at least 450, but on average over 1000) and asked volunteers to manually label whether the trajectories are anomalous or not. On average, about 5.1% of the trajectories are labeled as anomalous. The statistical information of each dataset is shown in Table 6.1.

6.5.2 Evaluation Criteria

A classified trajectory will fall into one of four scenarios: (1) True positive (TP), when an anomalous trajectory is correctly classified as anomalous; (2) false positive (FP), when a normal trajectory is incorrectly classified as anomalous; (3) false negative (FN), when an anomalous trajectory is incorrectly classified as normal; and (4) true negative (TN), when a normal trajectory is correctly classified as normal.

[1] The S and D areas are twice as big as the regular grid cells.

Table 6.1 Datasets used in our experiments

	# Trajectories	# Anomalousness (%)
T-1	453	15 (3.3%)
T-2	1494	57 (3.8%)
T-3	528	43 (8.1%)
T-4	946	58 (6.1%)
T-5	1018	68 (6.7%)
T-6	1369	72 (5.3%)
T-7	1310	67 (5.1%)
T-8	1216	71 (5.8%)
T-9	1254	24 (1.9%)

The TP rate (TPR) measures the proportion of correctly labeled anomalous trajectories, and it is defined as

$$TPR = \frac{TP}{TP + FN} \tag{6.7}$$

The FP rate (FPR) measures the proportion of false alarms (i.e., normal trajectories that are labeled as anomalous) and is defined as

$$FPR = \frac{FP}{FP + TN} \tag{6.8}$$

A perfect classifier will have TPR = 1 and FPR = 0. In a receiver operating characteristic (ROC) [18] curve, we plot FPR on the x-axis and TPR on the y-axis, which indicates the tradeoff between false alarms and accurate classifications. By measuring the area under a curve (AUC), we can quantify this tradeoff.

6.5.3 Results

To test *iBOAT*, we selected trajectory t as an ongoing trajectory from data set T and used both *iBOAT* and fixed window approaches with $\theta = 0.05$. Parameter sensitivity experiments can be found in [19], revealing our method's performance (in terms of running time and accuracy) under different parameter settings. In the left panel of Fig. 6.8, we display the output of our method for a test trajectory from T-6, where we plot the set of trajectories $T - \{t\}$ in light blue; for the test trajectory (t), the anomalous points are drawn in red and the rest (normal points) in dark blue. As shown, our method can accurately detect which parts of a trajectory are anomalous and which are normal. In the middle and right panels of Fig. 6.8, we plot support $(T - \{t\}, t)$ [see Eq. 6.3] and the score [see Eq. 6.6] for the ongoing trajectory t. We can see that the value of support is a clear indication of when trajectories become anomalous and that there is little difference between the different variants of *iBOAT*.

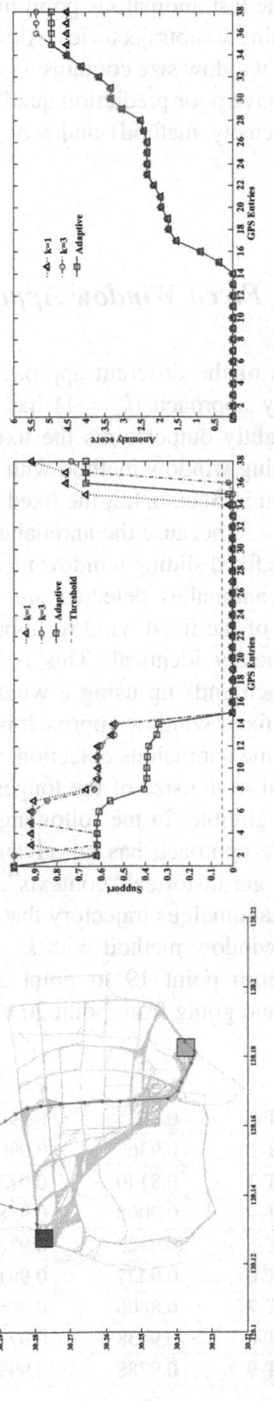
k=1
k=3
Adaptive
Threshold
Support
GPS Entries
Anomaly score

Fig. 6.8 (Left) Detected anomalous subtrajectories from T-6 using *iBOAT*. (Middle) Plot of ongoing support. (Right) Plot of ongoing score

Note, however, that there is a trailing lag for the fixed window approach, which is equal to k. This is because the last anomalous point in an anomalous subtrajectory will be included in the following k subtrajectories. Although setting $k = 1$ will solve the lag problem, this minimal window size contains no contextual information on the trajectory and will therefore have poor prediction quality. This was observed in [15] (therein referred to as the density method) and will be evident in the following figures.

6.5.4 Adaptive Versus Fixed Window Approach

We display the AUC values of the different approaches on the nine data sets in Table 6.2. While the density approach ($k = 1$) has the worst performance, our proposed *iBOAT* method slightly outperforms the fixed sliding window approach with $k = 2$, and the fixed sliding window method with $k = 2$ is better than that with $k = 1$ and $k > 3$. As explained in Sect. 6.4.2, the fixed sliding window method with $k = 1$ is worse than that of $k = 2$ because the anomalies detected are fewer than the actual anomalies, whereas the fixed sliding window method with $k = 3$ is worse than that of $k = 2$ because the anomalies detected are much more than the actual anomalies. The performance of the fixed window approach with $k = 2$ and that of the adaptive approach are nearly identical. This is because, for the anomalous sections, the adaptive approach ends up using a window with a size of 2, just as $k = 2$. The advantage of the fixed window approach is that it requires a very small amount of memory for real-time anomalous detection, whereas the adaptive method requires memory proportional to the size of the longest "normal" subtrajectory. In practice, this difference is negligible. In the following, we will use an example to demonstrate that the adaptive approach has an advantage over the fixed window approach due to its use of longer historical "contexts".

In Fig. 6.9, we display an anomalous trajectory that "switches" from one normal route to another. The fixed window method with $k = 2$ is not able to detect this anomalous switch. Going from point 19 to point 20 seems normal since this sequence occurs in route A, and going from point 20 to point 21 also seems normal

Table 6.2 AUC values of the different algorithms

	$k = 1$	$k = 2$	$k = 3$	Adaptive
T-1	0.9635	0.9904	0.9811	0.9985
T-2	0.9367	0.9902	0.9887	0.9952
T-3	0.8140	0.9733	0.9152	0.9962
T-4	0.9005	0.9586	0.9575	0.9890
T-5	0.9323	0.9885	0.9821	0.9967
T-6	0.9227	0.9912	0.9840	0.9952
T-7	0.8806	0.9853	0.9849	0.9937
T-8	0.9438	0.9739	0.9724	0.9937
T-9	0.9788	0.9991	0.9987	0.9995

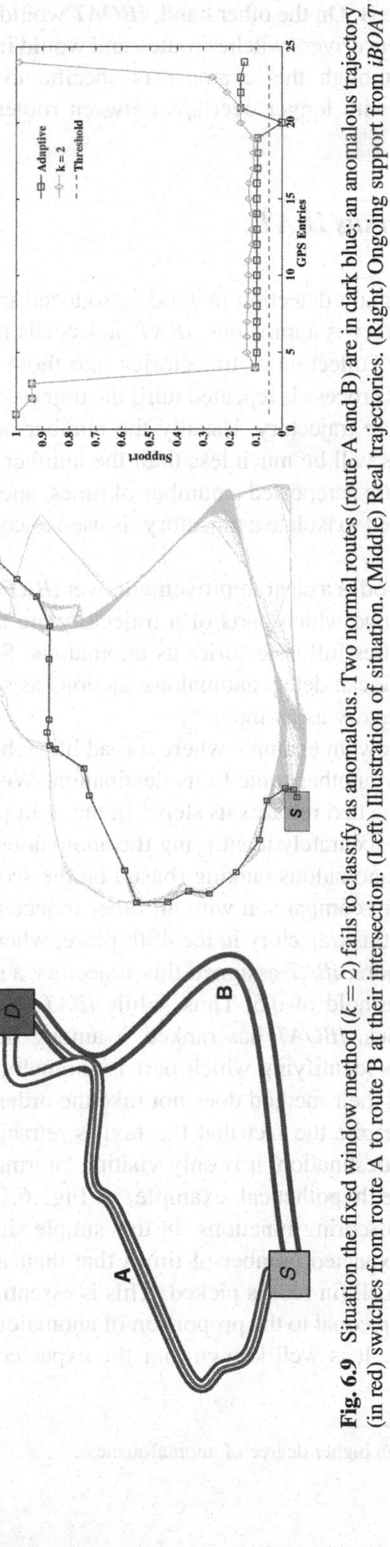
A
B
S
D
S
D
Support
GPS Entries
Adaptive
k = 2
Threshold

Fig. 6.9 Situation the fixed window method ($k = 2$) fails to classify as anomalous. Two normal routes (route A and B) are in dark blue; an anomalous trajectory (in red) switches from route A to route B at their intersection. (Left) Illustration of situation. (Middle) Real trajectories. (Right) Ongoing support from *iBOAT* since it occurs in route B. On the other hand, *iBOAT* would maintain the entire route up to the point when the driver switches routes and would immediately detect it as an anomalous point. Although this example is specific to window sizes equal to 2, similar situations (with longer overlaps between routes) will produce a similar effect.

6.5.5 iBOAT Versus iBAT

iBAT is a similar anomaly detection method introduced in [15] recently. To determine whether a trajectory is anomalous, *iBAT* picks cells from the testing trajectory at random to split the collection of trajectories into those that contain the cell and those that do not. This process is repeated until the trajectory is isolated or until there are no more cells in the trajectory. Usually the number of cells required to isolate anomalous trajectories will be much less than the number of cells in the trajectory. This isolation procedure is repeated a number of times, and $\mathbb{E}(n(t))$, i.e., the average number of cells required to isolate a trajectory, is used to compute the score, which is proportional to $2^{-\mathbb{E}(n(t))}$.

Our proposed method is a clear improvement over *iBAT* on two levels. First of all, we are able to determine which parts of a trajectory are anomalous, in contrast to iBAT that only classifies full trajectories as anomalous. Second of all, our method works in real time. We can detect anomalous sections as soon as they occur and do not require a full trajectory as an input.

In Fig. 6.10, we show an example where a road block has forced a taxi to retrace its path and search for another route to its destination. We focus on the first part of the trajectory where the taxi retraces its steps. In the right panel of Fig. 6.10, we can see that the support is accurately identifying the anomalous section of the trajectory. We determined what anomalous ranking (based on the scores) both methods assign this partial trajectory in comparison with all other trajectories.[2] Out of 1418 trajectories, *iBOAT* ranked this trajectory in the 48th place, whereas *iBAT* ranked it in the 831st place. Furthermore, *iBAT* assigned this trajectory a score of 0.4342, which is below their usual threshold of 0.5. Thus, while *iBAT* is unable to detect that this trajectory is anomalous, *iBOAT* has ranked it among the top 3% of anomalous trajectories, as well as identifying which part is anomalous. The reason *iBAT* fails in this example is that their method does not take the order that the points appear in into consideration; despite the fact that the taxi is retracing its steps and actually going away from the destination, it is only visiting "normal" grid cells.

Now, consider the hypothetical example in Fig. 6.11, which highlights the differences in the two scoring functions. In this simple situation, the value $\mathbb{E}(n(t))$ for *iBAT* is just the expected number of times that their algorithm must pick cells before an anomalous cell (in red) is picked. This is essentially a Bernoulli trial with "success" probability p equal to the proportion of anomalous cells to total number of cells in the trajectory. It is well known that the expected number of trials before

[2]A higher ranking means a higher degree of anomalousness.

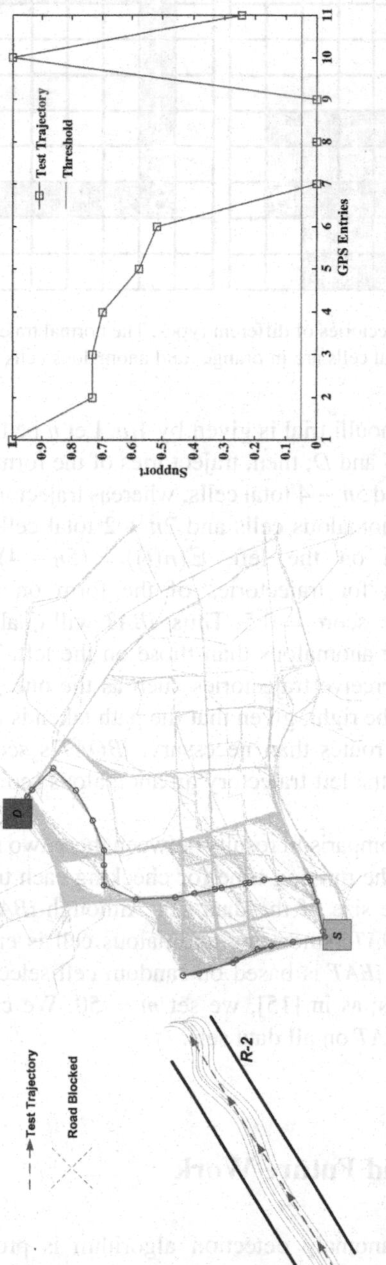

Test Trajectory
Road Blocked
R-1
R-2
S
D
Test Trajectory
Threshold
Support
GPS Entries

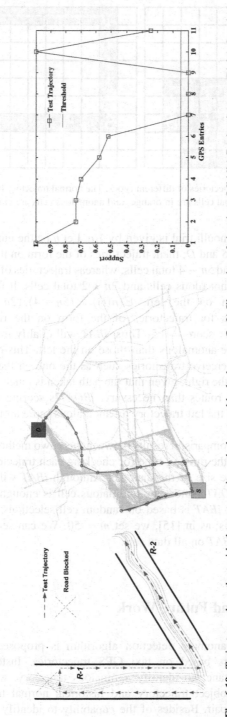

Fig. 6.10 Trajectory where the taxi had to retrace its path due to a blocked route. (Left) Illustration of situation (Middle) Real trajectory. (Right) Ongoing support from *iBOAT*

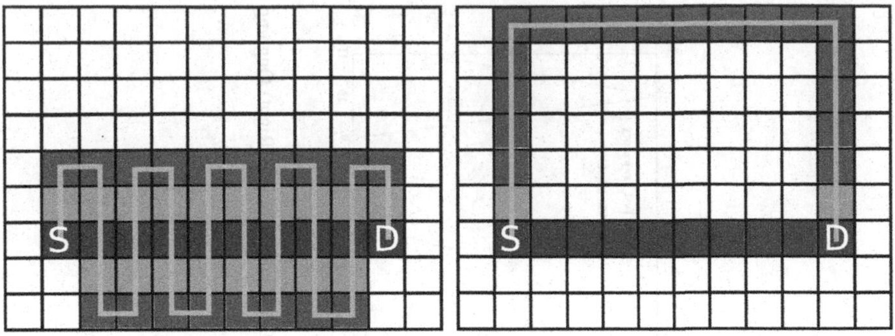
S
D
S
D

Fig. 6.11 Two anomalous trajectories of different types. The normal trajectory between *S* and *D* is in blue, cells adjacent to normal cells are in orange, and anomalous cells are in red

reaching success in a Bernoulli trial is given by $1/p$. Let n be the number of cells in the straight line between *S* and *D*; then, trajectories of the form on the left will have $2n - 2$ anomalous cells and $5n - 4$ total cells, whereas trajectories of the form on the right will have $2n - 2$ anomalous cells and $2n + 2$ total cells. It follows that, for trajectories of the form on the left, $\mathbb{E}(n(t)) = (5n-4)/(2n-2) \to 5/2 \Rightarrow$ score ≈ 0.1768 , whereas for trajectories of the form on the right, $\mathbb{E}(n(t)) = (2n+2)/(2n-2) \to 1 \Rightarrow$ score $= 0.5$. Thus, *iBAT* will qualify trajectories of the form on the right as more anomalous than those on the left. This runs contrary to intuition, which would perceive trajectories such as the one on the left at least as anomalous as the one on the right, given that the path taken is much longer and they are clearly taking longer routes than necessary. *iBOAT*'s scoring method, on the other hand, would assign the left trajectory an anomalous score around 33% higher than the one on the right.

Finally, the time cost comparison results between these two methods are shown in Fig. 6.12. We computed the running time for checking each trajectory in each data set and averaged over the size of the data set. Although *iBAT* will usually check fewer grid cells than *iBOAT* (since one anomalous cell is enough to classify the trajectory as anomalous), *iBAT* is based on random cell selections; therefore, they must average over m runs; as in [15], we set $m = 50$. We can see that *iBOAT* is consistently faster than *iBAT* on all data sets.

6.6 Conclusions and Future Work

In this chapter, a new anomaly detection algorithm is proposed to find fraud behaviors of taxi drivers based on taxi GPS trajectories. Instead of directly employing time and distance to identify anomalous trajectory, we focus on the difference between the object trajectory and historical normal trajectories with regard to the same S-D pair. Besides of the capability to identify anomalousness

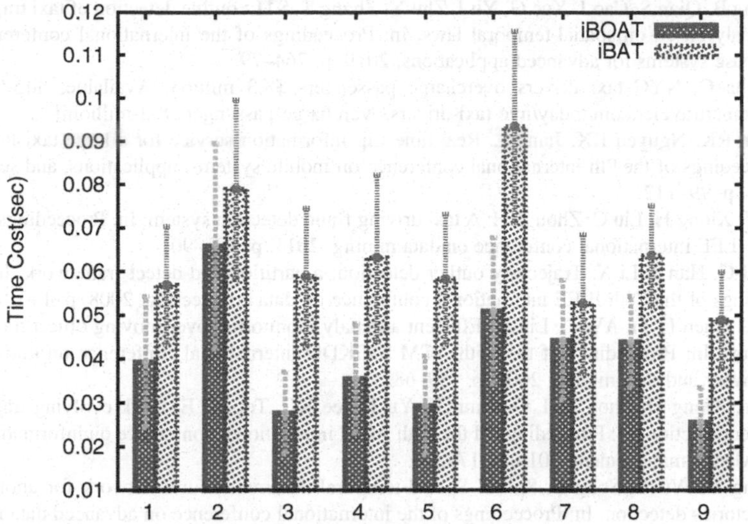
Time Cost(sec)
0.12
0.11
0.1
0.09
0.08
0.07
0.06
0.05
0.04
0.03
0.02
0.01
1
2
3
4
5
6
7
8
9
iBOAT
iBAT

Fig. 6.12 Running times of *iBOAT* and *iBAT* on all the data sets

of completed trips, *iBOAT* is able to be used to the unfinished trajectories and find out which segments of a trajectory cause the anomalousness. At last, we evaluated *iBOAT* on the real-world taxi GPS trajectories. Results show that the proposed algorithm can achieve superior performance (AUC $\geq$ 0.99 for all data sets), and it is similar to *iBAT* algorithm. However, we demonstrated a number of examples that highlight *iBOAT*'s advantage over *iBAT* and the sliding window method. Furthermore, we presented how to employ *iBOAT* to analyze dishonest behaviors of taxi drivers. Results indicate the fact that most anomalous taxi trajectories are by the reason of dishonest behaviors. In addition, we offered evidence to reject potential excuses for fraud.

In the future, we plan to broaden and deepen this study in several directions. First, we intend to try some statistical approaches to improve the performance and quicken the trajectory processing. Second, we also intend to develop a real-world dishonest driving behaviors detection system. Third, regarding the fact that not all S-D pairs have enough samples, we intend to group S and D areas to "merge" different trajectories, or use grid cells with different sizes to divide the map.

References

1. Rong H, Teixeira AP, Guedes Soares C. Data mining approach to shipping route characterization and anomaly detection based on AIS data. Ocean Eng. 2020;198:106936.
2. Liu Y, Zhao K, Cong G, Bao Z. Online anomalous trajectory detection with deep generative sequence modeling. In: Proceedings of the 36th IEEE international conference on data engineering (ICDE), 2020. p. 949–60.

3. Cheng B, Qian S, Cao J, Xue G, Yu J, Zhu Y, Zhang T. STL: online detection of taxi trajectory anomaly based on spatial-temporal laws. In: Proceedings of the international conference on database systems for advanced applications, 2019. p. 764–79.
4. Leocha C. NYC taxi drivers overcharge passengers $8.3 million. Available: http://www.consumertraveler.com/today/nyc-taxi-drivers-overcharge-passengers-8-3-millioin/
5. Balan RK, Nguyen KX, Jiang L. Real-time trip information service for a large taxi fleet. In: Proceedings of the 9th international conference on mobile systems, applications, and services, 2011. p. 99–112.
6. Ge Y, Xiong H, Liu C, Zhou Z-H. A taxi driving fraud detection system. In: Proceedings of the 11th IEEE international conference on data mining, 2011. p. 181–90.
7. Lee J-G, Han J, Li X. Trajectory outlier detection: a partition-and-detect framework. In: Proceedings of the 24th IEEE international conference on data engineering, 2008. p. 140–9.
8. Bu Y, Chen L, Fu AW-C, Liu D. Efficient anomaly monitoring over moving object trajectory streams. In: Proceedings of the 15th ACM SIGKDD international conference on knowledge discovery and data mining, 2009. p. 159–68.
9. Ge Y, Xiong H, Zhou Z-H, Ozdemir H, Yu J, Lee KC. Top-EYE: top-k evolving trajectory outlier detection. In: Proceedings of the 19th ACM international conference on information and knowledge management, 2010. p. 1733–6.
10. Cheng Y, Wu B, Song L, Shi C. Spatial-temporal recurrent neural network for anomalous trajectories detection. In: Proceedings of the international conference on advanced data mining and applications, 2019. p. 565–78.
11. Morais R, Le V, Tran T, Saha B, Mansour M, Venkatesh S. Learning regularity in skeleton trajectories for anomaly detection in videos. In: Proceedings of the IEEE conference on computer vision and pattern recognition, 2019. p. 11996–12004.
12. Li X, Li Z, Han J, Lee J-G. Temporal outlier detection in vehicle traffic data. In: Proceedings of the 25th IEEE international conference on data engineering, 2009. p. 1319–22.
13. Liu FT, Ting KM, Zhou Z-H. Isolation-based anomaly detection. ACM Trans Knowl Discov Data (TKDD). 2012;6(1):3:1–3:39.
14. Chen C, Zhang D, Castro PS, Li N, Sun L, Li S. Real-time detection of anomalous taxi trajectories from GPS traces. In: Proceedings of the international conference on mobile and ubiquitous systems: computing, networking, and services, 2012. p. 63–74.
15. Zhang D, Li N, Zhou Z-H, Chen C, Sun L, Li S. iBAT: detecting anomalous taxi trajectories from GPS traces. In: Proceedings of the 13th international conference on ubiquitous computing, 2011. p. 99–108.
16. Zobel J, Moffat A. Inverted files for text search engines. ACM Comput Surv (CSUR). 2006;38(2):1–55.
17. Liao Z, Yu Y, Chen B. Anomaly detection in GPS data based on visual analytics. In: Proceedings of the IEEE symposium on visual analytics science and technology, 2010. p. 51–8.
18. Fawcett T. An introduction to roc analysis. Pattern Recogn Lett. 2006;27(8):861–74.
19. Chen C, Zhang D, Castro PS, Li N, Sun L, Li S, Wang Z. iBOAT: isolation-based online anomalous trajectory detection. IEEE Trans Intell Transp Syst. 2013;14(2):806–18.

Chapter 7
Real-Time Imputing Trip Purpose Leveraging Heterogeneous Trajectory Data

7.1 Introduction

Travel behavior understanding is an important yet challenging research topic in the area of smart cities [1]. The underlying travelling knowledge can enable urban planners and policy makers to increase their abilities in addressing urban planning, management and operating issues [2]. As one kind of the semantic information regarding travel behaviors, trip purpose (i.e., why people take taxis) indicates the type of activity that passengers will take after drop-offs. Knowing such information is quite essential in developing smart passenger-centered services in Intelligent Transportation Systems, such as enabling personalized recommendations and advertisements for passengers.

Although the taxi trip purpose inference has been a long-standing research topic for many years [3, 4], rare efforts have been devoted to addressing the following two problems: (1) Obtain the trip purpose at an individual level. To be more specific, previous studies mainly devoted to inferring trip purposes at a city-wide scale, so as to support the smart urban services at the macro level. On the contrary, the imputation of the trip purpose at the individual level is strongly necessary for the micro smart urban services (e.g., recommendation services to each passenger); (2) Require the real-time response, i.e., returning the corresponding purpose as soon as the trip ends. With the real-time identification of passengers' travel purposes, taxis are able to present timely recommendations, and passengers are able to make an efficient and economic schedule. There is one point we need to clarify that the inference of trip purposes is triggered when the taxi arrives at the destination. The reasons are twofold: (1) Although taxi drivers have known the destinations in advance, the

---

Part of this chapter is based on a previous work: C. Chen, S. Jiao, S. Zhang, W. Liu, L. Feng and Y. Wang, "TripImputor: Real-Time Imputing Taxi Trip Purpose Leveraging Multi-Sourced Urban Data," in IEEE Transactions on Intelligent Transportation Systems, vol. 19, no. 10, pp. 3292–3304, Oct. 2018, doi: 10.1109/TITS.2017.2771231.

C. Chen et al., *Enabling Smart Urban Services with GPS Trajectory Data*,
https://doi.org/10.1007/978-981-16-0178-1_7

drop-off information usually cannot be automatically recorded by the embedded GPS systems until taxi drivers manually change the passenger status from occupied to free. (2) The accurate prediction of the taxi destinations by leveraging their unfinished trajectories is extremely challenging, which is also a research problem in the Intelligent Transportation Systems [5]. There are two challenges for the real-time individual taxi trip purpose:

- **Lack ground-truth**. The collection of the ground-truth of trip purpose is often based on the proactively prompted recall. In this way, only a small percentage of people are asked to annotate their traces with corresponding trip purposes. To make matters worse, such obtained ground truth is generally contaminated due to some people's incorrect recall.
- **Real-time response**. On one hand, previous trip purpose inference methods are not able to be used directly due to the incapacity of real-time responses. On the other hand, it's very challenging to enable the real-time response based on the continuously and intensively generated taxi trip.

For the accurate trip purpose imputation, a few meaningful information should be taken into consideration, including the time and location characteristics of the drop-off event and the neighboring geographical context. Specifically, around the drop-off location, the distribution of different categories of human activities at the drop-off time is a meaningful clue for the trip purpose imputation. Fortunately, the users' check-in records at POIs from the LBSNs (e.g., Foursquare), can tell detailed information of POIs, such as the hierarchical category and opening/closing time) [6]. Such check-in information can provide a great opportunity for the understanding of passengers' activities at a given area during a specific time. For example, people visit a shopping mall for shopping and visit a restaurant for dining. Thus, imputing taxi trip purpose *can be converted to inferring the probabilities of performing activity in different kinds of POIs when the taxi arrives at the destination.* In summary, the main contributions of this chapter are as follows.

- Based on the Bayes' theorem, a new two-phase framework named TripImputor is proposed to obtain individual taxi trip purpose in real time. In phase I, a two-stage clustering algorithm is proposed to obtain urban activity regions (UARs) and candidate activity areas (CAAs) based on the POI data. Then, we extract fine-granularity spatial and temporal patterns regarding human behaviors inside the CAAs from Foursquare check-in data to approximate the prior probability for each category of human activity, and compute the posterior probabilities using the Bayes' theorem. In phase II, in order to enable the real-time response, a procedure is proposed to reduce the online computation time, which includes the clustering of historical drop-off points and the matching between drop-off clusters and CAAs.
- Systematical experiments are conducted to evaluate the efficiency and effectiveness of TripImputor based on real-world datasets in Manhattan, NYC. The datasets consist of check-in records from Foursquare, taxi GPS trajectories and the road network. Since the ground truth of trip purposes is unavailable, the

effectiveness is evaluated in an indirect manner. We exploit a comparison between the inferred results and travel survey records at the region scale from the perspective of statistical analysis. At last, the results show that the proposed method achieves superior performance compared to baselines methods. In addition, the time cost for an individual imputation is about 1.588 s, and the shortest time cost is nearly 40 ms while the longest time (7.54 s) is still applicable in the real world scenario.

7.2 Related Work

In recent years, many efforts of researches have been devoted to using the raw trajectory data to explore the latent motivations behind human mobility in a data mining manner [7]. However, rare methods are designed for the inference of passengers' activities after taxi trips namely the trip purposes. Existing methods include machine learning, statistics based data mining algorithms, and deterministic/heuristic rule based methods [3, 4, 8]. To name a few, deterministic rules were proposed to identify trip purpose based on the land use data [8].

Based on the GIS information and respondents' social-demographics data, a decision tree was proposed to identify trip purposes [3]. Similar to our work, Bayes' rules were used to derive the probabilities of POIs to be the final activity place of passengers [4], so as to obtain taxi trip purposes. Although extensive works have been done regarding the semantics enrichment of the raw trajectory, there was no requirement of the real-time response for the trip purpose inference, thus the recommendations cannot be timely presented to passengers. Recent years also witness the trend that incorporating taxi trajectory data with multi-sourced data (e.g., Flickr images and POI check-ins) for the smart urban service development. The smarter services include the building usages interpreting, personalized routing, hitchhiking package delivering [9, 10]. It worth noting that Check-in data has been employed in the development of various smart urban services. For example, the underlying knowledge of check-in data can be used to enable the personalized advertising and location searching, POI recommendation and the visualization of POI popularity evolving patterns, and so on [6, 9].

7.3 Basic Concepts and Problem Statement

7.3.1 Basic Concepts

Definition 7.1 (Road Network). A road network is a graph G (N, E), consisting of a node set N and an edge set E, where each element n in N is an intersection with a pair of longitude and latitude coordinates (x, y) representing its spatial location. Edge set E is a subset of the Cartesian product N × N. Each element e (u, v) in E is a street

Table 7.1 Nine trip purposes and the corresponding primary POI categories

Trip purposes	Primary POI categories
Work-related	Office building; government; scientific research institution
In-home	Residential subdivisions; neighbourhood
Transportation transfer	Airport; railway station; long-coach station; subway station
Dining	Restaurant
Shopping	Shopping mall; supermarket; store
Recreation	Culture facilities (museum, science center, etc.) recreational facilities (movie theater, park, etc.) sports facilities; coffee house; bar
Schooling	Primary school; high school; university
Lodging	Hotel; hostel
Medical	Hospital

connecting node u and node v, which has several attributes including speed limit, number of lanes.

Definition 7.2 (A Taxi Drop-Off Point). A taxi drop-off point (pi) is defined as a time-stamped location where the passenger was dropped off, denoted by $((x_i, y_i), t_i)$.

Definition 7.3 (POI Category). A POI category is a semantic label for a place, indicating the correlation between the place and potential human activities.

Foursquare maintains a three-level ontology structure for category description [9]. In the first level, it has 9 categories in total. In the second and third levels, it has 412 sub-/sub-subcategories in total. Table 7.1 shows the trip purposes (travel activities) and the corresponding primary POI categories [4].

Definition 7.4 (A Check-In). A check-in is represented by a triple $c_k = (u_{id}, v_{id}, t_i)$, indicating a user with id u_{id} checked-in at a venue (i.e. POI) with id v_{id} at time t_i using Foursquare.

Generally, popular POIs would be frequently visited by many citizens or tourists. Additionally, the Foursquare platform also records the longitude&latitude, tags, and the opening&closing time of POIs.

Definition 7.5 (Response Time). The response time is defined as the time difference between the drop-off time when the passenger gets off taxis and the time when the passenger receives the recommendation services.

7.3.2 Problem Statement

Inferring the taxi trip purposes leveraging multi-sourced urban data can be viewed as predicting the probabilities of taking one of the nine activities, which can be formulated as:

Given:

1. A drop-off point $((x_r, y_r), t_r)$, which is generated in real-time;
2. A set of historical check-ins $\{u_{id}, v_{id}, t_i\}$ (e.g., the last month), together with check-ins accumulated several hours before t_r in the designated city;
3. POIs in the designated city, which can be obtained from the check-in data;
4. A road network G (N, E) of the designated city.

Predict the probabilities of taking each of the nine activities respectively for the drop-off point (the objective of Phase I), and **provide** timely service recommendations related to the top-ranked trip purposes (activities with top probabilities) for the passenger (the objective of Phase II).

7.4 Phase I: Imputing Trip Purposes

7.4.1 *Urban Activity Region Identification*

Human beings are known as collective people (i.e., most of people live, work together with others in nature), thus it is highly likely that people take activities in a small and scattered fraction of the whole city space. A preliminary step for inferring the travel purpose of passengers is to identify all the scattered activity regions in the whole urban space. To ease the presentation, we name these regions as *Urban Activity Regions (UARs)*.

Urban activity regions are bounded and separated by some physical barriers such as main roads, rivers, and mountains, as can be witnessed in the human civilization and urbanization process in history [11]. Each separated UAR is isolated and bounded by main road segments (or rivers), covering several neighborhoods and narrow streets. Inside each UAR, passengers can easily reach between two points if they are located to each other. Usually, passengers who get off taxis at one side of the primary way will not cross it (i.e., go to the other side) to take activities due to the huge barrier. On the contrary, when getting off taxis at small and narrow streets, the passengers can easily walk towards another direction. Based on the above observations, in this study, we mainly rely on the road network data to identify the UARs in the target city. We propose a two-step procedure to divide the whole city into a number of disjointed UARs.

- *Step 1*: We extract the road network data including coordinates of nodes, edges, as well as the attributes of edges (e.g., number of lanes, speed limits, road levels/types) from an open crowdsourced platform, i.e., the OpenStreetMap. With the information of road level/type attributes, we are able to keep high-level road segments that are only tagged as 'motorway', 'trunk', or 'primary'.
- *Step 2*: For the trimmed road network only consisting of high-level road segments, we apply the image-processing-based map segmentation algorithm in [12] to obtain connected components. Each connected component is just a piece of the separated urban activity region (UAR, R_1–R_5 in Fig. 7.1).

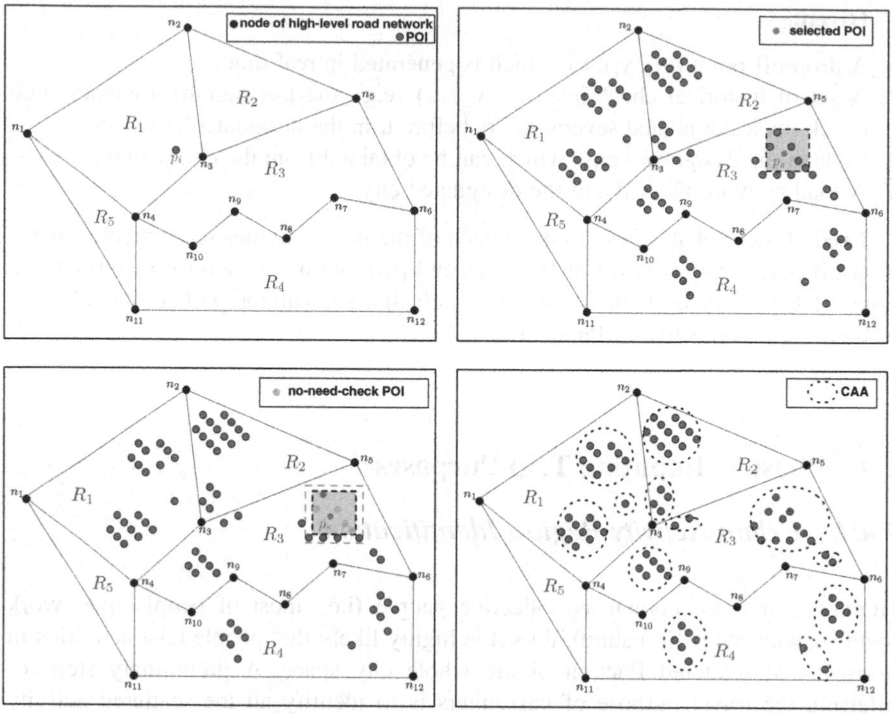
node of high-level road network
POI
selected POI
no-need-check POI
CAA
R_1
R_2
R_3
R_4
R_5
n_1
n_2
n_3
n_4
n_5
n_6
n_7
n_8
n_9
n_{10}
n_{11}
n_{12}
p_i

Fig. 7.1 Illustrative example of determining the region that a given POI belonging to (top left); the illustrative examples of assigning a huge number of POIs to regions (top right and bottom left); the identified CAAs for the illustrative example (bottom right)

7.4.2 Candidate Activity Area Identification

As we all known, POIs are the basic human activity units in the city. In the case of people taking taxi to travel, on one hand, they always prefer to get off as close to the true destination as possible. On the other hand, in the modern city, there are usually many different categories of POIs located in a same building (e.g., a shopping mall). In this respect, people are more likely to be attracted by the nearby one or two buildings after getting off taxis. Hence, we propose the concept of candidate activity area (CAA) in which different POIs locate close to each other. The CAAs correspond to small areas, and we use CAA as the activity unit for taxi passengers.

To identify such a CAA, we first determine which UAR a given POI belongs to. Then, we aggregate the POIs belonging to the same UAR to several clusters based on the spatial proximity. Finally, we identify each cluster as a CAA. In this sense, a UAR contains serval CAAs. However, the assignment of POIs to UARs is quite challenging since we have to address the following two issues:

1. Each UAR is usually of an arbitrary shape, thus we cannot simply compare the POI locations to the locations of the UAR boundaries. A simpler but essential

problem is the point-in-polygon problem [13]. More specifically, it's the problem of determining whether a given point is inside/outside a given closed polygon (i.e., region), which is proved to be hard [14].

2. The number of POIs is huge (e.g., the number of POIs in the Manhattan of NYC is more than 10k), and how to efficiently determine which UAR each POI locates at is also a challenging issue.

Algorithm 7.1 Algorithm for Determining the Region That a Given POI Belongs to

Input: a given of POI (p_i): the trimmed road network and the identified UARs in the target city;
Output: the UAR that the given point is located, denoted by $R_i = PinR(p_i)$.

- Step 1: Based on the location of the given point (p_i), we can find its nearest node n_i;
- Step 2: According to the identified node n_i and the topology of the high-level road network, we can easily identify all the regions that share the node n_i. We denote these regions by $\{R_i\}$;
- Step 3.a: For each region in the set of $\{R_i\}$, we apply ***ifPinR*** algorithm to check whether p_i is inside that region;
- Step 3.b: Loop ends when ***ifPinR*** returns 1.

Without loss of generality, to deal with the first issue, we apply a popular and mature algorithm to determine the relationship (i.e., inside or outside) between a given point and a given region [15]. For simplicity, we denote the algorithm as *ifPinR* (p_i, r_i). If the point p_i is inside region r_i, *ifPinR* (p_i, r_i) returns 1; otherwise, it returns 0. To determine which region that a given point belongs to, we propose the algorithm by recalling *ifPinR* repeatedly. The pseudo-code of this algorithm is presented in Algorithm 7.1. For the given point, Step 1 and Step 2 identify all the possible regions that it may belong to, according to the geometrical relationship in the space. Note that a region is represented by a sequence of nodes in the clockwise direction. For instance, the possible regions for p_i in the illustrative example (as shown in top left of Fig. 7.1) are marked as R_1, R_2, and R_3. Step 3 shows the repeated recalling procedure of algorithm *ifPinR*. The number of loops is usually small since the possible region set contains few and limited regions. In the best case, the number of loops is 1, while in the worst case, the number of loops is just equal to the size of the possible region set. The loop number is 1 for the illustrative example since *ifPinR* returns 1 when checking R1 at the first loop.

To deal with the second issue, a straightforward but computationally expensive method is to check each POI based on Algorithm 7.1. In theory, the computation complexity is O (N × M × C), where N is the number of POIs; M is the average number of possible regions for a given POI, which is usually small and O(C) is the complexity of *ifPinR* algorithm. Therefore, in order to accelerate the computation process, we should reduce the number of POIs to be checked. Actually, it is unnecessary to check some POIs. More specifically, if we have determined the region where a given POI locates at, then we can directly infer that its 'nearby' POIs should also be located inside the same region with high confidence level. Inspired by this observation, we propose a novel and efficient algorithm to determine the regions of the POIs. Briefly speaking, the algorithm mainly consists of *POI random selection, point in which region determination and cell growing, as illustrated in Algorithm 7.2.*

Algorithm 7.2 Algorithm for Determining Regions That a Huge Number of POIs Belong to

---

Input: a pool of POIs ($\{\boldsymbol{p_i}\}$) and a set of UARs ($\{\boldsymbol{R_i}\}$) in the target city;
Output: $\{\boldsymbol{R_i}\} = \boldsymbol{PinR}(\{\boldsymbol{p_i}\})$.

- Step 1: Randomly select a POI from {pi} (e.g., $\boldsymbol{p_s}$);
  Step 1.1: $\boldsymbol{P_s} = \boldsymbol{PinR}(\boldsymbol{p_s})$;
- Step 2: Take $\boldsymbol{P_s}$ as the center, get a grid cell with equal width and length ($\boldsymbol{g_0}$);
  Step 2.1: $\boldsymbol{g_i} = \boldsymbol{g_0}$;
- Step 3: If $\boldsymbol{g_i}$ has no intersection with $\boldsymbol{R_s}$, then
  Step 3.1: Identify all POIs inside the grid based on the geometric relationship (denoted by $\boldsymbol{P_{sub}}(\boldsymbol{g_i})$); $\boldsymbol{P_{sub}}(\boldsymbol{g_i})$) should be all located at $\boldsymbol{R_s}$;
  Step 3.2: $\{\boldsymbol{p_i}\} = \{\boldsymbol{p_i}\} - \boldsymbol{P_{sub}}(\boldsymbol{g_i})$);
  Step 3.3: Increase the grid cell size by **50%**, $\boldsymbol{g_{i+1}} = \mathbf{1.5} \times \boldsymbol{g_i}$;
- Step 4: Repeat Step 1~3 until $\{\boldsymbol{p_i}\}$ is empty.

In the first step, we randomly pick up a POI from the pool and determine which region the selected POI belongs to (Step 1.1) based on Algorithm 7.1. In the second step, we determine a grid cell with the selected POI as the center. Figure 7.1 (top right) demonstrates the result after the first two steps. All POIs inside the grid cell should be located at the same region of the selected POI if there is no inter-section between the grid cell and the region boundaries (Steps 3.1 and 3.2 respectively). Thus, there is no need for us to check for those POIs and we can remove them from the POI pool directly (Step 3.3). With the objective of further increasing the number of no-need-check POIs, the grid cell will grow bigger to contain more POIs (Step 3.4), as demonstrated in Fig. 7.1 (bottom left). In the case that the grid cell (g_i) crosses over the region, the algorithm will restart the whole procedure from the first step by selecting a new POI randomly again. The process will terminate until there is no POI in the pool (Step 4). Finally, each POI will be associated with a label of the region that it belongs to.

For POIs inside the same UAR (POIs with the same region label), we apply the popular DBSCAN algorithm to get clusters since the algorithm can identify clusters with different density and shape [16]. POIs that are close to each other and within the same UAR would be identified as a Candidate Activity Area (CAA). However, as demonstrated in Fig. 7.1 (bottom right), POIs scattering at different UARs are grouped to different CAAs, even if they are close to each other.

Remark Although the clustering and identification of CAAs can be done offline, it should be a plus if we can accelerate the procedure, since we have a huge number of points of interests and dozens of regions in the city. What is more, POIs in the city are dynamic, for instance, some POIs are disappearing while some POIs are emerging, necessitating the regular update of CAAs. Thus, it is desirable if we have an efficient algorithm for clustering and identification of CAAs.

7.4.3 Trip Purpose Imputation

The objective of the trip purpose imputation is to predict the POI category that the passenger intends to visit after getting off the taxi, given the drop-off point location and the drop-off time. We denote the drop-off information of the passenger by ((x, y), t). To infer the trip purpose correctly, several factors need to be considered. The first is the distance from passenger's final destination to the drop-off location. In more detail, the closer is the POI to the drop-off point, the more likely would the POI be visited, since taxis offer door-to-door services to passengers. Under such circumstance, most passengers prefer to get off taxis as close as possible to the final destination. The second factor that needs to be considered is the distribution of nearby POI categories to the drop-off point. Heading to an area mostly covered by Restaurants, the trip purpose would probably be the dining activities. Last but not the least, the alighting time of the passenger from the taxi is also vital as people take different activities at different time.

To integrate the above three factors comprehensively, we mainly take the following three major steps. First, given the location of the drop-off point, we select the top-k nearest CAAs within the walkable distance (e.g., 500 m). We note that passengers will visit the top-k CAAs with different probabilities. That is, the closer is the CAA to the drop-off point, the higher is the probability that the CAA will be visited, which exhibits the distance decay effect. Specifically, the probability that a CAA will be visited can be determined by Eq. (7.1).

$$\begin{aligned} & P(CAA_i \mid (x,y)) \propto (d_i)^{-\beta} \\ s.t. \quad & \sum_{i=1}^{k} P(CAA_i \mid (x,y)) = 1 \end{aligned} \tag{7.1}$$

where d_i refers to the Euclidean distance from the center of CAA_i to the drop-off point (x, y) of the passenger; k is the number of the nearby CAAs considered, which is set to 3 in our study; β is the distance decay parameter. We set $\beta = 1.5$, which is also consistent with existing findings in [1, 9].

Second, even if the visited CAA has been determined, because there are different POIs, each with a unique category and visiting popularity, the prediction of the POI categories for passengers is still challenging [4]. To alleviate the issue, inside a determined CAA (e.g., CAA_i), we compute the probability for visiting each POI category (i.e., taking activity) based on Bayes' theorem [17], as shown in Eqs. (7.2) and (7.3).

$$P\big(a_j \mid (x,y),t,CAA_i\big) = \frac{P\big((x,y) \mid a_j,t,CAA_i\big) \times P\big(a_j \mid t,CAA_i\big) \times P(t,CAA_i)}{P\big((x,y),t,CAA_i\big)} \tag{7.2}$$

$$P\big((x,y),t,CAA_i\big) = \sum_{j=1}^{n} P\big((x,y) \mid a_j,t,CAA_i\big) \times P\big(a_j \mid t,CAA_i\big) \times P(t,CAA_i) \tag{7.3}$$

n is the number of total activities considered in the study; $p((x,y)| a_j, t, CAA_i)$ represents the probability that a passenger gets off the taxi at location (*x*, *y*) if he/she has decided to take the activity a_j at CAA_i at time t. Gong et al. [4] simply assume that the location and the time of the drop-off point are *conditionally independent*, given the activity type (a_j), i.e., the following equation can be satisfied.

$$P\big((x,y) \mid a_j,t,CAA_i\big) = P\big((x,y) \mid a_j,CAA_i\big) \tag{7.4}$$

However, we argue that Eq. (7.4) does not hold for most cases, since where passengers select to get off taxis does not only depend on the nearby land use (i.e., spatial context) [3, 8], but also the alighting time. On one hand, passengers may get off taxis near a shopping plaza to shop; while on the other hand, passengers might get off taxis at places in a business district to have meal in the evening. In other words, the locations and the times of the drop-off point are *interdependent*. Here, we use the following equation to approximate the true value of $P((x,y) \mid a_j, t, CAA_i)$ by considering the attractiveness and the POI distribution on categories of the CAA collectively, as shown in Eq. (7.5).

$$P\big((x,y) \mid a_j,t,CAA_i\big) \propto \frac{\text{numberofPOIs}(a_j,CAA_i)}{\text{numberofPOIs}(CAA_i)} \times A_i(t) \quad \text{s.t.} \sum_{j=1}^{n} P\big((x,y) \mid a_j,t,CAA_i\big) = 1 \tag{7.5}$$

numberofCheckins (CAA_i) and *numberofPOIs* (a_j, CAA_i) in Eq. (7.5) refer to the number of POIs and the number of POIs related to a_j within the CAA_i respectively; Ai (t) refers to the attractiveness of the CAA_i at the given time slot, which can be measured by the popularity of CAA_i at that time, compared to the rest of other CAAs among the top-k list. In more detail, we calculate the value of A_i (t) by dividing the number of check-ins of CAA_i by the total number of check-ins of all top-k CAAs during the given time slot in the historical days (e.g., last month), as can be seen in Eq. (7.6). Note that it is easy to extract the information about the check-ins and categories of POIs from the Foursquare platform.

$$A_i(t) = \frac{\text{numberofCheckins}(CAA_i, t, \text{days})}{\sum_{i=1}^{k} \text{numberofCheckins}(CAA_i, t, \text{days})} \tag{7.6}$$

P (a_j|t, CAA_i) in Eq. (7.2) is the probability of taking activity a_j if the passenger is located in CAA_i at time t. The distribution of P (a_j|t, CAA_i) depends on the spatial and temporal patterns of human activity in that area. It has been well recognized that human behaviours in terms of taking activities present strong and regular patterns. For instance, with respect to the time dimension, the probability of visiting work-related places during 8:00 am–10:00 am is generally much higher than that of visiting shopping malls. With respect to the space dimension, the case may vary depending on geographical areas. To capture such temporal and spatial regularities in a fine granularity, again in this study, we rely on the check-ins from Foursquare. Given the time t and candidate activity area CAA_i, we approximate the probability of visiting a certain POI category (i.e., taking the activity of a_j) by the ratio of the number of check-ins on the given POI category to the total number of check-ins in CAA_i during the given time slot in the historical days (e.g., last month), as shown in Eq. (7.7).

$$P(a_j \mid t, CAA_i) = \frac{\text{numberofCheckins}(a_j, CAA_i, t, \text{days})}{\text{numberofCheckins}(CAA_i, t, \text{days})} \tag{7.7}$$

Although strong and regular patterns (i.e., regularity) of human behaviours are frequently observed, dynamic is also an another salient feature. For instance, human behaviours are interrupted and changed when encountering unexpected sudden and big social events. To capture such changes, we propose to combine the freshest check-ins in the studied area since the live data may reflect the affected human activities timely. Therefore, the probability can be updated by Eq. (7.8).

$$\begin{aligned} \text{P}(a_j|\text{t}, CAA_i) \propto \alpha \times & \underbrace{\frac{numberofCheckins(a_j, CAA_i, t, days)}{numberofCheckins(CAA_i, t, days)}}_{regularity} \\ + (1-\alpha) \times & \underbrace{\frac{numberofCheckins(a_j, CAA_i, t, 4\,h)}{numberofCheckins(CAA_i, t, 4\,h)}}_{dynamic} \end{aligned} \tag{7.8}$$

where *numberofCheckins* (a_j, CAA_i, t, 4 h) refers to the number of check-ins in the given POI category and *numberofCheckins* (CAA_i, t, 4 h) indicates the total number of check-ins in the area of $CAA_i$by counting the check-ins accumulated in the most

	tr_1	tr_2	tr_3	tr_4	tr_5	tr_6
CAA_1		0.3		0.1		0.25
CAA_2			0.3		0.1	
CAA_3	0.2	0.1	0.2		0.1	
CAA_4		0.3			0.4	
CAA_5	0.2	0.3	0.4		0.4	
CAA_6	0.3			0.3		0.25
CAA_7			0.1	0.3		0.25
CAA_8	0.3			0.3		0.25

Fig. 7.2 Illustration for the computation of P (t, CAA_i). Value in the grid cell refers to the probability of taking activity in the corresponding CAA after the ending of the corresponding trip

recent 4 h just before time t, respectively. α is a weighting factor (we set α = 0.9 in this study). We note that the probability obtained by Eq. (7.8) needs to be normalized, i.e., $\sum_{j=1}^{n} P(a_j|t, CAA_i) = 1$ with n being the total number of activities considered in the study.

P (t, CAA_i) in Eq. (7.2) is the probability of taking activities in CAA_i after the passengers gets off taxis at time t, which can be computed by Eq. (7.9), as follows.

$$P(t, CAA_i) = P(t) \times P(CAA_i \mid t) \tag{7.9}$$

The probability of the passenger getting off taxis at time t (i.e., P(t)) is different at different times of the day, since human activity has strong time regularity. The probability P(t) can be estimated by the ratio of the number of drop-offs during the given time slot to the number of drop-offs during the whole day. The computation of P(CAA_i| t) is a bit more complicated. In the following, to better understand how to compute the value of P(CAA_i| t), we use an example to illustrate the basic idea, as shown in Fig. 7.2. We suppose that there are 6 taxi trips occurred during the given time slot and there are 8 CAAs that have been identified. For each taxi trip, passengers would choose one of the CAAs to take activities after getting off taxis. Furthermore, as discussed earlier in the section, for each trip, we assume the passenger would take activities in one of the top-k CAAs within the walkable distance. In the example, the value of the grid cell (e.g., g_{ij}) refers to the probability of passengers from taxi trip tri taking activity in area CAA_i, which can be computed based on Eq. (7.1). For each time slot, the probability of taking activity in a given CAA (CAA_i) is just the average value of the corresponding row values, i.e.,

$$P(CAA_i \mid t) = \frac{\sum_{m=1}^{N} g_{im}}{N} \tag{7.10}$$

where N is the number of taxi trips occurred in the studied time slot.

In summary, for the taxi trip (x, y, t), the probability of passengers taking a given activity a_j after getting off the taxi can be approximated by the following equation.

$$P\left(a_j \mid (x,y),t\right) \propto P(CAA_i \mid (x,y)) \times P\left(a_j \mid (x,y),t,CAA_i\right)$$
$$\text{s.t.} \sum_{j=1}^{n} P\left(a_j \mid (x,y),t\right) = 1 \tag{7.11}$$

7.5 Phase II: Enabling Real-Time Response

In order to enable the real-time response for each drop-off event (i.e., compute the posterior probability of taking each activity for each drop-off point using Bayes' theorem in real-time), we need to identify the most time-consuming component. As discussed in the Sect. 7.3, the posterior probability calculation mainly consists of four components, the details of which are shown in Table 7.2.

As shown in the table, the first component is related to the probability of visiting a given candidate activity area (CAA_i) if the passenger was dropped off at point (x, y). The probability is computed online because the distance to each top-k nearest CAAs varies if the passengers get off taxis at different points. However, we argue that two drop-off points that are close to each other would have similar value of P (CAA_i| (x, y)), i.e., P (CAA_i | (x_1, y_1)) ≈ P (CAA_i| (x_2, y_2)) if (x_1, y_1) is close to (x_2, y_2). Hence,

Table 7.2 Details on each component of inferring trip purpose

	Component	Notion	Calculation formula	Calculation manner[a]
1	P(CAA_i|(x, y))	The probability of visiting CAA_i if getting off taxis at (x, y)	Eq. (7.1)	Online
2	P((x, y)|a_j, t, CAA_i)	The probability of getting off taxis at (x, y) if taking activity a_j in CAA_i at t	Eqs. (7.5) and (7.6)	Offline
3	P(a_j|t, CAA_i)	The probability of taking activity a_j if visiting CAA_i at t	Eq. (7.8)	Online (partially)
4	P(t, CAA_i)	The joint probability of visiting after getting off taxis at t	Eqs. (7.9) and (7.10)	offline

[a]It has two options, i.e., online and offline, respectively. "Online" indicates that the value of the corresponding component can be only computed after the arrival of drop-off points, while "offline" means it can be pre-computed in advance before the arrival of drop-off points

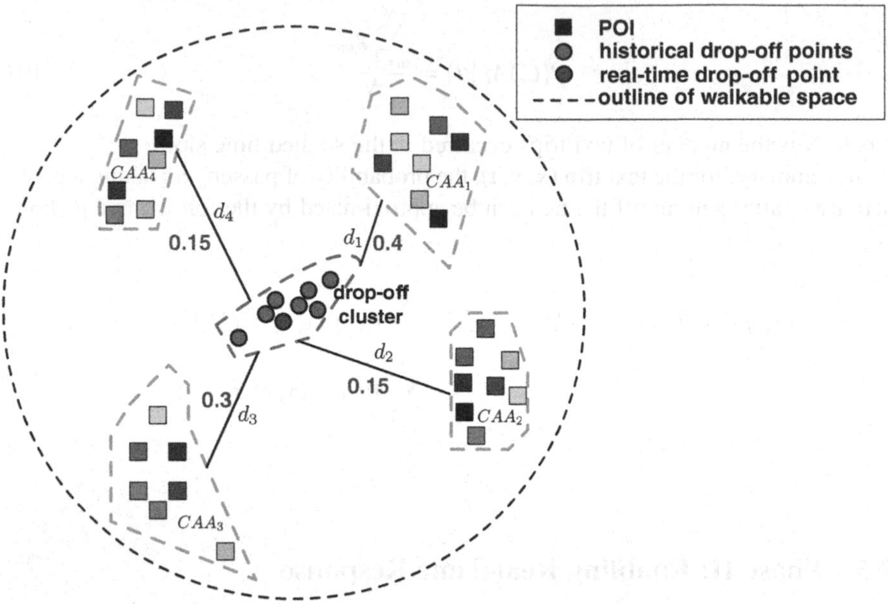
POI
historical drop-off points
real-time drop-off point
outline of walkable space
CAA4
CAA1
CAA2
CAA3
d4
d1
d2
d3
0.15
0.4
0.15
0.3
drop-off
cluster

Fig. 7.3 A schematic diagram of reducing the time complexity of the first component. The value on the edge carries the information about the visiting probability to the corresponding CAA

we aggregate historical information on drop-off points to drop-off cluster and assume all drop-off points in the same cluster would have equal value of P (CAA_i| (x, y)). In such way, the value of the first component can be pre-computed offline. The only online job is to identify which drop-off clusters that it should belong to. Once receiving a real-time drop-off point, this online job is quite efficient. In this manner, the computation time can be reduced significantly. As shown in Fig. 7.3, the top-k CAAs of the drop-off cluster can be identified and the distance to each CAA can be measured by the one between the centroid of drop-off cluster and the centroid of each CAA. Thus, the probability of visiting CAA_i from a drop-off point inside the drop-off cluster can be calculated offline efficiently. We note that many drop-off clusters can be obtained in advance, given the historical taxi trip data. Each of the drop-off clusters is associated with k visiting probabilities to its nearby top-k CAAs.

The second component is related to the probability of getting off taxis at point (x, y) if the passenger walks to area CAA_i and intends to take activity a_j at time t. As discussed earlier, two factors are considered. The first is the attractiveness of CAA_iat the given time slot, which is measured by the popularity of that area. Note that the popularity of a CAA at a given time slot can be calculated in advance, using the historical check-in data contributed by mobile users. The second factor is the POI category distribution in the CAA_i, which remains relatively stable. Thus, it is obviously that the value of the second component can be pre-computed offline.

The third component is the conditional probability of taking a given activity (e.g., a_j) if the passenger is at CAA_i at the time t. To approximate the true value of this component, both the "regularity" and "dynamic" patterns of the area are taken into consideration. As shown in the formula, the "regularity" pattern is based on the historical check-in data, and the "dynamic" pattern is captured by the most recent check-in data just before the drop-off time. Thus, the former part can be pre-processed offline, while the latter part can only be computed online.

The fourth component is about the joint probability of visiting the area of CAA_i at the time of t. As can be seen, the value is determined by two parts. One is the frequency of getting off taxis at the given time slot, and the other is the spatial distribution of the drop-off pints. Both parts are quantified using the historical taxi trip data. Thus, the value can be pre-computed offline.

In summary, two online jobs, identifying the drop-off clusters and extracting the "dynamic" patterns of the top-k CAAs, are required when receiving a streaming drop-off point (x_r, y_r, t_r). With the other components computed and structured offline purposely, the whole process can be quite efficient. We will validate this in the experiments.

7.6 Evaluations

In this section, we provide an evaluation of the proposed approach using a real-world taxi trajectory dataset, Foursquare check-in dataset and road network dataset in Manhattan area, the city of New York (NYC). For more details about the experimental setup, please refer to the article [18].

7.6.1 Evaluation on Candidate Activity Area Identification

Figure 7.4 presents the clustering results (i.e., the identification of UARs and CAAs) of our two-stage clustering algorithm. In total, we have identified 30 UARs, all of which are based on the road network data. As shown in Fig. 7.4a, most POIs are located at midtown and downtown of Manhattan, while only very are scattering at the upper town. A close view of some selected regions is shown in Fig. 7.4b to highlight the advantages of our proposed clustering algorithm. For example, due to the physical barriers (i.e., wide roads), POIs in purple color at Region 6 are not grouped together with their nearby POIs at Region 5, and several POIs at Region 4 are not merged with their neighbors at Region 5 either. Each UAR contains different number of CAAs, depending on the spatial distribution of the POIs inside. Figure 7.4c shows the number of CAAs for each UAR. The x coordinate corresponds to the region number and the y coordinate is the number of CAAs in that region. As shown in the figure, region 17 contains the maximal number of CAAs, while most of regions have a number of CAAs less than 20.

The size of the identified CAA is also an important metric to evaluate the clustering algorithm. The size of each CAA should be within a region of the

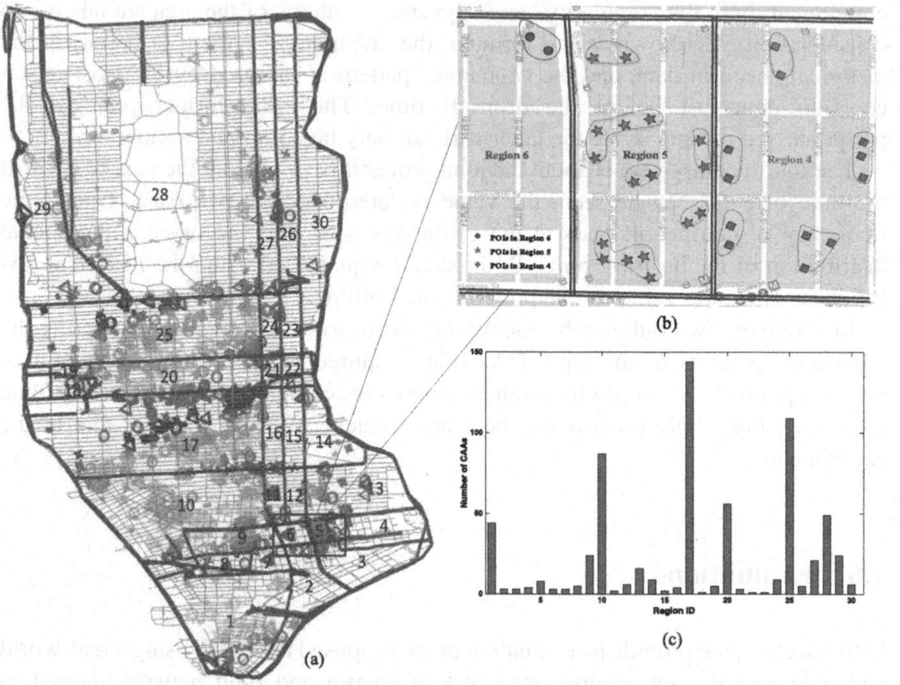
Region 6
Region 5
Region 4
POIs in Region 6
POIs in Region 5
POIs in Region 4
(a)
(b)
(c)
Number of CAAs
Region ID

Fig. 7.4 Results of UARs and CAAs identification in Manhattan, NYC. A full-view of clustering results (**a**); a close-view of some selected regions (**b**); the number of CAAs in the UARs (**c**). (Best viewed in an enlarged digital version)

walkable distance. Here the size of a CAA is defined as the minimal rectangle which covers all POIs in the CAA. If the CAA size is too big, then the POIs in the CAA are difficult to be reached by foot. Figure 7.5 shows the Cumulative Distribution Function (CDF) of the size of all CAAs. As can be seen from the figure, the size of over 96% of CAA are less than 10,000 m^2, showing the effectiveness of our proposed two-stage clustering algorithm.

7.6.2 Evaluation on Trip Purpose Imputation Algorithm

As discussed earlier, due to the lack of ground-truth of the taxi trip purpose, it is impossible to calculate the inference accuracy directly. Fortunately, we are provided with the travel purpose survey data at the regional scale (e.g., Manhattan) [19], which motivated us to evaluate the system accuracy indirectly. The rationale here behind is: if the distribution of the trip purposes inferred by our proposed method is close to the one obtained by the survey data in the statistical sense at the regional scale, our proposed method should be reliable. Since the survey data classifies the

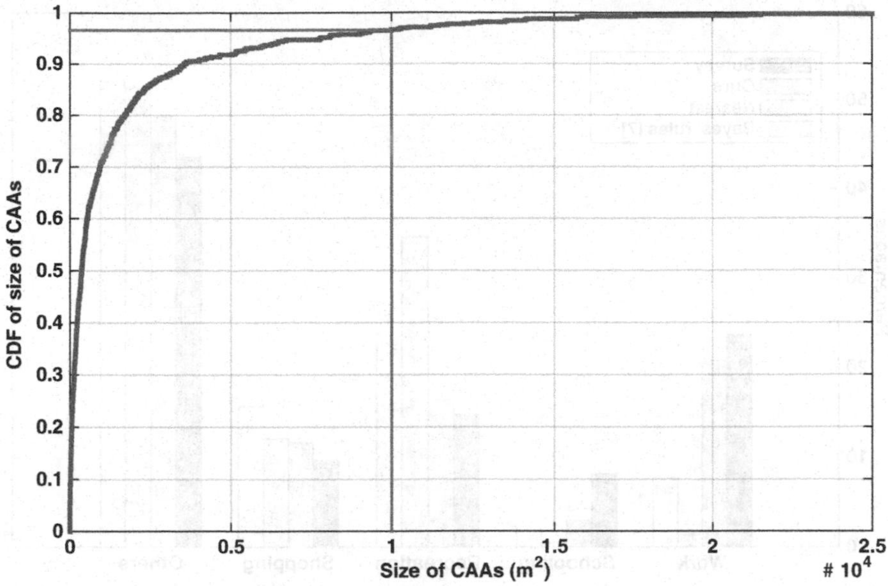
CDF of size of CAAs
1
0.9
0.8
0.7
0.6
0.5
0.4
0.3
0.2
0.1
0
0
0.5
1
1.5
2
2.5
Size of CAAs (m²)
10⁴

Fig. 7.5 The CDF distribution of the size of CAAs

travel purposes into four categories, i.e., work, education, recreation, shopping and others, to make the results comparable, we manually put 'dining', 'In-home', 'Transportation transfer', 'Lodging' and 'Medical' into the 'Others' category. Next, for each taxi trip, with the proposed inference algorithm, we are able to get five probabilities of five new trip purposes. Finally, for each trip purpose, we average the probabilities of all taxi trips generated in 1 month, and use the average value as the percentages of the travel for that trip purpose.

We show the comparison between our inference results to the travel survey data in Fig. 7.6. Besides, the results obtained by the other two baselines are also plotted for comparison. It is easy to understand that, the closer the percentage value on each category to the corresponding survey data value, the better performance our algorithm achieves. As can be seen from the results, our proposed algorithm achieves the best performance, while the Nearest algorithm achieves the worst performance and the Bayes' Rules [4] achieves the performance in-between.

Our proposed inference algorithm also enables us to gain insights on trip purpose in a much finer resolution. We thus select a representative urban activity region (UAR) to investigate the trip purpose trend at different time of the work day. The selected UAR together with inside distributed POIs is shown in Fig. 7.7, where only four POI categories can be found. Figure 7.8 shows the trip purpose inference results of the selected region across the whole day (top chart). We also show the corresponding results in other regions of Manhattan for comparison (bottom chart). As shown in the figure, travel for shopping and dining in the selected region is more common since it is a well-known shopping and dinner center in NYC.

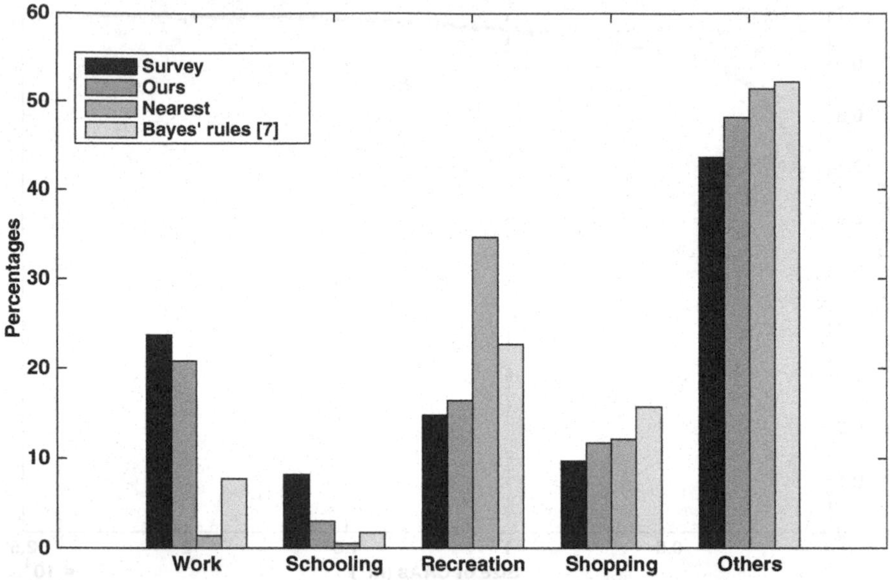

60
50
40
30
20
10
0
Percentages
Survey
Ours
Nearest
Bayes' rules [7]
Work
Schooling
Recreation
Shopping
Others

Fig. 7.6 Comparison results to baseline algorithms and survey data

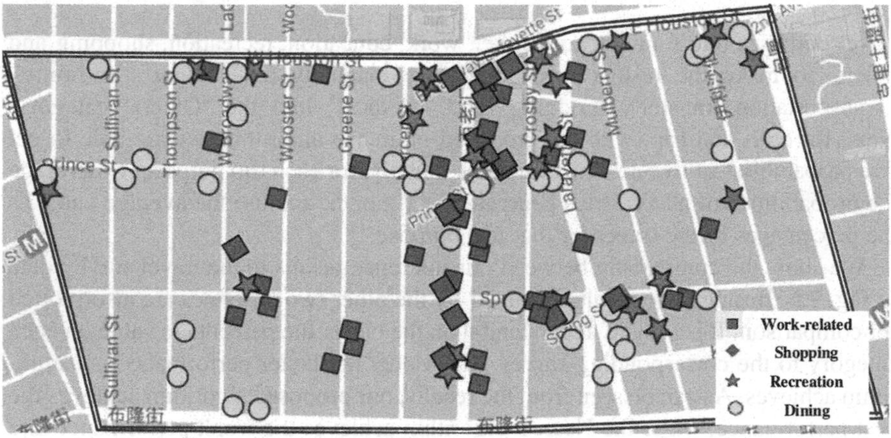

Work-related
Shopping
Recreation
Dining
Sullivan St
Thompson St
Wooster St
Greene St
Crosby St
Mulberry St
Prince St
E Houston St

Fig. 7.7 A selected UAR with four kinds of POIs. (Best viewed in an enlarged digital version)

Moreover, the number of trips for shopping purpose keeps increasing and remain high in the daytime, even in the work days. In both selected UAR and other regions in Manhattan, the number of trips for recreation purpose climbs after the work title me.

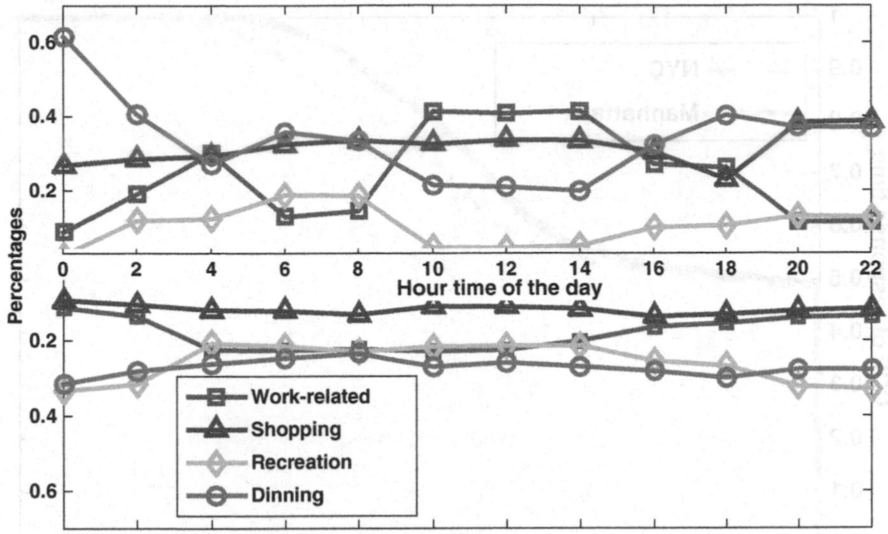
0.6
0.4
0.2
Percentages
0
2
4
6
8
10
12
14
16
18
20
22
Hour time of the day
0.2
0.4
0.6
Work-related
Shopping
Recreation
Dinning

Fig. 7.8 Trip purpose imputation results for a given day in the selected UAR and in Manhattan, respectively

Table 7.3 Response time in the worst case in Manhattan and NYC, respectively

Longest response time	Avg. ± std. (s)
Manhattan	7.54± 0.4187
NYC	8.15 ± 0.4808

7.6.3 Evaluation on Response Time

Another key system metrics is how long a passenger can get the recommendation services after getting off the taxi. Because all the requests are processed sequentially in one machine following the First-Come-First-Out (FIFO) rule, when a request arrives, one of the following two situations may occur. (1) There are no other requests are being processed or waiting to be processed in the system; (2) There are other requests in the system, being processed or waiting to be processed. In the first case, the request can enter service immediately upon arrival. In the second case, the request has to wait in queue until the server has finished processing other requests that arrive earlier. Thus, the response time for a request is the time from the request arrives till the time the request has been processed. In other words, the response time includes the wait time and the process time.

We are more interested in the longest response time that a request needs to spend during a day, i.e., the longest time that a request (or a taxi trip) needs to wait before being proceeded. The logic is that if the longest response time is acceptable for most users, then the system is useful in practice. The longest response time corresponds to the worst case during a day. Table 7.3 shows the average of the longest response time and its standard deviation values in Manhattan and the whole NYC respectively.

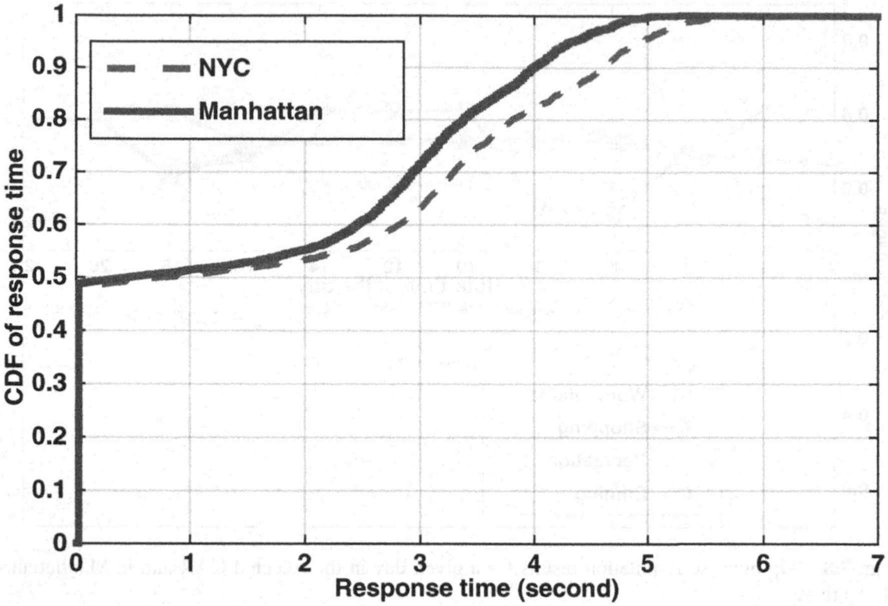
NYC
Manhattan
CDF of response time
Response time (second)
0
0.1
0.2
0.3
0.4
0.5
0.6
0.7
0.8
0.9
1
0
1
2
3
4
5
6
7

Fig. 7.9 The CDF distributions of the response times at a day in Manhattan and the whole NYC, respectively

Note that the observation days is 15. On average, the worst case takes 7.54 and 8.15 s to respond requests from Manhattan and from the whole NYC, respectively, which are acceptable in our application scenarios. Hence, we conclude that our proposed TripImputor is not only able to process requests from the whole NYC with a single normal PC, but also provide timely recommendation services.

We are also interested at the distribution of all response times in Manhattan and the whole NYC, as shown in Fig. 7.9. As can be observed, although Manhattan contributes 90% trip inferring requests of the whole city, it still takes more time to respond to a request from the city, because the more requests come per unit time, the longer the waiting time and so is the response time. Moreover, almost half of requests can be responded within 50 ms in both Manhattan and whole NYC. As shown in the figure, although in the worst case it takes up to around 7.54 s to process a request, 80% of the requests from Manhattan can be responded within 4.5 s and that from the NYC can be responded within around 5 s. On average, it takes only 1.588 and 1.812 s to respond for Manhattan and the whole NYC respectively. The above results demonstrate the efficiency of our system.

The previous experimental results ensure the efficiency of our proposed system in handling requests from the whole NYC. We are also aware that it takes more time to respond to a request when there are more requests arrive (as in the NYC). Going a step further, we intend to investigate how many cities (like NYC) can a single normal PC support and return a timely response. As shown in Fig. 7.10, x-axis refers to the number of requests per hour and y-axis refers to the longest response time of all

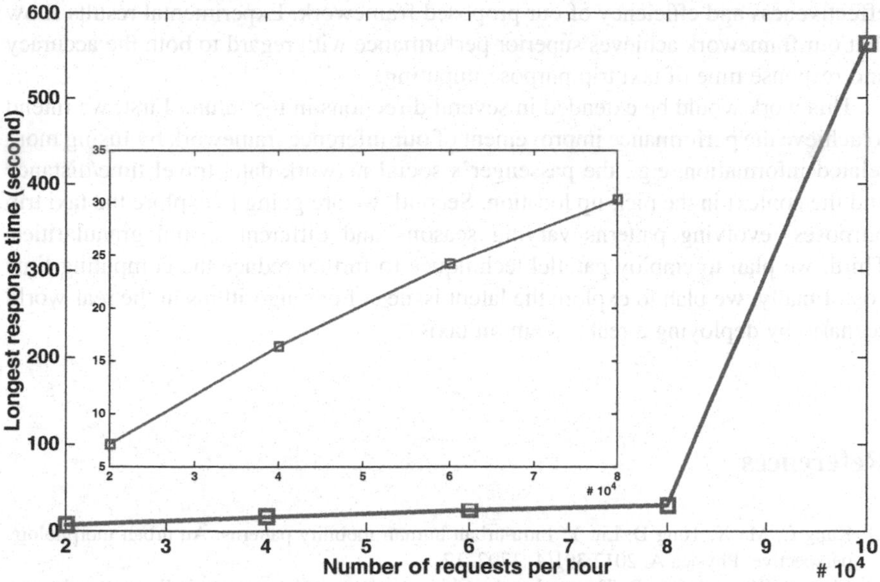
Longest response time (second)
Number of requests per hour
10^4

Fig. 7.10 The longest response time (corresponds to the worst case) under different number of requests per hour

requests. As can be seen, it takes around 7, 16, 24, 30 s at most to process 20,000, 40,000, 60,000, 80,000 requests, respectively. When the number of requests received during 1 h keeps increasing, the total processing time will increase exponentially, because all the requests are processed sequentially in one PC. The longest response time is more than 9 min if the number of requests per hour is 100,000. Note that there are around 20,000 requests arriving in 1 h in the whole NYC during the peak hours. Thus, facilitated by our method, we are capable of taking care of requests for four cities like NYC by just using one normal PC, if users can accept the maximal response time as around 30 s.

7.7 Conclusions and Future Work

In this chapter, a new two-phase framework named TripImputor is proposed to enable the real-time taxi trip purpose inference. Specifically, in the first phase, a two-stage clustering algorithm is proposed to obtain candidate activity areas (CAAs) based on the POI data. Then the posterior probabilities of different activities being taken by the passenger would be computed by employing the Bayes' theorem. In the second phase, in order to enable the real-time response, a sophisticated process is employed to reduce the online computation time, which includes the historical drop-off points clustering and the matching between drop-off clusters and CAAs. At last, extensive experiments are conducted on the real world datasets to evaluate the effectiveness and efficiency of our proposed framework. Experimental results show that our framework achieves superior performance with regard to both the accuracy and response time of taxi trip purpose imputing.

This work would be extended in several directions in the future. First, we intend to achieve the performance improvement of our inference framework by fusing more related information, e.g., the passenger's social network data, travel time/distance and the context in the pick-up location. Second, we are going to explore the taxi trip purposes' evolving patterns varying seasons and different spatial granularities. Third, we plan to employ parallel techniques to further reduce the computing time cost. Finally, we plan to explore the latent issues of our algorithms in the real-world scenario by deploying a real system on taxis.

References

1. Kang C, Ma X, Tong D, Liu Y. Intra-urban human mobility patterns: An urban morphology perspective. Physica A. 2012;391(4):1702–17.
2. Chen C, Wang Z, Guo B. The road to the Chinese smart city: progress, challenges, and future directions. IT Prof. 2016;18(1):14–7.
3. Deng Z Ji M. Deriving rules for trip purpose identification from GPS travel survey data and land use data: a machine learning approach. In: Traffic and transportation studies, 2010. p. 768–77.
4. Gong L, Liu X, Wu L, Liu Y. Inferring trip purposes and uncovering travel patterns from taxi trajectory data. Cartogr Geogr Inform Sci. 2016;43(2):103–14.
5. Li X, Li M, Gong Y-J, Zhang X-L, Yin J. T-DesP: Destination prediction based on big trajectory data. IEEE Trans Intell Transp Syst. 2016;17(8):2344–54.
6. Yang D, Zhang D, Zheng VW, Yu Z. Modeling user activity preference by leveraging user spatial temporal characteristics in LBSNs. IEEE Trans Syst Man Cybernet Syst. 2015;45 (1):129–42.
7. Krumm J, Rouhana D. Placer: semantic place labels from diary data. In: Proceedings of the ACM international joint conference on pervasive and ubiquitous computing, 2013. p. 163–72.
8. Wolf J. Using GPS data loggers to replace travel diaries in the collection of travel data. PhD dissertation, School of Civil and Environmental Engineering, Georgia Institute of Technology, Atlanta, GA, USA, 2000.
9. Chen C, Zhang D, Guo B, Ma X, Pan G, Wu Z. TripPlanner: personalized trip planning leveraging heterogeneous crowdsourced digital footprints. IEEE Trans Intell Transp Syst. Jun. 2015;16(3):1259–73.
10. Chen C, et al. CrowdDeliver: planning city-wide package delivery paths leveraging the crowd of taxis. IEEE Trans Intell Transp Syst. Jun. 2017;18(6):1478–96.
11. Newman D, Paasi A. Fences and neighbours in the postmodern world: Boundary narratives in political geography. Progr Human Geogr. 1998;22(2):186–207.
12. Yuan NJ, Zheng Y, Xie X. Segmentation of urban areas using road networks. Microsoft Corp., Redmond, WA, Tech. Rep. MSR-TR-2012-65, 2012.
13. Shimrat M. Algorithm 112: position of point relative to polygon. Commun ACM. 1962;5 (8):434.
14. Hormann K, Agathos A. The point in polygon problem for arbitrary polygons. Comput Geom. 2001;20(3):131–44.
15. Huang J, Li Y, Crawfis R, Lu S-C, Liou S-Y. A complete distance field representation. In: Proceedings visualization, 2001 (VIS'01), San Diego, CA, USA, 2001. p. 247–54.

16. Ester M, Kriegel H-P, Sander J, Xu X. A density-based algorithm for discovering clusters in large spatial databases with noise. Kdd. 1996;96:226–31.
17. Koch K-R. Introduction to Bayesian statistics, Springer, Berlin, 2007. Available: http://www.springer.com/gp/book/9783540727231#aboutBook.
18. Chen C, Jiao S, Zhang S, Liu W, Feng L, Wang Y. TripImputor: real-time imputing taxi trip purpose leveraging multi-sourced urban data. IEEE Trans Intell Transp Syst. 2018;19(10):3292–304.
19. Chen C, Gong H, Lawson C, Bialostozky E. Evaluating the feasibility of a passive travel survey collection in a complex urban environment: lessons learned from the New York City case study. Transp Res A Pol Pract. 2010;44(10):830–40.

Part IV
Enabling Smart Urban Services: Urban Planners

Chapter 8
GPS Environment Friendliness Estimation with Trajectory Data

8.1 Introduction

The error of GPS positioning has negative impacts on location-based services (LBS) users and even leads to decision-making mistakes. Although some methods can improve the accuracy of GPS positioning (e.g., using more satellites and the vector tracking based on Kalman filtering), they are still not sufficient enough to reduce errors due to multipath effects [1, 2].

Concretely, the multipath effect refers to the phenomenon in urban canyons (urban areas surrounded by tall buildings creating a canyon-like environment); Instead of reaching the receiver directly through the line of sight (LoS), the GPS signals are reflected via the surface of the building or ground. Recent works [3, 4] confirmed that the multipath effect is the main factor affecting GPS positioning accuracy. Besides, according to the statement of the National Marine Electronics Association, the locating information included in the GPS raw data can be used to measure satellite constellations' geometry errors and receivers' instrumental errors to some extent. However, the work in [3] indicated that the extra distance travelled due to the reflection of the primary signal can inflate the pseudo-range estimate, which cannot be reliably distinguished by GPS receivers. In conclusion, it is not enough to measure the multipath error or GPS positioning error definitely based on such information alone [4, 5].

To measure the impact of the urban environment on GPS positioning accuracy, we define GPS environment friendliness (GEF) as the metric: for a specific area, the more negative the effect of the building layout environment on GPS accuracy, the lower the GEF. The estimation of GEF information in different areas is a

Part of this chapter is based on a previous work: Ma L, Zhang C, Wang Y, Peng G, Chen C, Zhao J and Wang J. "Estimating Urban Road GPS Environment Friendliness with Bus Trajectories: A City-Scale Approach," *Sensors*, vol. 20, no. 6, pp. 1580, Jan 2020, doi: https://doi.org/10.3390/s20061580.

C. Chen et al., *Enabling Smart Urban Services with GPS Trajectory Data*,
https://doi.org/10.1007/978-981-16-0178-1_8

fundamental work: First, it helps to improve the user experience of LBS in a situation where GPS accuracy is limited. Second, estimating GEF is also helpful to improve the accuracy of GPS in urban areas. There are many methods to improve the accuracy of GPS by combining GPS samples with other complementary information, such as Wi-Fi fingerprints, street-view media, and 3D-maps. However, implementing those solutions usually introduces extra costs. The GEF can serve as a reminder of where the deployment of assisted positioning solution is most needed, to minimize the overall cost while achieving satisfactory positioning accuracy.

There are many interesting studies on GPS accuracy in different urban areas. For example, Schipperijn [6] recorded 68,000 GPS points on four routes to test the dynamic accuracy of GPS positioning. Drawil [4] proposed a scheme that employs the GPS dataset collected by a vehicle and the knowledge of the surrounding environment to solve the estimation of localisation accuracy. However, most of these apply to small-scale and road-by-road field studies. It is not practical to obtain a comprehensive city-scale evaluation.

To this end, we propose an approach for estimating the city-scale GEF based on the historical GPS trajectory data of buses. The basic idea is first to divide the urban road network into short road segments of equal length, with the GEF being the same at different locations within a single road segment. We then utilise historical bus GPS data and the bus routes' information to estimate the error of each GPS localisation record. Finally, we statistically analyse the positioning error of the buses on different road segments and calculate their GEF level.

Although the basic idea above is easy to understand, our proposed approach is not straightforward on account of the following challenges. All bus routes have a high coverage for roads in a city, while the trajectory data of a single bus can only cover a small part of the road network. It is impossible to assess the GEF of all roads by merely using the GPS data of a single bus. Moreover, the quality of GPS receivers varies significantly among different buses. If we estimate the GEF of a road only on the bus routes that cover it, the conclusion reached may be incorrect. We cannot solve the problem by simply aggregating GPS samples from different buses. Therefore, when integrating the GPS data of different buses, we need to develop more sophisticated mechanisms to eliminate the effects of GPS receiver quality differences in order to compare the GEF of different road segments. The main contributions of the study are:

- We use the historical GPS trajectories of buses to estimate the GEF of city-scale roads, which makes our approach more scalable and less expensive and is easier to transfer to other cities, as it only uses existing bus trajectory data. In addition, we employ a priori knowledge that buses run on a fixed route many times a month to improve the accuracy and efficiency of the map matching process.
- We propose a novel three-phase framework for estimating the GEF of urban roads. First, we employ the map matching algorithm to map bus routes' data and the historical bus GPS data into the road network, and then calculate the errors of each bus on road segments it passes through. Secondly, we infer the GPS errors of buses on all road segments adopting our proposed matrix completion-based

method. Finally, we integrate the errors of buses on all road segments to estimate the GEF.
- We evaluate and validate our estimated GEF by comparing it with the ground truth collected through field study and the street views on some road segments. The results confirm the effectiveness of our proposed approach.

8.2 Related Work

8.2.1 GPS Error and Calibration

GPS data have attracted a great deal of attention in the field of data mining. Most of the research combines GPS data with multiple sources of heterogeneous data, such as POI data and crowd movement data, using a variety of machine learning techniques to analyse and discover knowledge about the city and further resolve problems in smart city construction [7–13].

However, most works referring to GPS data are seriously misguided by GPS positioning error. GPS errors consist of three main components, including Satellite clock error, signal transmission error, and the GPS terminal device's error. Recent work [4] derived from experiments that the multipath effect, especially in built-up urban areas, has a significant impact on the precision of GPS positioning. Although many methods can improve the accuracy of GPS positioning, they are not sufficient enough to reduce multipath errors [1, 2]. Wu et al. [14] proposed a novel error reduction system for trajectories. However, this method cannot figure out the actual position of any single GPS positioning record as it is suitable for sequential localisation trajectories.

To obtain a reliable position in urban areas, some existing positioning techniques fusing GPS data with extra information, such as Wi-Fi fingerprints, street view videos/images, and 3D maps. For the sake of maximising the benefits and minimising the total cost, decision-makers should select where to deploy expensive devices cautiously to collect such complementary information. The introduction of GEF provides economic guidance on where to map out those devices.

8.2.2 Measuring GPS Positioning Performance

GPS positioning errors and their causes have been the subject of intensive research by researchers. GPS receivers cannot reliably distinguish between reflected and direct signals [3]. In addition, according to a statement by the National Marine Electronics Association, there is an effect of satellite geometry on accuracy in GPS measurement data, known as the dilution of precision (DOP). However, the work in [5] indicated that the DOP of the site varies throughout the day. The work in [4] also showed that although the DOP can offer some power to figure out the positioning performance of a given measurement, it cannot be relied upon to classify the accuracy of the measurement.

Data from various scenarios are collected to measure the precision of the GPS records. The work in [15] evaluated GPS positioning accuracy based on the vehicle trajectory data only. The work in [4] used a vehicle equipped with a standard GPS receiver to collect 6520 real-life GPS measurements and further proposed a method for localisation accuracy estimation. Some studies have also employed knowledge of the surrounding environment to optimise classification performance. The work in [6] collected information only from a closely spaced bodybuilding apparatus in an outdoor fitness area. These existing studies are not desirable with limited budgets due to the amount of effort they consume. Therefore, they were unable to estimate the GPS positioning performance on the city scale.

8.3 Basic Concepts

Definition 8.1 Bus Route *The bus route BR_i is a subgraph of the road network graph RN. In this study, there were 184 different bus lines in Chengdu that covered n = 8831 road segments in road network RN. There was always more than one bus running on the same route. For example, the red lines in Fig. 8.1 denote a part of the bus line route.*

Definition 8.2 Bus Trajectory *The trajectory $G_i = \{g_{i,t}\}(i = 1, \ldots, m)$ of bus_i is a sequence of GPS points $g_{i,t}$. We used m to denote the number of buses. m is equal to 4835 in this work. The GPS point $g_{i,t} = (time_{i,t}, latitude_{i,t}, longitude_{i,t})$ consists of a time-stamp $time_{i,t}$, a latitude record $latitude_{i,t}$, and a longitude record $longitude_{i,t}$. For example, the black points in* Fig. 8.1 *denote the GPS trajectory data of buses.*

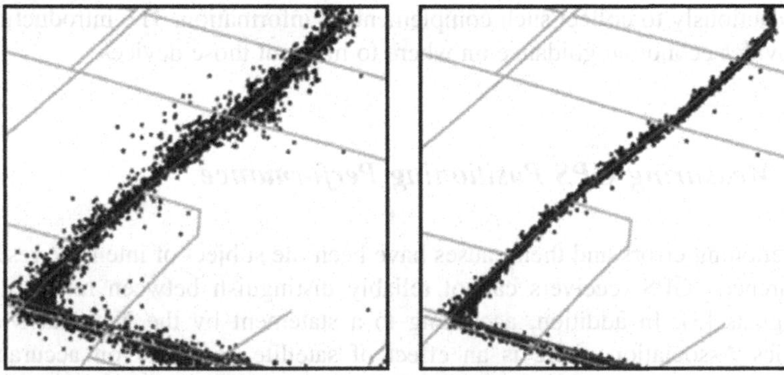

Fig. 8.1 GPS trajectory data of two buses on the same roads

Definition 8.3 POI Information of the Road Segment *The POI information of the road segment is depicted by several different POI categories from the online map. For road*$_i$*, we constructed a POI feature vector* $c_i = (cnt_1, \ldots, cnt_{num})$*, where num denotes the number of different POI categories and* $cnt_j(j = 1, \ldots, num)$ *denotes the number of nearby (within 200 m) POI, which belong to category* poi_j*. Concretely in this study, there were num* $= 17$ *different POI categories according to the Gaode Online Map: catering services, traffic infrastructures, government agency, vehicle sales, corporations, scenic spots, sports services, science education services, shopping services, accommodation services, vehicles services, serviced apartment, finance insurance services, life services, vehicle maintenance, and medical care services.*

Definition 8.4 Tags of the Road Segment *According to the OpenStreetMap, road segments could be categorized by tags: PrimaryLink, LivingStreet, service, residential, SecondaryLink, primary, MotorwayLink, unclassified, motorway, trunk, TrunkLink, tertiary, secondary. Each road segment is labelled with only one tag.*

Definition 8.5 Layout Information of the Road Segment *The layout information of the road segment is depicted by several different floors. For road*$_i$*, we constructed a layout feature vector* $h_i = (height_1, \ldots, height_{num})$*, where num denotes the number of different floors and* $height_j(j = 1, \ldots, num)$ *denotes the number of nearby (within 200 m) buildings with j floors. Concretely in this study, there were num* $= 60$ *different floors within the second-ring road in Chengdu, China.*

Definition 8.6 GPS Positioning Bias *The GPS positioning bias refers to the linear distance between the GPS positioning record and the real position of the bus. It ranges from a few meters in open sky environments to over 80 m in urban canyons* [4]. *The positioning bias of a bus on the road could be divided into two orthonormal parts. One is vertical to the road, while the other is parallel with the road. The vertical component is much greater than the parallel component, which can be ignored* [4]. *In this study, such bias is measured as the vertical distance between the GPS positioning point and the real road where the bus is running.*

Definition 8.7 GPS Positioning Error *The real horizontal position of the bus along the roads can be figured out based on map-matching algorithms. However, the width of the actual road cannot be ignored about the GPS positioning bias. It is difficult to tell on which lane the bus is running. As a result, we utilized the standard deviation (std) of the GPS positioning biases to measure the buses' GPS positioning errors on roads, instead of the mean values of the biases. In this way, the GPS positioning error is defined as the standard deviation (std) of the GPS positioning biases. Such error is affected by satellite ephemeris error, receiver clock error, multipath error, spherical error, receiver measurement noise, and so on. Multipath error is the major component when locating in urban areas. The concepts above are shown in* Fig. 8.2.

Definition 8.8 GPS Environment Friendliness (GEF) *Multipath error is caused by the delay of the signal arrival due to its reflection off building surfaces in the area.*

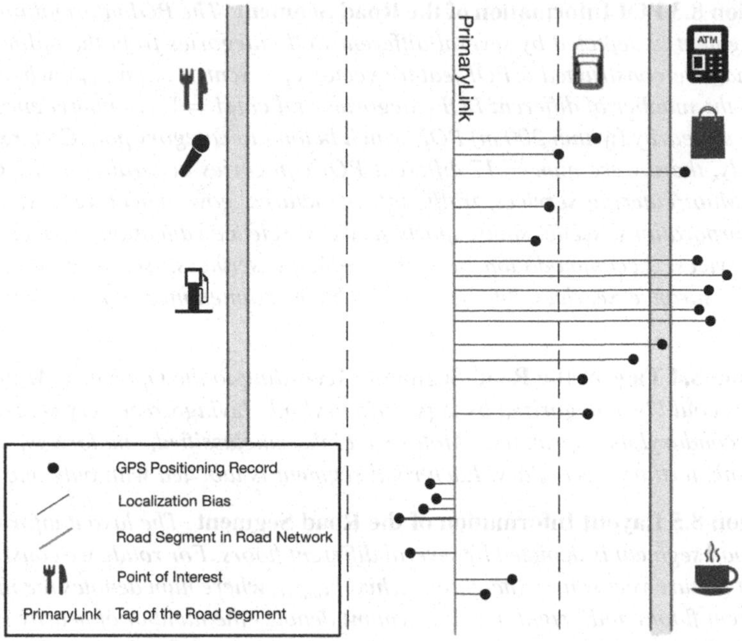
PrimaryLink
ATM
GPS Positioning Record
Localization Bias
Road Segment in Road Network
Point of Interest
PrimaryLink Tag of the Road Segment

Fig. 8.2 Meta information of the data

GPS environment friendliness defines the degree to which the multipath phenomenon affects the GPS performance. The GEF depends on the surrounding environment. It is independent of time, weather, the quality of GPS positioning terminal device, and the number of visible GPS satellites. We assumed that different locations within the same road segment shared a similar environment and the same GEF.

For a specific road segment, the GEF is considered poor if the std is high, while a lower std indicates that the GEF is better. To understand the GEF introduced in this study intuitively, we show the GPS trajectory data of one bus on different roads in Fig. 8.3. Yellow lines denote the road network, and black points denote the GPS records of the bus. The GPS points in the green circle are densely distributed, which means that their variance is small. It is indicated that the std of the GPS positioning error of the bus on this road is small and the GEF here is good. On the contrary, the GEF of the road marked by the red circle is poor.

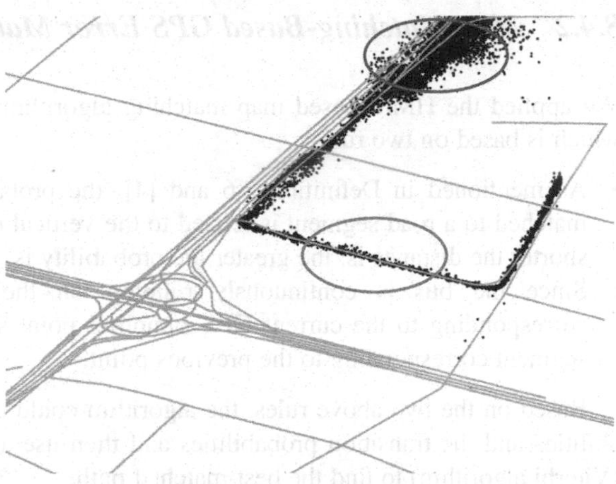
Fig. 8.3 GPS trajectory data of one bus on different roads

8.4 Methodology

8.4.1 Overview of the Framework

We developed an urban road GPS environment friendliness estimation approach based on the historical bus GPS trajectory data. The whole framework of the GEF evaluation was composed of the following main components:

1. We utilized the hidden Markov model (HMM)-based map matching algorithm [16] to map the bus trajectories' data to the roads. Then, we constructed a matrix, where the element of the matrix represented the positioning error standard deviation of each bus on each road segment. Note that the route of one bus only covered a small portion of the roads in the city. Few buses were running on any given road. Thus, the matrix to be completed was very sparse.
2. We estimated the positioning errors of each bus on each road segment based on the matrix completion algorithm, considering the nearby environment information. Due to the variance of the quality of the GPS receivers, an incorrect conclusion would be drawn if we estimated the GEF of a road only depending on the buses whose routes covered the road. Therefore, we needed to complete the matrix that was constructed in the first phase.
3. The GEF of each road segment was estimated based on the completion result. The buses whose GPS terminal device had a higher quality would have more weight on the evaluation of the GEF.

The details will be presented in the following subsections.

8.4.2 Map Matching-Based GPS Error Matrix Construction

We applied the HMM-based map matching algorithm as an algorithm prototype, which is based on two rules:

- As mentioned in Definition 8.6 and [4], the probability that a GPS point is matched to a road segment is related to the vertical distance between them. The shorter the distance is, the greater the probability is.
- Since the bus is continuously running on the road, the road segment corresponding to the current GPS sampling point should be close to the road segment corresponding to the previous point.

Based on the two above rules, the algorithm could calculate the emission probabilities and the transition probabilities and then use a dynamic planning strategy (Viterbi algorithm) to find the best-matched path.

However, if we applied the above algorithm prototype to our bus trajectory dataset directly, the amount of calculation would be relatively large. It is difficult for the map matching algorithm to achieve good performance when applied directly to trajectories with large errors [14]. In fact, fixed bus routes data can provide important supplementary information for map matching, which can reduce the number of candidate roads and improve the computational efficiency and matching accuracy. Therefore, the real position of each GPS point was believed to be located on the nearest road segment that was covered by the route of this bus, and the GPS positioning error of the record could be calculated. Besides, several buses were running on the same fixed line, and each of them went through the routes many times under different weather conditions and in time periods. As a result, the error produced by map matching and accidental factors (e.g., the position of satellites, weather) was reduced to some extent.

Concretely, we divided roads in the road network into short equal-length road segments. On the one hand, we assumed that the GEF at different locations within the same road segment was the same. On the other hand, the GEF of a segment could be estimated only if there were enough GPS record points.

As the bus route data provided by the public transport company were also designated by GPS points, we then needed to match bus route data to the road network. After that, the bus routes were designated by road segments, to which bus trajectory data would be mapped (i.e., the red lines in Fig. 8.4). Figure 8.4 shows the map matching result of bus route data of Line 1022. The yellow lines denote roads in Chengdu. The black points denote the GPS point of line 1022 route data. The red lines denote the map matching results of the black points.

Some GPS record points could not be mapped to the bus routes within the threshold distance. The main reason for the failure of map matching was that the bus did not travel exactly on the given route. Bus route data provided by the bus company may not be entirely accurate because of a temporary road diversion for construction or the delayed update of bus route data after route adjustment. The bus may not travel on a given route because of repair or refuelling. What is more, the bus

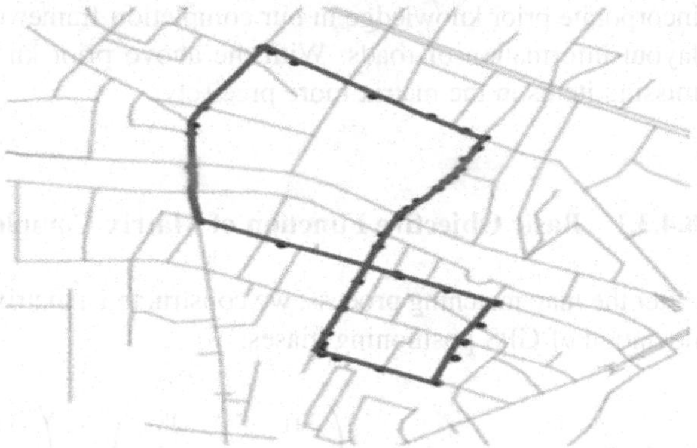

Fig. 8.4 Map matching result of Bus Line 1022 route data

may be temporarily scheduled to travel on another route. If there were more than five consecutive points far away (more than 50 m) from the given route, the bus was believed to be veering off its route. Those points would be mapped to other nearby segments on the road network until the bus returned to its given routes.

8.4.3 GPS Error Estimation with Additional Environment Information Integration

After mapping each GPS record to the corresponding road segment, we calculated the positioning biases of each bus on every passing road segment. The standard deviation of biases was utilized to measure the error. Matrix **Var** was then constructed, where the entry v_{ij} denotes the error of bus_i on $road_j$. However, there existed no bus that could pass all roads, making this matrix very sparse.

The routes of all buses in total had a high coverage for roads in the city, while the bus trajectory data of a single bus was quite sparse in the city. For example, within the second-ring of Chengdu, a single bus' coverage was only about 2.5%. The GEF of roads in the city had to be estimated based on the GPS records of many buses instead of a single bus.

However, there was a significant variance in the quality of GPS receivers among different buses, which meant that the GPS positioning performance of different buses on the same road segment may differ from each other. Therefore, a high GPS error may be caused by GPS receivers with low quality, even on the road segment with good GEF. To reduce such negative influence and estimate the GEF of road segments fairly, we needed to estimate the error of each bus on each road segment.

To complete a matrix, compressive sensing is widely applied [17]. Given a sparse matrix for which most of its items are missing, compressive-based matrix completion will estimate those missing items according to the specific cost function and optimization algorithm. In addition to the common cost function, we tried to incorporate prior knowledge in our completion framework, i.e., the nearby building layout information of roads. With the above prior knowledge, we could estimate missing items in the matrix more precisely.

8.4.3.1 Basic Objective Function of Matrix Completion

After the map matching process, we constructed a matrix **Var** recording the standard deviation of GPS positioning biases:

$$\mathbf{Var} = \begin{pmatrix} v_{11} & \cdots & v_{1n} \\ \vdots & \ddots & \vdots \\ v_{m1} & \cdots & v_{mn} \end{pmatrix} = \begin{pmatrix} \overrightarrow{v_1} \\ \vdots \\ \overrightarrow{v_m} \end{pmatrix}_{m \times n}$$

v_{ij} denotes the bias std of bus_i on road segments $road_j$. If the number of GPS records of bus_i on $road_j$ was less than 20, the v_{ij} would be set as a missing value. The basic objective function of matrix completion was set as [17]:

$$F(\mathbf{Sign}, \mathbf{Var}, \mathbf{L}, \mathbf{R}) = \|\mathbf{Sign} \cdot \mathbf{LR}^T - \mathbf{Var}\|_F^2 + \lambda(\| \mathbf{L}_F^2 \| + \| \mathbf{R}_F^2 \|) \tag{8.1}$$

The size of the binary identification matrix **Sign** was the same as matrix **Var**. s_{ij} equalled 1 if v_{ij} was known. Otherwise, s_{ij} equalled 0. $s_{ij} = 1_{\{(i,j)|v_{ij} \text{ is known.}\}}$. The result of matrix completion was $\mathbf{LR}^T$. The size of matrix **L** was $m \times a$, and the size of matrix **R** was $n \times a$. a was a hyper-parameter of matrix completion. The penalty term $\|$ Sign $\cdot$ LR^T $-$ Var $\|_F^2$ measured the similarity between the completion result and original matrix. $\| \mathrm{L}\|_F^2 + \| \mathrm{R}\|_F^2$ was the regularization term. λ was the hyper-parameter denoting the importance of the penalty term.

8.4.3.2 Measure the Relative Advantage of GPS Receivers' Qualities

To measures the relative advantage of GPS positioning terminals' qualities between each of two buses, an $m \times m$ matrix **Qua** was constructed. To test the equality of variations, we used the F-test [18].

$$\mathbf{Qua} = \begin{pmatrix} k_{11} & \cdots & k_{1m} \\ \vdots & \ddots & \vdots \\ k_{m1} & \cdots & k_{mm} \end{pmatrix}_{m \times m}$$

where k_{ij} measures the relative advantage of bus_i over bus_j.

$$k_{ij} = \begin{cases} 1 & \text{if the quality of } bus_i \text{ is better than } bus_j \\ -1 & \text{if the quality of } bus_i \text{ is worse than } bus_j \\ 0 & \text{if the relative advantage cannot be determined, or } i = j \end{cases}$$

Concretely, bus_i and bus_j only compared with each other on $road_r$, which has the most GPS points of them. It was assumed that GPS errors followed a Gaussian distribution. Thus, we performed an F-test between the GPS error sequences of bus_i and bus_j on $road_r$, while the confidence coefficient was 95%. As a result, the quality of bus_i was considered as better than the quality of bus_j, if $v_{ir} < v_{jr}$.

However, if there was not any road that had been travelled by both bus_i and bus_j, we would try to find another intermediate bus_q. The quality of bus_i was considered as better than the quality of bus_j, if $k_{iq} = 1$ and $k_{qr} = 1$, while both of the confidence coefficients should be higher than 97.5%; or the relative advantage between bus_i and bus_j was considered not able to be determined. In order to make the matrix completion result meet the relative advantage between different buses, we constructed an $m \times m$ matrix **Tran** based on matrix **Qua**.

$$\mathbf{Tran} = \begin{pmatrix} \Sigma_{j=1}^{m} k_{1j} & -k_{12} & \cdots & -k_{1m} \\ -k_{21} & \Sigma_{j=1}^{m} k_{2j} & \cdots & -k_{2m} \\ \vdots & \vdots & \ddots & \vdots \\ -k_{m1} & -k_{m1} & \cdots & \Sigma_{j=1}^{m} k_{mj} \end{pmatrix}_{m \times m}$$

Consider the transformation of matrix **Var**:

$$\text{Tran} \cdot \text{Var} = \begin{pmatrix} \sum_{j=1}^{m} k_{1j}\overrightarrow{v}_1 - k_{12}\overrightarrow{v_2} - \cdots - k_{1m}\overrightarrow{v_m} \\ \vdots \\ -k_{m1}\overrightarrow{v_1} - k_{m2}\overrightarrow{v_2} - \cdots - +\Sigma_{j=1}^{m} k_{mj}\overrightarrow{v_m} \end{pmatrix}_{m \times n}$$

To get a better insight into this transformation, consider Row_1 of **Tran** · **Var**.

$$Row_1(\text{Tran} \cdot \text{Var}) = \left(\sum_{j=2}^{m} k_{1j}(v_{11} - v_{j1}), \cdots, \Sigma_{j=2}^{m} k_{1j}(v_{1m} - v_{jm}) \right)_{1 \times m}$$

Recall the construction of k_{ij}; ideally, the value of $k_{ij}(v_{il} - v_{jl})$, $l = 1, \ldots, m$ should be a negative value for all (i, j) pairs. Thus, every input of **Tran** · **Var** should be a negative value in the ideal case, while a positive value is an inappropriate input.

Due to the lack of a zero lower bound of the **Tran** · **Var** F-norm, the cost function would not converge if we added this matrix into the cost function directly. Consider a monotone matrix operation $E_\theta(\cdot)$, $\theta > 0$. θ is a predetermined positive number controlling the absolute values of $e^{\theta y}s$ to avoid overflow while processing the algorithm. Here, $\theta = max\,(y_{ij})^{-1}$, $(i = 1, 2, \ldots, m; j = 1, 2, \ldots, n)$.

$$E_\theta(\mathbf{Y}) = \begin{pmatrix} e^{\theta y_{11}} & \cdots & e^{\theta y_{1n}} \\ \vdots & \ddots & \vdots \\ e^{\theta y_{m1}} & \cdots & e^{\theta y_{mn}} \end{pmatrix}$$

where Y is an arbitrary $m \times n$ matrix:

$$\mathbf{Y} = \begin{pmatrix} y_{11} & \cdots & y_{1n} \\ \vdots & \ddots & \vdots \\ y_{m1} & \cdots & y_{mn} \end{pmatrix}$$

The preferred properties of matrix operation $E_\theta(\cdot)$:

1. The elements in $E_\theta(Tran \cdot Var)$: inherit the relative magnitudes of the elements in **Tran** · **Var**, small values for the ideal case, large values for an inappropriate case.
2. It guarantees a lower bound of $\| E_\theta(\mathbf{Tran} \cdot \mathbf{LR}^T) \|_F^2$, so that the objective function below has a lower bound. As a result, we added the penalty below to the objective function:

$$\lambda_2 \left\| E_\theta\left(\mathbf{Tran} \cdot \mathbf{LR}^T\right) \right\|_F^2$$

8.4.3.3 Measure the POI Information of Road Segments

The POI information of road segments was able to characterize the nearby building layout environment. For example, there may be more POI of catering services and shopping services on the road segments with tall buildings. We assumed that the GEF of two roads was similar to each other when the Euclidean distance between two POI vectors annotating two roads was small. According to Gaode Map, the road was depicted by 17 different POI categories. For $road_i$, we constructed a POI feature vector $\vec{c_i}$:

$$\vec{c_i} = (cnt_1 \quad \cdots \quad cnt_{17})$$

where $cnt_q (q = 1, \ldots, 17)$ is the number of nearby (within 200 m) POI, which belongs to $category_q$. Then, compute the Euclidean distance between each POI vector of roads segments:

$$\text{Dist} = \begin{pmatrix} d_{11} & \cdots & d_{1n} \\ \vdots & \ddots & \vdots \\ d_{m1} & \cdots & d_{mn} \end{pmatrix}$$

where d_{ij} denotes the Euclidean distance between the POI vector of c_i and c_j. Thus, we can construct matrix **Poi** to describe the similarity of the POI distribution between each of two roads.

$$\text{Poi} = \begin{pmatrix} p_{11} & \cdots & p_{1n} \\ \vdots & \ddots & \vdots \\ p_{n1} & \cdots & p_{nn} \end{pmatrix} \text{where:}$$

$$p_{ij} = \begin{cases} 0 & \text{EuclideanDistance}\left(r_i, r_j\right) > \varepsilon \\ \dfrac{1/d_{ij}}{\Sigma_k 1/d_{ik}} & \text{Euclidean Distance}\left(r_i, r_j\right) < \varepsilon \end{cases}$$

k denotes the number of road segments, which had a similar POI distribution as $road_i$. In the Experiment Section, ε was set to 250, tuned by three-fold cross-validation. According to our assumption, the objective function should be penalized if there was a big difference between GPS errors of buses on roads, whose POI distributions were similar to each other. As a result, we added the penalty below to the objective function:

$$\lambda_3 \cdot \Sigma_{i=1}^{n} \sum_{j=1, j \neq i}^{n} \frac{1}{d_{ij}} \left\|\mathbf{LR}^T p_{ij}\right\|_F^2$$

8.4.3.4 Measure the Tag Information of Road Segments

According to the OpenStreetMap, road segments could be categorized by tags. Similar to the POI distribution, it was also assumed that the GEF of roads would be similar, if they had the same tag. As a result, we constructed the matrix:

$$\text{Tag} = \begin{pmatrix} t_{11} & \cdots & p_{1n} \\ \vdots & \ddots & \vdots \\ t_{n1} & \cdots & t_{nn} \end{pmatrix}$$

$$t_{ij} = \begin{cases} -1 & i = j, \\ 1/(k-1) & \text{if road}_i \text{ and } road_j \text{ have the same tag,} \\ 0 & \text{otherwise.} \end{cases}$$

k denotes the number of road segments that have the same tag as $road_i$ and $road_j$. The new regularization term was designed as:

$$\lambda_4 \cdot \left\|\mathbf{LR}^T\mathbf{Tag}\right\|_F^2$$

8.4.3.5 Measure the Layout Information around Road Segments

It was assumed that the layout information around road segments was able to characterize the nearby environment. The GEF of two roads should be similar to each other if the nearby building environments were similar as well. We assumed that the GEF of two roads was similar to each other when the Euclidean distance between two layout vectors annotating two roads was small. For example, there may be more urban canyons or other terrains that lead to poor GEF on the road segments with tall buildings.

The number of floor levels in Chengdu ranged from 1 to 60. Therefore, the layout of each road segment was depicted as a 60-dimensional vector, which meant the number of buildings (within 200 m) of each corresponding height. For $road_i$, we constructed a layout feature vector $\overrightarrow{h_i}$ to depict its nearby building layout:

$$\overrightarrow{h_i} = (height_1 \quad \cdots \quad height_{60})$$

where $height_q(q = 1, \ldots, 60)$ is the number of nearby (within 200 m) buildings that belongs to *q floors*. Then, we computed the Euclidean distance between each height vector of road segments: $\mathbf{Dist} = (d_{ij})_{m \times n}$. d_{ij} denotes the Euclidean distance between the layout vector of h_i and h_j. Thus, we could construct a matrix $\mathbf{Layout} = (l_{ij})_{n \times n}$ to describe the similarity of the layout between each of the two segments.

$$l_{ij} = \begin{cases} 0 & \text{EuclideanDistance}\left(h_i, h_j\right) > \varepsilon \\ \dfrac{1/d_{ij}}{\Sigma_k 1/d_{ik}} & \text{EuclideanDistance}\left(h_i, h_j\right) < \varepsilon \end{cases}$$

k denotes the number of road segments that have a similar layout as $road_i$. According to our assumption, the objective function should be penalized if there was a big difference between the GPS errors of buses on roads, whose layouts were similar to each other. As a result, we added the penalty to the objective function, and the final objective function of matrix completion was:

$$\begin{aligned} &F(\text{Sign}, \text{Var}, \text{L}, \text{R}, \text{Tran}, \text{Poi}, \text{Tag}) = \| \text{Sign} \cdot \text{LR}^T - \text{Var} \|_F^2 + \lambda_1 \left(\|\text{L}\|_F^2 + \|\text{R}\|_F^2 \right) \\ &+ \lambda_2 \left(\| E_\theta\left(\text{Tran} \cdot \text{LR}^T\right) \|_F^2 \right) + \lambda_3 \Sigma_{i=1}^n \sum_{j=1, j \neq i}^{n} \frac{1}{d_{ij}} \| \text{LR}^T p_{ij} \|_F^2 + \lambda_4 \| \text{LR}^T \text{Tag} \|_F^2 \\ &+ \lambda_5 \Sigma_{i=1}^n \sum_{j=1, j \neq i}^{n} \frac{1}{d_{ij}} \| \text{LR}^T l_{ij} \|_F^2 \end{aligned} \tag{8.2}$$

8.4.3.6 Optimization of the Objective Function

The general objective function of matrix completion can be solved iteratively, where each iteration consists of two steps [17]. First, for a given fixed **L**, update **R** element-wisely in the gradient descent direction of the objective function. Second, fixing the updated **R**, update **L** element-wisely in the same manner.

However, it was intractable to adopt the gradient descent method because of the term $\lambda_2\|E_\theta(\mathbf{Tran}\cdot\mathbf{LR}^T)\|_F^2$ in our objective function. The computational complexity in a single iteration to update terms was $O(m^2n^2)$, and m × n was the size of the input matrix. If we took 50 m as the length of the road segment, then the complexity was about $O\ (4835^2 \times 8831^2)$, which was intolerably high.

Here, we took advantage of the simulated annealing algorithm to get a more effective solution. An empirically well-adopted initialization [17] of **L** and **R** is given by non-negative matrix factorization (NMF) [19].

8.4.4 Weighted Estimation of GEF

After completing the GPS positioning error matrix, we obtained the approximate GPS positioning error for each bus on each road segment. Our goal was to rank the road segments based on GPS environment friendliness. However, considering the different quality of GPS terminal devices on different buses, we needed to give buses different weights when estimating the GEF of road segments. The intuition was that the bus with a high-quality GPS terminal device could better distinguish between road segments. The quality of the GPS receivers of most buses was acceptable. The GPS receiver would be considered as unconvincing if its positioning performance was significantly different from other buses on the same road segment. We used $distinction_i$ and $consistency_i$ to measure the weight of bus_i. $distinction_i$ represents the capacity of bus_i to distinguish between road segments. $consistency_i$ represents the degree of bus_i's consistency with other buses on the same road segment.

$$distinction_i = std(v_{i,:})$$

$$consistency_i = \frac{1}{mean\left(\left[\frac{v_{ij}-mean(v_{:,j})}{std(v_{:,j})}\right]_{j=1,2,\ldots,n}\right)}$$

$$weight_i = distinction \times consistency_i$$

Given the weights of buses, we could calculate the average error of $road_j$ as follows:

$$GEF_j = \frac{\sum_{i=1}^{m}(weight_i \cdot v_{ij})}{\sum_{k=1}^{m} weight_k}$$

After that, we ranked road segments based on average errors. The smaller the average error was, the better the GEF was.

8.5 Experiments

In this section, we estimate the GEF of roads covered by bus routes within the second-ring road in Chengdu, China. The estimation results were compared with the baseline methods. We also selected several road segments to collect real-life GPS measurements as the ground-truth to verify the rationality of the results by a case study.

8.5.1 Dataset Description

The data we used were from a real-world dataset collected in Chengdu, China. The GPS points were recorded by the buses running on their fixed routes for 30 days (2015.11.01–2015.11.30), which meant that the GPS readings were recorded under different conditions. For each bus, it generated 2–4 records/min, and thus, the total number of GPS point records was about 62,783,000, which was far more than the existing field-test works. The basic statistics about the data are shown in Table 8.1.

The urban road network was obtained from OpenStreetMap (Please check the official site of OpenStreetMap for more details: http://www.openstreetmap.org/). The urban road network was divided into short and equal-length road segments, so that the GEF at different locations within the same segment could be treated as the same.

As mentioned above, there was a significant variance in the quality of GPS receivers among different buses. This was because buses were managed by different public transportation operating companies, and the time of GPS installation and update varied from each other, which lead to the diversity in GPS receivers' brands and models. Taking the city of Chengdu as an example, there were more than 80 different types of GPS receivers in 4835 buses. For different GPS receivers, the quality varied obviously. To understand the difference intuitively, we show the GPS trajectory data of two buses on the same roads in Fig. 8.1. Yellow lines denote the

Table 8.1 Dataset description

Bus line number	184
Bus number	4835
Duration	30 days
GPS point record number	62,783,000
Sampling rate of GPS receiver	2–4 points/min
Number of types of GPS receivers	>80
Length of road segment	50 m
Road segment number	8831
Average of buses running on each segment	121
Average of segments covered by each bus line	171
Number of GPS points a bus recorded on a segment	>20

road network. Black points denote the GPS records of the buses. Red lines denote the route where the buses are running. Obviously, the GPS positioning accuracy of the first bus was worse than the second bus.

8.5.2 *Result of Map-Matching*

About 80.90% (50,789,815) of the GPS points were mapped to their given bus routes under the distance threshold. About 81.09% of such remaining (19.10%) points were mapped to nearby road segments under the distance threshold. As a result, 96.39% (60,517,304) of all points were mapped successfully. Other points were abandoned as accidental outliers.

8.5.3 *Evaluation of the Matrix Completion Result*

We employed k-fold cross-validation to evaluate the precision of the completed results of our completion algorithm. Concretely, k was set as 3 in this experiment. We used estimate error to measure the accuracy of matrix completion.

In detail, all non-zero positions of matrix **Var** were equally divided into k parts $(P_1, P_2, \ldots, P_k)$. For each part P_i, we covered it and preserved the remaining $k-1$ parts. We applied our completion algorithm to matrix **Var** and obtained the completed matrix $\mathbf{LR}^T$. We calculated the estimate-error [20] according to $\mathbf{LR}^T$ as follows: $\xi_i = \frac{\sum_{r,t:v_{r,t}\in \mathbf{P}_i} |v_{r,t} - \mathbf{LR}^T_{r,t}|}{\sum_{r,t:v_{r,t}\in \mathbf{P}_i} |v_{r,t}|}$. Enumerate the covered part from P_1 to P_k, and calculate the final estimate-error as: $\xi = \frac{\sum_{i=1}^{k} \xi_i}{k}$. Repeat the above operations t times, and calculate the average estimate error as the evaluation result of the completion algorithm. The rank comparing result is shown in Table 8.2, and we can see that our method outperformed the following baseline methods:

Table 8.2 The estimate error of our method and baseline methods. NMF, negative matrix factorization

Method		Matrix completion error
Baseline approaches	NAKNN	0.37242
	CBKNN	0.32951
	NMF	0.31883
Our approach	Basic method 1	0.29371
	Integrating layout information	0.29348
	Integrating layout and tag and POI	0.29311
	Integrating all penalty terms 2	**0.29220**

Naive KNN: For each empty entry in one row (column), we searched the k nearest rows (columns) whose corresponding entry was not null according to the Euclidean distance. Then, KNN used these non-empty entries to do the estimation.

Correlation-based KNN: This was similar to naive KNN. The only difference was that it used the correlation to measure the similarity instead of the Euclidean distance.

Non-negative matrix factorization (NMF) [19]: The matrix was factorized into two matrices, with the property that all matrices had no negative elements. Matrix multiplication of the factorized matrices was the completion result.

Our proposed algorithm consistently outperformed the baseline methods, which showed the superiority of our approach over other methods. The layout information, as well as the POI information represented the arrangement of the buildings at both sides along the road, and the tag information indicated the width of the road. They measured the signal occlusion effect to some extent. When integrating this prior information as additional penalty terms into the algorithm, the matrix completion performance was improved. Besides, it was also necessary to consider the variance between receivers' qualities when estimating the error.

8.6 Conclusions and Future Work

We presented in this chapter an approach for the GPS environment friendliness estimation of urban road segments. The method begins by constructing a mapping from GPS data to road sections using the unique feature of bus fixed routes. Secondly, we completed the missing data based on the inherent correlation among GPS errors and the environment information. Finally, a weighted evaluation strategy is proposed to estimate the GEF, taking fully into account the influence of the quality of different GPS devices. We evaluated the GEF of 8,831 different road segments using 1-month trajectory data of 4835 buses within the Second Ring Road of Chengdu, and we verified the rationality of the results through satellite maps, street view and field tests.

Since bus routes are difficult to cover all segments of a city, in the future, we plan to apply our approach to taxi trajectory data to tackle the bypasses whose GEF could not be estimated. What's more, we intend to use the methods in this chapter to location-based services to improve the user experience. As an example, the GEF evaluation method can provide services for real-time bus location estimation and arrival time estimation of buses.

References

1. Rezaei S, Sengupta R. Kalman filter-based integration of DGPS and vehicle sensors for localization. IEEE Trans Control Syst Technol. 2007;15(6):1080–8.

2. Hsu L-T, Jan S-S, Groves PD, Kubo N. Multipath mitigation and NLOS detection using vector tracking in urban environments. GPS Solutions. 2015;19(2):249–62.
3. Liu X, Nath S, Govindan R. Gnome: a practical approach to NLOS mitigation for GPS positioning in smartphones. New York, NY, USA: Proceedings of the 16th Annual International Conference on Mobile Systems, Applications, and Services; 2018. p. 163–77.
4. Drawil NM, Amar HM, Basir OA. GPS localization accuracy classification: a context-based approach. IEEE Trans Intell Transp Syst. 2013;14(1):262–73.
5. Renfro BA, An analysis of global positioning system (GPS) standard positioning system (SPS) performance for 2014, p. 105
6. Schipperijn J, Kerr J, Duncan S, Madsen T, Klinker CD, Troelsen J. Dynamic accuracy of GPS receivers for use in health research: a novel method to assess GPS accuracy in real-world settings. Front Public Health. 2014;2:21.
7. Wang J, Wang Y, Zhang D, Lv Q, Chen C. Crowd-powered sensing and actuation in smart cities: current issues and future directions. IEEE Wirel Commun. 2019;26(2):86–92.
8. Wang J, et al. HyTasker: hybrid task allocation in mobile crowd sensing. IEEE Trans Mob Comput. 2020;19(3):598–611.
9. Wang Y, Zheng Y, Xue Y. Travel time estimation of a path using sparse trajectories. New York, NY, USA: Proceedings of the 20th ACM SIGKDD international conference on Knowledge discovery and data mining; 2014. p. 25–34.
10. Zhou P, Jiang S, Li M. Urban traffic monitoring with the help of bus riders. New York, NY: 2015 IEEE 35th international conference on distributed computing systems; 2015. p. 21–30.
11. Chen P, Chen F, Qian Z. Road traffic congestion monitoring in social media with hinge-loss markov random fields. New York, NY: 2014 IEEE International Conference on Data Mining; 2014. p. 80–9.
12. Chen C, Ding Y, Wang Z, Zhao J, Guo B, Zhang D. VTracer: when online vehicle trajectory compression meets mobile edge computing. IEEE Syst J. 2020;14(2):1635–46.
13. Guo S, et al. ROD-revenue: seeking strategies analysis and revenue prediction in ride-on-demand service using multi-source urban data. IEEE Trans Mob Comput. 2020;19(9):2202–20.
14. Wu H, Sun W, Zheng B, Yang L, Zhou W. CLSTERS: a general system for reducing errors of trajectories under challenging localization situations. Proc ACM Interact Mobile Wearab Ubiquit Technol. Sep. 2017;1(3):1–28.
15. Liantao M, Yasha W, Guangju P, Yuxin Z, Yuanduo H, Jingyue G. Evaluation of GPS-environment friendliness of roads based on bus trajectory data. J Comput Res Developm. 2016;53(12):2694.
16. Raymond R, Morimura T, Osogami T, Hirosue N. Map matching with Hidden Markov Model on sampled road network. New York, NY: Proceedings of the 21st International Conference on Pattern Recognition (ICPR2012); 2012. p. 2242–5.
17. Dax A. Imputing missing entries of a data matrix: a review. J Adv Comput. 2014;
18. Casella G, Berger RL. Statistical inference, 2nd edn. Australia, Pacific Grove, CA: Thomson Learning; 2002.
19. Lee DD, Seung HS. Algorithms for non-negative matrix factorization. In: Leen TK, Dietterich TG, Tresp V, editors. Advances in neural information processing systems 13. Cambridge: MIT Press; 2001. p. 556–62.
20. Zhu Y, Li Z, Zhu H, Li M, Zhang Q. A compressive sensing approach to urban traffic estimation with probe vehicles. IEEE Trans Mob Comput. 2013;12(11):2289–302.

Chapter 9
B-Planner: Planning Night Bus Routes Using Taxi Trajectory Data

9.1 Introduction

As a more environmentally friendly means of transportation than cars and taxis, buses are a popular and economical way for people to travel around the city. The daytime bus transportation systems in many cities are usually well designed. However, most of these systems are not available at night, so taxis have become the only option for travel within the city. With the aim of providing a cost-effective and environmentally friendly transport service to the public, many cities are beginning to plan night-through bus routes.

Previously, in the task of bus route planning, understanding the mobility patterns of citizens relied mainly on human surveys [1]. This approach has the obvious disadvantage of being costly and time-consuming. More importantly, the method based on human surveys cannot adapt to the frequent changes in road networks and traffic, especially for cities that are growing at high speeds. With the widespread deployment of GPS devices and wireless communication in taxis, we can obtain a wealth of information of taxis, including when and where passengers are picked-up or dropped-off, which route the taxi is taking on given a certain trip, etc. By knowing the origin-destination (OD) of each taxi trip, we can further extract valuable information from the crowd flow in a city at different times of the day. Finally, we can accurately plan new night bus routes to maximise the number of passengers along the routes.

However, identifying the candidate bus stops from taxi GPS data and enumerating the top-ranked bi-directional bus routes efficiently are not trivial and straightforward. For example, as illustrated in Fig. 9.1, given the taxi GPS trajectories of

---

Part of this chapter is based on a previous work: C. Chen, D. Zhang, N. Li and Z. Zhou, "B-Planner: Planning Bidirectional Night Bus Routes Using Large-Scale Taxi GPS Traces," in IEEE Transactions on Intelligent Transportation Systems, vol. 15, no. 4, pp. 1451–1465, Aug. 2014, doi: https://doi.org/10.1109/TITS.2014.2298892.

C. Chen et al., *Enabling Smart Urban Services with GPS Trajectory Data*,
https://doi.org/10.1007/978-981-16-0178-1_9

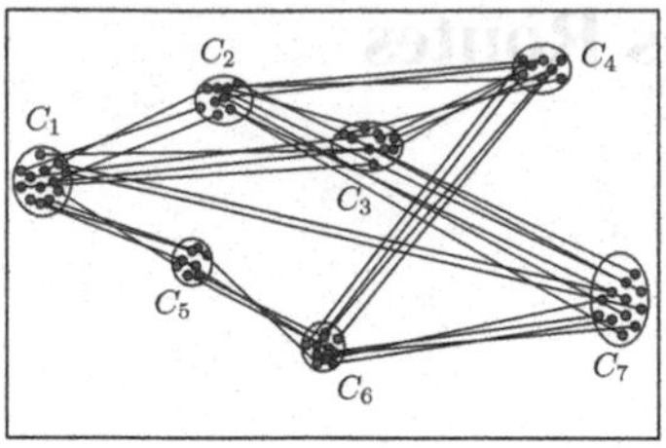
C_1
C_2
C_3
C_4
C_5
C_6
C_7

Passenger Flow

	C_1	C_2	C_3	C_4	C_5	C_6	C_7
C_1	0	15	7	0	9	0	5
C_2	18	0	12	8	0	0	12
C_3	14	3	0	20	0	0	9
C_4	0	5	10	0	0	22	0
C_5	12	0	0	0	0	9	0
C_6	0	0	0	13	0	0	10
C_7	3	17	13	0	0	4	0

Travel Time(in minute)

	C_1	C_2	C_3	C_4	C_5	C_6	C_7
C_1	0	8	10	15	9	15	25
C_2	9	0	4	10	16	19	22
C_3	11	3	0	6	7	8	14
C_4	16	12	7	0	29	27	10
C_5	8	14	6	25	0	5	12
C_6	14	18	7	26	4	0	10
C_7	27	21	15	12	14	9	0

Fig. 9.1 An illustrative example of the taxi GPS traces (left); the passenger flow (middle), and the travel time among bus stops (right)

night time for a certain period of time, there are seven dense taxi pick-up/drop-off locations (i.e. $C_1 - C_7$) have been identified as candidate bus stops, where C_1 and C_7 are the bus origin and destination, respectively, and the corresponding passenger flow and travel time among stops are shown in the right panel of Fig. 9.1. The objective of bidirectional bus route design is to find a bi-directional bus route ($C_1 \longrightarrow C_7$ and $C_7 \rightarrow C_1$) with the maximum number of passengers expected given the bus operation time constraints.

For the sake of design an effective bus route, we need to deal with the following challenges:

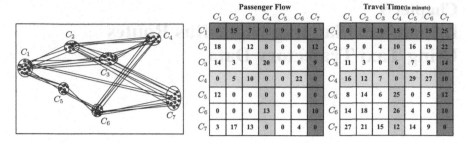

- ***Candidate bus stop identification***: The pick-up and drop-off points for taxi passengers are distributed throughout the city. Some areas having more pick-up/drop-off records (PDRs) than other areas, but there is no clear guideline about where the bus stops should be put.
- ***Tradeoff between the number of passengers and travel time***: A bus route that passes through more stations can transport more passengers but will increase travel time accordingly. Hence, a non-trivial trade-off is needed.
- ***Passenger flow accumulation effect***: Assuming there is no taxi passenger travelling from C_4 to C_7, if we plan the route as $C_1 \rightarrow C_2 \rightarrow C_3 \rightarrow C_7$, then the significant passenger flow in $C_2 \rightarrow C_4$ and $C3 \rightarrow C4$ cannot be accommodated. Alternatively, by including C_4 in the route as $C_1 \rightarrow C_2 \rightarrow C_3 \rightarrow C_4 \rightarrow G_7$, this passenger flows can be accommodated with the cost of adding one stop. Therefore, we need to consider this accumulation effect, which tends to lead to a globally better solution.
- ***Dynamic passenger flow***: The passenger flows are usually different from time to time, for example, the passenger flow during 23:00–24:00 can be very different from that during 3:00–4:00, thus we need to consider this dynamics when planning bus routes.
- ***Asymmetry of passenger flow and travel time***: It is easy to see that the best route in terms of passenger flow and travel time for one direction (from C_1 to C_7) is probably not the best one for the opposite direction (from C_7 to C_1), we thus need to select the bus route with maximum accumulated number of passengers in two directions.

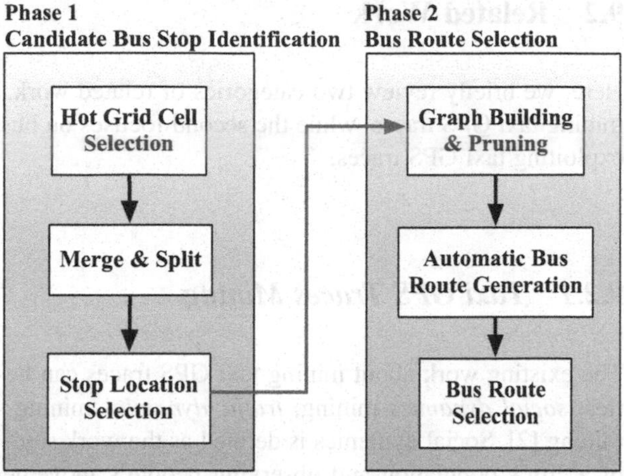
Phase 1
Candidate Bus Stop Identification
Hot Grid Cell Selection
Merge & Split
Stop Location Selection
Phase 2
Bus Route Selection
Graph Building & Pruning
Automatic Bus Route Generation
Bus Route Selection

Fig. 9.2 The two-phase bus route planning framework

In this chapter, we address the challenges mentioned above through a two-phase approach, and the process shown in Fig. 9.2. Generally speaking, in the first phase, we identify candidate bus stops by clustering and splitting hot areas, and then we propose several strategies to find the best bus routes. Specifically, the main contributions of the study are:

- A two-phase approach is presented to tackle the problem of bi-directional night bus route design leveraging the taxi GPS traces. To the best of our knowledge, this is the first work on bi-directional night bus route design exploiting the taxi travel speed, time, pick-up and drop-off information in large-scale, real-world taxi GPS traces.
- We develop a novel process with effective methods to cluster "hot" areas with dense passenger pick-up/drop-off, split big "hot" areas into walkable size ones and identify candidate bus stops. Moreover, we study how different thresholds in the merge and split algorithms affect bus stop identification results and final selected bus routes.
- We propose rules to build and prune the directed bus route graph. Based on the graph, we design a new heuristic algorithm, named Bi-directional Probability based Spreading (*BPS*) algorithm, to select the best bi-directional bus route which can achieve the maximum number of passengers expected in two directions. We also investigate the impact of different bus stop distances on the final bus routes selection.
- We determine the night bus capacity by computing the maximum number of passengers on buses for the selected bus route at different stops and different bus frequencies. To understand the impact of the new opened bus route on taxi services, we further report the passenger flow change along the bus route before and after the new bus route opened date.

9.2 Related Work

Here, we briefly review two categories of related work. The first category is about mining taxi GPS traces, while the second focuses on bus network design other than exploiting taxi GPS traces.

9.2.1 Taxi GPS Traces Mining

The existing work about mining taxi GPS traces can be grouped into three categories: *social dynamics* mining, *traffic dynamics* mining, and *operational dynamics* mining [2]. Social dynamics is defined as the work studying the collective behavior of a city's population and observing people's movement in the city which are is motivated by diverse needs and influenced by external factors. Common research problems in this subcategory using taxi GPS traces include: where do people go throughout the day [3, 4], what are the "hottest" spots around a city [5], what are the "functions" of these hotspots [6], how strongly connected are different areas of the city [7], etc. Moreover, social dynamics also aims to understand movement patterns of people, and traffic dynamics studies the resulting flow of the population through the city's road network. Most work in this subcategory aimed at uncovering the root cause of traffic outliers [7], predicting traffic conditions which are useful for providing real-time traffic forecast [8], and travel time estimation for drivers [9]. The aim of operational dynamics is to learn from taxi drivers' expert knowledge of the city and detect abnormal or effective driving behaviors. The last two sub-categories used mainly the origin-destinations (OD) of a taxi trip trajectory; in the study of operational dynamics, researchers make use of full trajectories, as the routes taken by drivers are of utmost importance. Researchers have mined these trajectories to suggest strategies for quickly finding new passengers/taxis [10], recommend time-dependent navigational routes for reaching a destination quickly [11], plan flexible bus routes [12], and suggest driving routes to achieve dynamic taxi ride-sharing [13]. Additionally, new trajectories can be compared against a large collection of historical trajectories to automatically detect abnormal behavior [14]. Before taxi GPS trace mining, data clean and repair may be required since the data can be noisy.

Among all the papers related to taxi GPS trace mining, the work involving "hot spots" and frequent travel OD patterns in the social dynamics sub-category is relevant to our work. Still, there are not many papers except two [12, 15] using those data for bus route planning. The goal of [12] is to suggest a flexible bus route by mining historical taxi GPS trips. Unlike our goal, this goal selects only the route that maximizes the sum of each connected trip cluster. The research objective of [15] is to find an optimal bus route for a given OD pair in a single direction. However, the one-direction optimal route obtained by [15] is generally not the optimal route in both directions due to the asymmetry of passenger flow. Therefore, planning the bi-directional bus route is an ideal choice, which can achieve the maximum number of passengers in both directions. In this chapter, we try to solve the problem of bi-directional bus route planning. Besides conducting bus route planning, the work of this chapter is also different from [16] in the following two aspects. First, we provide a mechanism to determine the maximum number of passengers at different bus operation frequencies. Second, we employ an empirical study to investigate the impact of adding new bus routes on taxi services.

9.2.2 Bus Network Design

The design of the bus network is an intensively studied area in the field of urban planning and transportation. The aim is to determine bus routes and operation frequencies in order to achieve certain objectives, subject to the constraints and passenger flows. The popular objectives include the shortest route, shortest travel time, lowest operation cost, maximum passenger flow, maximum area coverage, and maximum service quality while the constraints have time, capacity, and resources.

However, the selection of the objectives should take into account the requirements of the operator and user. These two often conflicting requirements may lead to design trade-off rather than the a best solution.

As pointed out in [16], the early design of the bus network was mainly based on the human survey to get passenger flows and user requirements, and it relied heavily on heuristics and intuitive principles formed by the experience and practice of designers. Recent work on bus network design is also based on the assumption that passenger flow is obtained from user surveys or population estimation. There are many complex optimization approaches. The best solving algorithms among them are based on heuristic procedures [17] to find near-optimal solutions. The reader can find a detailed review of the route network design in [18].

Despite the renewed attention for bus network design, the use of taxi passenger OD flow data to solve the problem of bi-directional night-bus route design has not yet started. Different from existing research, the work in this chapter aims to find a bi-directional bus route with a fixed frequency, maximizing the number of passengers expected along the route subject to the total travel time constraint. This problem is different from the traditional Travelling Salesman Problem (TSP) in nature, which aims to find the shortest path that visits each given location (node) exactly once. TSP evaluates different routes with exact locations, which means the route includes all candidate stops. Our problem is also different from the shortest path finding problem, and we have to consider the accumulated effect (passenger flows) from all previous stops to the current stop for choosing the bi-directional bus route.

9.3 Candidate Bus Stop Identification

In the proposed two-phase bus route planning framework, the objective of the phase one is to identify candidate bus stops by exploiting the taxi PDRs. As shown in Fig. 9.2, the whole process consists of three steps: (1) Divide the whole city into small equal-sized grid cells, mark those "hot" grid cells with high taxi passenger PDRs for further processing; (2) Merge the adjacent "hot" grid cells to form "hot" areas, divide each big area into "walkable size" cluster; (3) Choose one grid cell as the candidate bus stop location in each walkable size "hot" cluster.

9.3.1 *Hot Grid Cells and City Partitions*

In this work, we first divide the city into equal-sized grid cells, with each cell about 10m $\times$ 10m in size. In such a way, the whole city is partitioned into 5000 $\times$ 2500 cells in total. Out of all the grid cells, over 95% of them contain no taxi passenger *PDRs* as they are either lakes, mountains, buildings, and highways that cannot be reached by taxis, or suburb areas that people seldom travel to. Only 0.11% of the grid cells have more than 0.2 *PDRs* per hour on average if we only count the *PDRs* in late night. And we name these grid cells as "hot" ones.

As each grid cell has maximum eight neighbors, if we define the connectivity degree (*CD*) of a "hot" grid cell as the number of "hot" neighboring cells, the *CD* of any grid cell will range from 0 to 8, where the "hot" grid cell with *CD* equals to 0 is called an isolated cell. As the city is composed of mixed hot grid cells and common grid cells, both hot cells and common cells form irregular "hot areas" and "common areas" as a consequence of the same type of cells being adjacent to each other. These "hot areas" are also called city partitions, as shown in Fig. 9.3. Apparently, some small partitions (e.g., the green one in Fig. 9.3) can be very close to some big ones (e.g., the red one in Fig. 9.3). It would be necessary to consider all the city partitions globally in order to plan the bus stop locations, thus city partitions close to each other had better merge to form big clusters for better overall bus stop distribution. In the next section, we propose a simple strategy to merge the close partitions into bigger clusters.

9.3.2 *Cluster Merging and Splitting*

We present the cluster merging and splitting approach in Algorithm 9.1. We first sort them in descending order according to the number of *PDRs* (Line 1). To merge the partitions, we propose to use the hottest partition to *absorb* its nearby partitions according to the descending order of *PDRs*, until no more nearby partitions meet the merging criterion (Line 8). Then we choose the next hottest partition to repeat the

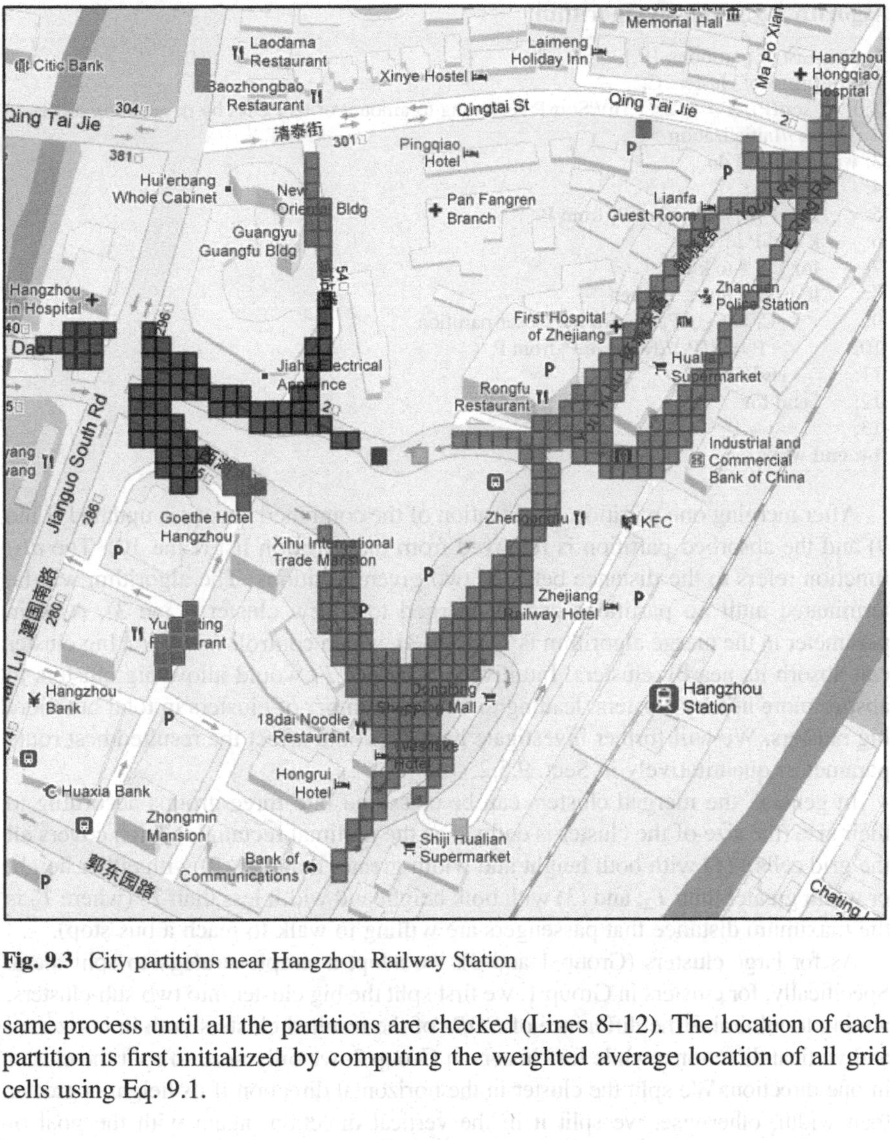
Memorial Hall
Laodama Restaurant
Laimeng Holiday Inn
Ma Po Xiang
Citic Bank
Hangzhou Hongqiao Hospital
Xinye Hostel
Baozhongbao Restaurant
Qing Tai Jie
Qingtai St
Qing Tai Jie
清泰街
Pingqiao Hotel
Hui'erbang Whole Cabinet
Pan Fangren Branch
Lianfa Guest Room
Guangyu Guangfu Bldg
Hangzhou Hospital
First Hospital of Zhejiang
Police Station
Rongfu Restaurant
Supermarket
Industrial and Commercial Bank of China
Jianguo South Rd
Goethe Hotel Hangzhou
KFC
Zhejiang Railway Hotel
建国南路
Hangzhou Bank
Hangzhou Station
18dai Noodle Restaurant
Hongrui Hotel
Huaxia Bank
Zhongmin Mansion
Shiji Hualian Supermarket
Bank of Communications
郭东园路

Fig. 9.3 City partitions near Hangzhou Railway Station

same process until all the partitions are checked (Lines 8–12). The location of each partition is first initialized by computing the weighted average location of all grid cells using Eq. 9.1.

$$\mathrm{loc}(P) = \frac{\sum_{i=1}^{N} (\mathrm{PDRs}(g_i) * \mathrm{loc}(g_i))}{\sum_{i=1}^{N} \mathrm{PDRs}(g_i)} \tag{9.1}$$

where $loc(g_i)$ refers to the longitude/latitude of the member grid cell g_i.

Algorithm 9.1 Merge algorithm

Input: List of partitions $\{P_i\}$
Output: List of clusters C_i
1: $P \leftarrow$ sort(P), $(i = 1, 2, \ldots, n)$//Sort P according to amount of its PDRs by descending order.
2 : $i = 1$; //Initialization
3: **wihle** $P \neq \varnothing$ **do**
4: $C_i = P_1$;
5: $P = P\{P_1\}$//Remove P_1 from P
6: $k = |P|$;
7: **for** $j := 1$ to k **do**
8: **if** $dist(C_i, P_j) < T_1$ **then**
9: $C_i = C_i \cup P_j$//absorb the closer partition
10: $P = P\{P_j\}$//Remove P_j from P
11: **end**
12: **end for**
13: $i = i + 1$
14: **end while**

After merging one partition, the location of the combined cluster is updated (Line 9) and the absorbed partition is removed from the partition list (Line 10). The *dist* function refers to the distance between two given partitions. The algorithm will be terminated until no partitions can be merged to a new cluster (Line 3). A main parameter in the merge algorithm is T_1 (Line 8), which controls *how far* a big cluster can absorb its nearby clusters. Intuitively, a bigger T_1 would allow big clusters to absorb more nearby clusters, leading to a fewer number of clusters in total but more big clusters. We will further investigate how T_1 would affect the resulted best route parameters quantitatively in Sect. 9.5.2.

In general, the merged clusters can be classified into three groups according to their size (the size of the cluster is defined as the minimal rectangle which covers all the grid cells): (1) with both height and width greater than T_2; (2) with either height or width greater than T_2; and (3) with both height and width less than T_2 (where T_2 is the maximum distance that passengers are willing to walk to reach a bus stop).

As for large clusters (Group 1 and 2), we adopt a simple strategy to split them. Specifically, for clusters in Group 1, we first split the big cluster into two sub-clusters, aiming to minimize the difference of *PDRs* of the resulted clusters both in horizontal and vertical directions; while for clusters in Group 2, we only need to split the cluster in one direction. We split the cluster in the horizontal direction if its height is greater than width, otherwise, we split it in the vertical direction, again with the goal of minimizing the number difference of *PDRs* of the split sub-clusters. With one split, one big cluster would produce two smaller sub-clusters. Thus, a smaller T_2 would need more splitting times, and also leads to more smaller clusters finally.

Figure 9.4 shows an illustrative example of splitting a cluster into four sub-clusters with the proposed splitting strategy. The initial cluster belongs to Group 1 (Fig. 9.4 (left)), the splitting is first done in the horizontal direction to produce two sub-clusters with similar *PDRs*. After the first splitting, two sub-clusters with a width greater than T_2 are generated (T_2 is set to 500 meters), thus both sub-clusters require a further splitting in the vertical direction. The final result with four split sub-clusters is shown in Fig. 9.4 (right). We will also study how T_2 would affect the resulted best route parameters in Sect. 9.5.2.

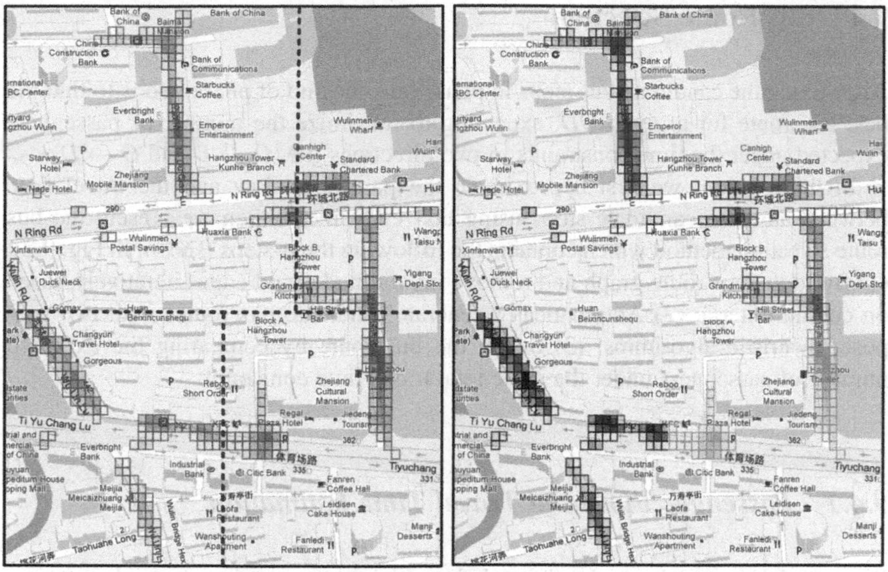
Bank of China
Bank of China
China Construction Bank
Bank of Communications
Starbucks Coffee
Everbright Bank
Emperor Entertainment
Wulinmen Wharf
Canhigh Center
Starway Hotel
Hangzhou Tower Kunhe Branch
Standard Chartered Bank
Zhejiang Mobile Mansion
N Ring Rd
环城北路
Wulinmen Postal Savings
Huaxia Bank
Xinfanwan
Juewei Duck Neck
Block B, Hangzhou Tower
Yigang Dept Sto
Gomax
Huan Beixincunshequ
Block A, Hangzhou Tower
Hill Street Bar
Changyun Travel Hotel
Gorgeous
Reboo Short Order
Zhejiang Cultural Mansion
Regal Plaza Hotel
Ti Yu Chang Lu
Everbright Bank
Industrial Bank
Citic Bank
体育场路
Tiyuchang
Fanren Coffee Hall
Meijia Meicaizhuang Meijia
Laofa Restaurant
Leidisen Cake House
Wanshouting Apartment
Fanledi Restaurant
Manji Desserts
Taohuane Long
Wulin Bridge
Wulin Rd

Fig. 9.4 Illustrative example of splitting. Big cluster formed via merging (left). Big cluster split into 4 walkable size clusters (right, in four different colors)

9.3.3 *Candidate Bus Stop Location Selection*

After *merging* and *splitting* operations, we obtain a big number of "hot" clusters with the a size smaller than $T_2 \times T_2$, scattered in the dynamic districts of the city during late night. The next step is to select a *representative* grid cell in each cluster to serve as the candidate bus stop.

To select this *representative* grid cell, both the connectivity degree (*CD*) and the number of *PDRs* of each cell in the cluster are taken into consideration. While the CD of a grid cell characterizes the accessibility of the cell, the number of *PDRs* is an indicator of its "hotness". The grid cell having the maximum value defined in Eq. 9.2 in each cluster is selected as the "center" of the cluster, marked as the location of the candidate bus stop.

$$\arg\max_{i} \left[w_1 \times \frac{CD(i)+1}{9} + w_2 \times \frac{\text{PDRs}(i)}{\sum_{i=1}^{n}(\text{PDRs}(i))} \right] \tag{9.2}$$

We set $w_1 = w_2 = 0.5$ in the evaluation, and totally we get 579 candidate bus stops in the city by using the taxi GPS data from Hangzhou, China. Note that different weight settings in Eq. 9.2 would only affect locations of the bus stop, and have no impact on the total number of bus stops.

9.4 Bus Route Selection

After fixing the candidate bus stops in phase one, the aim of phase two is to find the best bus route for a given *OD*, expecting to maximize the number of passengers expected under the time constraints in two directions (i.e. $O \rightarrow D$ and $O \rightarrow D$).

In this section, we first approximate the passenger flow and the travel time between any two candidate stops using taxi GPS traces, then we present the bus route selection method which contains the following three-steps (shown in Fig. 9.2): (1) Build the bus route graph and remove invalid nodes and edges iteratively based on certain criteria; (2) Automatically generate candidate bus routes with two proposed heuristic algorithms; (3) Select the bus route by comparing the expected number of passengers under the same total travel time constraint.

9.4.1 Passenger Flow and Travel Time Estimation

We record the travel demand and time information in two matrixes, named passenger flow matrix (*FM*) and bus travel time matrix (*TM*). Each element in a matrix refers to the number of passengers or the bus travel time from one stop (i th) to another stop (j th, $i \neq j$). We count the total taxi trips from *ith* cluster to j th cluster as each stop is responsible for its cluster. We set the maximum waiting time for passengers at the stop as 30 min (equal to the bus operation frequency), so any pick-up or drop-off events taking place in this time window are counted. We simply assume the passenger flows among candidate bus stops remain unchanged during each 30-min duration. The final FM is got by averaging all flow matrixces at different bus frequencies. We also assume TM keeps unchanged across the night time. $tm(s_i, s_j)$ is the average travel time multiplied by, which is a constant. We set $\alpha = 1:5$ to consider the speed difference between taxis and buses. For the paths having no taxi trip occurring in history (for instance, nobody travels by taxi due to too short distance), we use $Ddist(s_i, s_j)/v$ to approximate $tm(s_i; s_j)$, where $Ddist(s_i; s_j)$ is the driving distance between s_i and s_j, and v is a constant and is set to 50 km/h.

9.4.2 Bus Route Graph Building and Pruning

Selecting the best bus route is a very challenging problem as two conflicting requirements must be met: one is to ensure that the bus route would traverse intermediate stops and finally reach the destination within a limited time; the other is to maximize the number of passengers accumulated along the route from all previous stops to the destination. For example, if we choose the stop with the heaviest passenger flow from the origin as the first node, and then keep choosing the next stop following the heaviest passenger flow principle, then we might neither be able to reach the destination nor achieve the objective of having the maximum number of passengers accumulated along the route. To meet the above two requirements and follow the intuitive principles in bus route design, some basic criteria should be set for the building of the bus route graph and selection of the candidate bus route.

(1) *Route graph building criteria:* Obviously, there would be numerous stop combinations for a given OD pair, and only a small proportion of them meet the first or second requirement. In order to reduce the search space of possible stops and routes, we can build a bus route graph starting from origin to destination using heuristic rules. These rules are either derived from one of the above two requirements, or from the intuitive bus route design principle. For instance, from the shortest travel time perspective, the bus route should extend from the origin towards the direction of the destination, which can be further converted into three rules: each new selected stop should be farther from the origin, closer to the destination, and farther from previous stops. From the intuitive bus route design principle, the bus stops should not be too far from each other, also the bus route should not comprise sharp zig-zag paths. These can also be translated into two criteria in building the bus route graph. Specifically, given the OD pair (s_1, s_n) and the candidate route $R = < s_1, s_2 \ldots s_n >$, we should follow the following criteria when building the bus route graph with stops (nodes) and directed edges among nodes.

Criterion 9.1 Adequate Stop Distance

$$\mathrm{dist}(s_{i+1}, s_i) < \delta (i = 1, 2, \cdots, n-1)$$

where δ is a user-specified parameter. It means the maximum distance between two consecutive stops. We will study the effect of varying δ values on the best route parameters in Sect. 9.5.

Criterion 9.2 Move Forward

$$x_{new}(i+1) > x_{new}(i) (i = 1, 2, \cdots, n-1)$$

$$x_{new}(i) = x(i)\cos\theta + y(i)\sin\theta$$

$$\theta = \tan^{-1}\frac{y(n)}{x(n)}$$

$(x(i), y(i))$ of si is got by simply subtracting the longitude and latitude value to that of s_1. x_{new} is the X-axis value of stop in the new coordination which is with s_1 as the new origin, and from s_1 to s_n as the new direction of X-axis (see the left panel in Fig. 9.5). This criterion guarantees the bus will always move forward along the OD direction.

Criterion 9.3 Origin-Farther

$$\mathrm{dist}(s_{i+1}, s_1) > \mathrm{dist}(s_i, s_1) (i = 1, 2, \cdots, n-1)$$

This ensures that the bus will move away from the origin s_1 farther in each step.

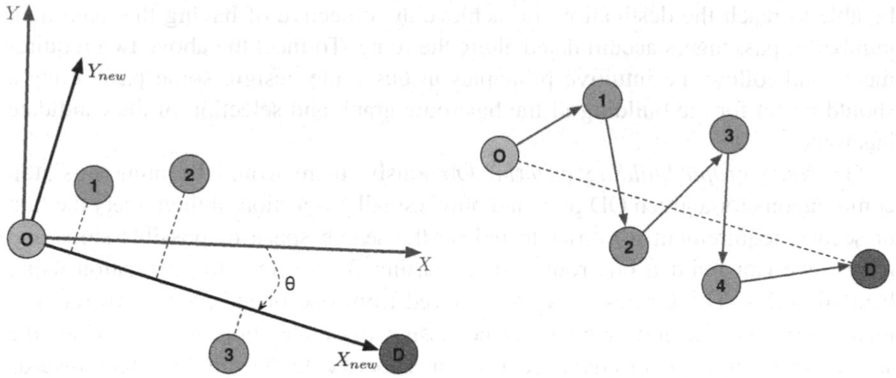
Y
Y_{new}
1
2
O
X
θ
3
X_{new}
D
1
3
O
2
4
D

Fig. 9.5 Demonstration of Criterion 9.2 (left) and Criterion 9.5 (right)

Criterion 9.4 Destination-Closer

$$\text{dist}(s_{i+1}, s_n) < \text{dist}(s_i, s_n)(i = 1, 2, \cdots, n-1)$$

This ensures the bus will move closer to the destination s_n in each step.

Criterion 9.5 No Zigzag Route

$$\arg\min_{s_j} \left(\text{dist}(s_{i+1}, s_j)\right) = s_i(\ j = 1, 2, \cdots, i)$$

Criterion 9.5 ensures the smoothness of the route. There would be no sharp zigzag path along the OD direction. The route demonstrated in the right panel of Fig. 9.5 should not happen, as it violates the no zigzag route criterion. We can see $\arg\min_{s_j} \left(\text{dist}(s_3, s_j)\right) = s_1 \neq s_2(\ j = 1, 2)$, also $\arg\min_{s_j} \left(\text{dist}(s_4, s_j)\right) = s_2 \neq s_3(\ j = 1, 2, 3)$.

(1) *Graph building and pruning*: The aim of graph building is to construct a directed graph with nodes and links given an OD pair, in which the nodes are the stops, and edges link the stop to its next possible stops, regardless of passenger flows among them. While the goal of graph pruning is to remove invalid edges and nodes according to the proposed criteria.

Graph Building: Given the bus route origin and destination, their locations are firstly used to narrow down the choice of valid candidate stops, only the candidate stops lying between them are under consideration. For each stop within the range, we determine links to its next possible stops according to the proposed Criterions 9.1–9.4. The process will terminate when all stops have been checked. At last, stops having no edges would be excluded.

As Criterion 9.5 is related to all stops in one bus route, so we use it to prune the route graph after it is built. Figure 9.6 (left) shows an illustrated example about a generated bus route directed graph. Note that the graph is built based on the geographical constraints, so the edge may have no taxi passenger flow on itself.

Graph Pruning: Some nodes and edges can be further pruned because they are not valid for candidate bus route selection. To be specific, nodes without in-coming

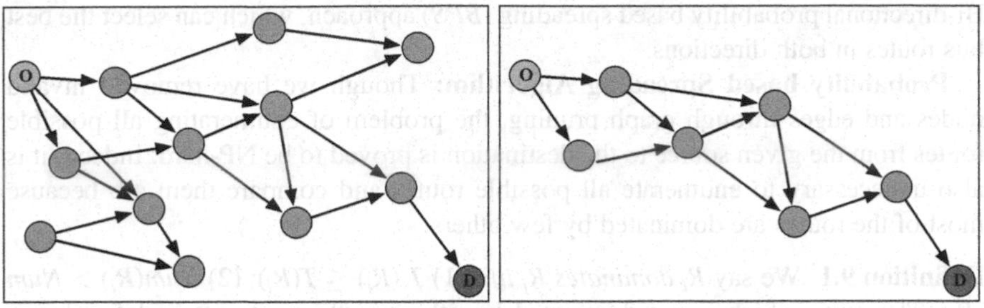
O
D
O
D

Fig. 9.6 A bus route directed graph for a given OD. The route graph is got by graph building algorithm (left) and its corresponding graph after applying graph pruning (right)

edges (if not the origin) or out-going edges (if not destination) should be deleted as they will not form any valid routes with the bus route OD pair.

We first calculate all the nodes' in-coming and out-going degrees. Afterward nodes (exclude the given OD) together with related edges would be iteratively deleted from the graph if their in-coming or out-going degree is zero. At last, a graph with only one zero in-coming degree node (i.e. the given origin) and one zero out-going degree node (i.e. the given destination) would be generated. After graph pruning, all the bus routes starting from the source and following the edges in the graph would eventually reach the destination. Figure 9.6 (right) displays the resulted graph after applying pruning to the graph in Fig. 9.6 (left).

Graph for $D \rightarrow O$: An intuitive way of building a route graph for $D \rightarrow O$ is to run the previous two steps again, with the D as the new origin and O as the new destination. However, Theorem 9.1 below ensures that the route graph from D to O is just the same as that from O to D, with all the edges having opposite directions.

Theorem 9.1 If $R = \langle s_1, s_2, \ldots, s_n \rangle$ is a candidate bus route for pair (s_1, s_n), then its reversed route $R = \langle s_n, s_{n-1}, \ldots, s_1 \rangle$ will be the candidate bus route for (s_n, s_1) pair.

Proof To prove R is the candidate bus route for (s_n, s_1) pair, we just need to check whether it meets all the five criteria. It is obviously that R meets the first four criteria. For Criterion 9.5, given a particular node $si(1 < i < n-1)$ in R, we can derive its two closest nodes are s_{i-1} and s_{i+1}. Thus $\arg\min_{s_j}\left(\text{dist}\left(s_i, s_j\right)\right) = s_{i+1}(\,j = n, n-1, \cdots, i+1)$ will hold.

9.4.3 Automatic Candidate Bus Route Generation

Based on the graph constructed in the previous section, we first propose our probability based spreading algorithm for $O \rightarrow D$, then followed by the

Bi-directional probability based spreading (*BPS*) approach, which can select the best bus routes in both directions.

Probability based Spreading Algorithm: Though we have removed invalid nodes and edges through graph pruning, the problem of enumerating all possible routes from the given source to the destination is proved to be NP-hard. Indeed, it is also unnecessary to enumerate all possible routes and compare them all, because most of the routes are dominated by few others.

Definition 9.1 We say R_i *dominates* R_j *iif* : (1) $T(R_i) \leq T(R_j)$; (2) $Num(R_i) > Num(R_j)$. The route which is not dominated by others in the route set is defined as a skyline route.

where T and Num are the total travel time and number of expected delivered passengers. We compute them based on Eqs. 9.3 and 9.4. The *skyline route* definition is similar to that in [19], and the rationale behind is that only routes with less travel time but larger number of passengers should be selected. *Skyline* detector [20] will prune the routes which are dominated by skyline routes in the candidate set. Thus, the comparison can be done among detected *skyline routes*.

$$T = \sum_{i=1}^{n-1} tm(s_{i+1}, s_i) + (n-2) \times t_0 \tag{9.3}$$

$$Num = \sum_{i;j(\ j>i)}^{n} fm(s_i, s_j) \tag{9.4}$$

where t_0 is the average time needed to board at each stop, and we set it to 1.5 min.

Algorithm 9.2 Probability Based Spreading

Input: $G(S, E)$: Single directional graph for the given OD pair
 FM : Flow matrix
 TM : Travel time matrix
Output: $\mathcal{R}^*$: *the set of skyline routes*
1: $\mathcal{R} = \varnothing$
2: **Repeat**
3: $currentR = s_1$
//starts from the given origin s_1
4: Choose the next stop s_i^* with respect to *currentR* according to Eq. 9.5
5: $R = currentR \cdot s_1$
//operation append s_i to *currentR*
6: **Repeat** Line 4~5 Until $s_i^* = s_n$
//ends at the given destination s_n
7: $\mathcal{R} = \mathcal{R} \cup R$
8: Get corresponding *skylineroutes*$\mathcal{R}^*$
9: **Until** $\mathcal{R}^*$ *keep unchanged*

Our proposed probability based spreading algorithm is described in Algorithm 9.2. The spreading starts from the given source (Line 3). The next stop in the candidate route is chosen based on Eq. 9.5.

$$P(s_i^*|\langle s_1, s_2, \cdots, s_j \rangle) = \frac{\sum_{m=1}^{j} fm(s_m, s_i^*)}{\sum_{i=1}^{|S^*|} \sum_{m=1}^{j} fm(s_m, s_i^*)} \quad (9.5)$$

where $fm(s_m, s_i)$ is the passenger flow from s_m to s_i^* , and S^* contains the next possible stops of s_j (child nodes of s_j in the route graph).

We can see the selection of the next stop in the candidate route is not only determined by the current stop, but also all the previous stops. The output of this algorithm is one candidate bus route with the number of stops associated with the number of spreading steps. The spreading would be terminated when the given destination is reached (Line 6). For each run, we get either a repeated route or a new route, thus the candidate route set $\mathcal{R}$ would increase as the spreading algorithm is activated. Then a question arises: *how many running times are sufficient to get the best results*? Based on Definition 9.1 about the skyline routes, we should consider if the skyline route set $\mathcal{R}^*$ remains changed or unchanged.

Theorem 9.2 below ensures that when the skyline route set stays unchanged with the increase of spreading algorithm runs, then the best route has been discovered.

Theorem 9.2 R_1 and R_2 are the detected skyline routes from R_1 and R_2 respectively. If $R_1 \subseteq R_2$, then we have: $\forall R_i \in \mathcal{R}_1^*, \exists R_j \in \mathcal{R}_2^*; R_i = R_j \; or \; R_i$ is dominated by R_j.

In Algorithm 9.2, we have $\mathcal{R}_{t_1}$ $\mathcal{R}_{t_2}$ if the running time $t_1 < t_2$, and the algorithm would be stopped when no better skyline routes are returned with the increase of running times, that is $R_{t_1}^* = R_{t_2}^*$ (Line 9). The computation complexity of the algorithm is $O(N)$.

Instead of choosing only one stop randomly at each spreading step like in the *probability based spreading algorithm*, an intuitive way is to select top-k stops each time, where those k nodes should have the highest accumulated passenger flow with previous stops. In such a way, the first step selects top-k nodes, thus leading to k routes from the origin to those nodes. In the second step, each k nodes would select another top- k nodes, thus the total candidate routes would be k^2 . Assume that n steps are needed to the destination, then the total candidate routes generated would be kn in the end. Thus, the computation complexity of this algorithm is $O(k^n)$, which grows exponentially with the spreading step (n). We use this **top- k spreading** method as the baseline.

Bi-directional Probability based Spreading (BPS) Algorithm: In practice, for a particular bus line, buses can run on the same route in both directions. Algorithm 9.2 can get the best bus route in one direction (e.g. from ZJU to Railway Station), however, it cannot guarantee the same route in the opposite direction (i.e. from Railway Station to ZJU) would still expect the maximum number of passengers, as the passenger flows in two directions of the route are generally asymmetrical. To get a bus route which has overall maximum expected number of passengers in both directions, we propose the *BPS* algorithm, whose basic idea is to run the *probability based spreading algorithm* in both directions so that we generate one candidate "optimal" route in each direction, and the best route is selected by evaluating all the candidate routes in two directions. The procedure is illustrated in Algorithm 9.3.

Algorithm 9.3 BPS Algorithm

Input: $G_{O\to D}(S,E)$: Graph for $O \to D$
$G_{D\to O}(S,E)$: Graph for $D \to O$
FM : Flow matrix
TM : Travel time matrix
Output: $\mathcal{R}^*$: the set of skyline routes
1: $\mathcal{R} = \varnothing$
2: **Repeat**
3: Run Line 2~6 in Algorithm 9.2 for $G_{O\to D}(S,E)$, and the output is $R_{O\to D}$
4: Run Line 2~6 in Algorithm 9.2 for $G_{D\to O}(S,E)$, and the output is $R_{D\to O}$
5: $\mathcal{R} = \mathcal{R} \cup R_{O\to D} \cup R_{D\to O}$
6: Get corresponding *skyline routes* $\mathcal{R}^*$
7: **Until** $\mathcal{R}^*$ keeps unchanged

9.4.4 *Bus Route Selection*

Given the bus operation frequency (once every 30 min), the total travel time constraint, and the taxi passenger flow from 21:30 to 5:30, we obtain the candidate bus routes for a given OD pair using the two different heuristic spreading algorithms, and the skyline route which achieves the maximum expected number of passengers will be selected as the operating route.

With the planned bus route consisting of the selected bus stops, the next step is to find a physical bus route in the real setting, which consists of road segments corresponding to the planned route. The selection of each road segment is done by following the dense and fine trajectories of taxis if they allow buses to operate; Otherwise similar bus routes near the planned ones can be adopted as a refined solution.

9.5 Experimental Evaluations

In this section, we validate the proposed approach with a large-scale real-world taxi GPS dataset which is generated from 7600 taxis in a large city in China (Hangzhou) in 1 month, with more than 1.57 million of night passenger-delivering trips. All the experiments are run in MATLAB on an Intel Xeon W3500 PC with 12GB RAM running Windows 7.

9.5.1 *Evaluation on Bus Stops*

We compare the bus stop results generated with our proposed method with that generated by the popular *k*-means clustering method. We set $k = 579$, which is the same as our method. We adopt the Eulerian distance as the similarity metric. The centroid of each cluster is selected as the stop. Fig. 9.7 shows the comparison results. Comparing with the popular *k*-means approach, our proposed candidate bus stop identification method has two advantages: (1) the centroid of each cluster got by *k*-means is the average location of all its members, and it may fall into non-reachable places like river, as highlighted by the black circles in Fig. 9.7 (left). In our proposed method, both hotness and connectivity of each grid cell is considered for the bus stop location selection, and the selected bus stops are meaningful and stoppable places; (2) Several identified stops by *k*-means fall into a small area (highlighted by the blue circle) as the size of clusters got by *k*-means is very different, while our proposed method generates candidate bus stops that are evenly distributed in the hot areas, which better meets the commonsense design criteria of bus stops.

9.5.2 *Evaluation on Bus Route Selection Algorithm*

We first show the convergence of the proposed algorithm and followed by a parameter sensitivity study. Then we perform a quantitative statistical analysis of all the candidate routes generated for three given OD pairs. We also give the computed skyline route results. Finally, we validate that our proposed bus route generation approach outperforms the baseline approach. Table 9.1 shows the details of three OD pairs for the night-bus route design experiment, where more than 70 candidate bus stops are in the candidate bus route selection list.

1. *Convergence Study*: As illustrated in Algorithms 9.2 and 9.3, our proposed bus route generation process would be terminated if the resulted *skyline routes* keep unchanged. We study the similarity of consecutively generated *skyline routes* from 5000 to 150,000 runs, with a constant interval of 5000 runs. We measure the similarity (*sim*) of two sets A and B as follows

$$\text{sim}(A, B) = \frac{|A \cap B|}{|A \cup B|} \tag{9.6}$$

The similarity results of the consecutively generated skyline routes with a 5000-run interval are shown in Fig. 9.8, and the time cost is put in the diagram as

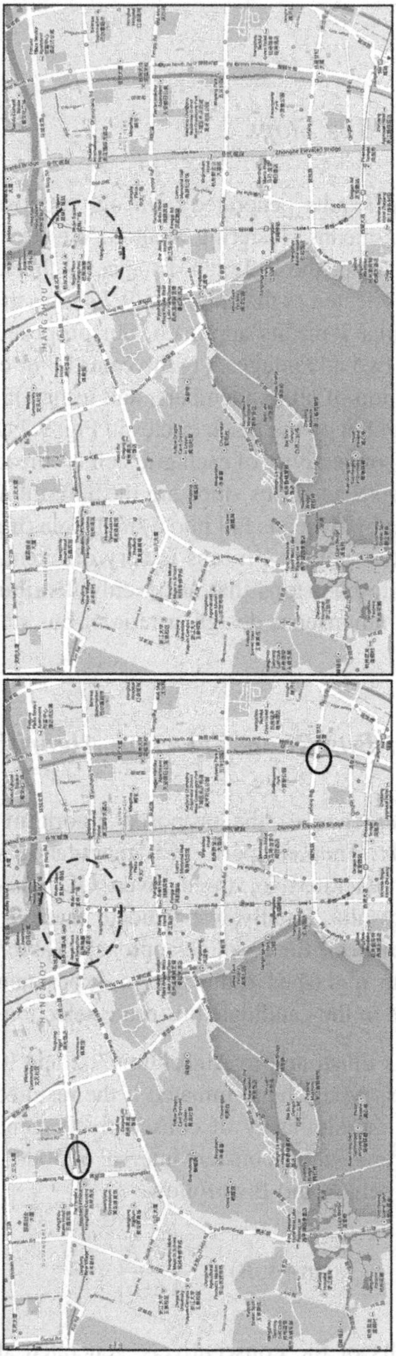

Fig. 9.7 Comparison results with *k*-means (best viewed in the digital version). Results got by *k*-means (left) and results got by our method (right)

Table 9.1 Detailed Information about Studied Odpairs

	OD Pairs	Distance(*km*)	Number of Stops
1	ZJU—Railway	5.70	104
2	Railway—East Railway	5.86	75
3	East Railway—ZJU	8.80	144

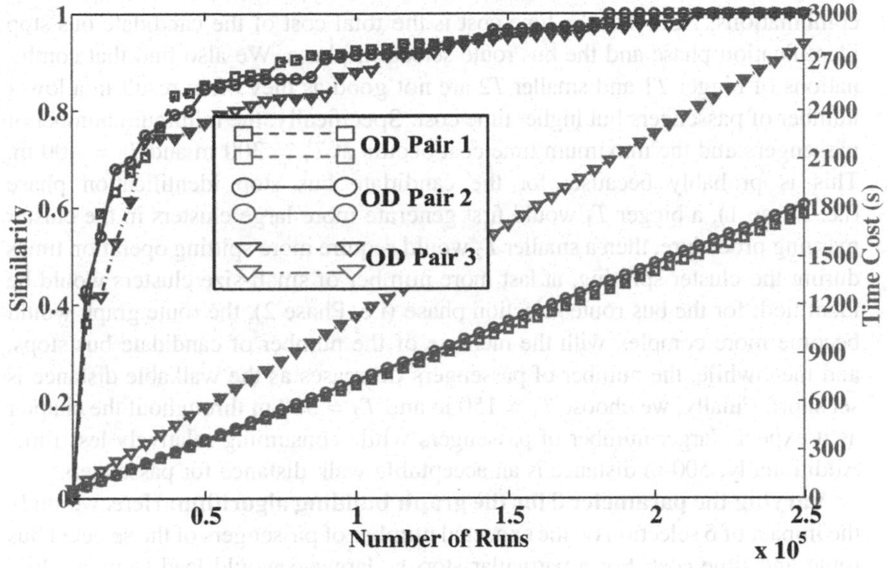

Similarity
Time Cost (s)
Number of Runs
x 10^5
OD Pair 1
OD Pair 2
OD Pair 3
0.5
1
1.5
2
2.5
0
0.2
0.4
0.6
0.8
300
600
900
1200
1500
1800
2100
2400
2700
3000

Fig. 9.8 Convergence study of the proposed *BPS* algorithm

well. In this study, we can see that *sim* values gradually reach 1 with the increase of runs for all three OD pairs, meaning that in all three cases the best bus route converges to one. Also, the time cost is almost linearly increased with the number of runs, suggesting that the spreading time cost at each run is almost constant. It is also noted that the three curves for three OD pairs have different slopes, the reason is probably that the bus routes corresponding to different ODs have different lengths and a varied number of candidate bus stops, thus the spreading time and candidate bus stop selection time should be also different.

2. *Parameter Sensitivity Study*: To better understand the bus stop identification and the bus route selection algorithms, we conduct experiments under different parameter settings to study how they affect the number of expected passengers of selected routes and running time. We examine three parameters in the process, while two of them are in the bus stop identification phase, the remaining one is in the route graph building algorithm.

Varying parameters (*T*1 and *T*2) for the cluster merge and split algorithms: As discussed in Sect. 9.3, a bigger *T*1 would produce more large clusters, and likewise, a bigger *T*2 would also generate more large clusters. Figure 9.9a, b show the *Cumulative Distribution Function* (CDF) of finally produced clusters in terms of size after cluster merging and splitting under various $T_1(\in[100:50:300]$ m) and $T_2(\in[400:50:650]$m) respectively. We also show the skyline route results under different T_1 and T_2 in Fig. 9.9c, d. From these results, we can see that choosing the relatively smaller T_1 and larger T_2 will lead to better skyline routes.

Figure 9.9e, f show the maximum number of expected passengers for the selected bus route and the time cost, respectively, under different T_1 and T_2 combinations. Note that the time cost is the total cost of the candidate bus stop identification phase and the bus route selection phase. We also find that combinations of bigger $T1$ and smaller $T2$ are not good as they often result in a lower number of passengers but higher time cost. Specifically, the minimum number of passengers and the maximum time cost occurs at $T_1 = 300$ m and $T_2 = 400$ m. This is probably because: for the candidate bus stop identification phase (i.e. Phase 1), a bigger T_1 would first generate more large clusters in the cluster merging procedure, then a smaller T_2 would require more spitting operation times during the cluster splitting, at last more number of small-size clusters would be identified; for the bus route selection phase (i.e. Phase 2), the route graph would become more complex with the increase of the number of candidate bus stops, and meanwhile, the number of passengers decreases as the walkable distance is set short. Finally, we choose $T_1 = 150$ m and $T_2 = 500$ m throughout the chapter as it expects larger number of passengers while consuming relatively less time. Additionally, 500-m distance is an acceptable walk distance for passengers.

Varying the parameter δ for the graph building algorithm: Here, we study the impact of δ selection on the expected number of passengers of the selected bus route and time cost. For a particular stop s_i, larger δ would lead to more child nodes. Mathematically, we have: $\forall s_i \in S, S'_{\delta_1}(s_i) \subseteq S'_{\delta_2}(s_i) \text{if} \delta_1 \leq \delta_2$, where $S'(s_i)$ is the child node of s_i in the route graph. And we also have $\mathcal{R}_{\delta_1} \subseteq \mathcal{R}_{\delta_2}$. Therefore, with the increase of δ value, better route can be obtained. Meanwhile, the route graph would become more complex, resulting in an increase of computation time.

We investigate different δ in the range of [1.0 km, 1.7 km] for OD pair 2, with a constant interval of 0.1 km. Figure 9.9g shows two metrics of the selected bus route under different values. One point on the plane stands for the selected route under a given δ. We can see that the selected route becomes steadily better with the increase of δ (deliver more passengers with less travel time). However, the difference is negligible after $\delta \geq 1.5$. We also show the complexity of the route graph and the time cost under different δ values in Fig. 9.9h. The complexity of the graph is simply quantified by the average Incoming/Out-going degrees. They are equal to the ratio of the total number of edges to the total number of nodes in the route graph. From the figure, we can see that the average Incoming/Out-going degrees under 1.7 km is twice more than that under 1.0 km. Furthermore, more computation time is needed when δ increases, because the route graph becomes more complex. We set $\delta = 1.5$ km throughout the chapter as it leads to good performance with low time cost.

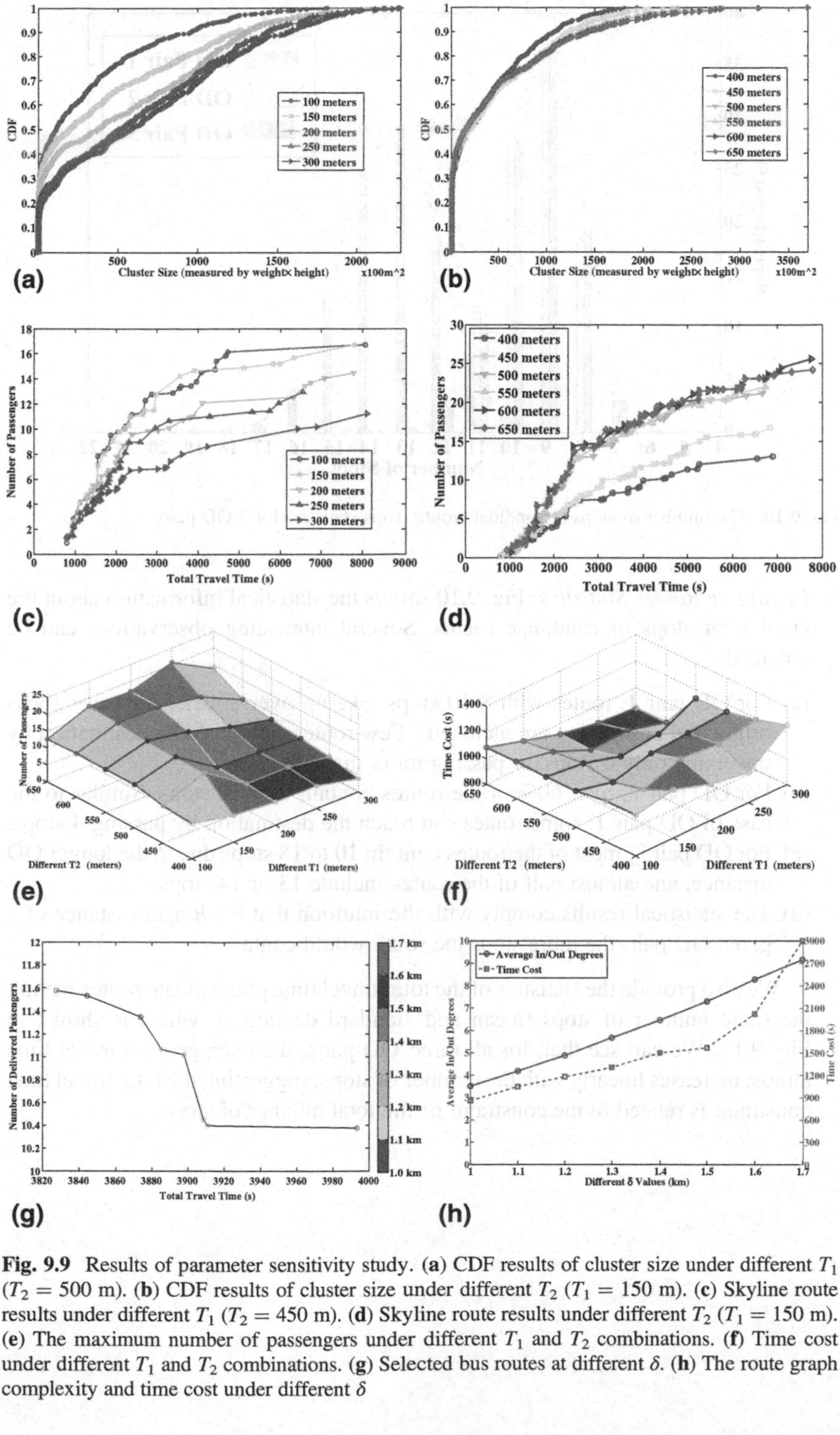
(a)
CDF
Cluster Size (measured by weight× height)
x100m^2
100 meters
150 meters
200 meters
250 meters
300 meters
(b)
CDF
Cluster Size (measured by weight× height)
x100m^2
400 meters
450 meters
500 meters
550 meters
600 meters
650 meters
(c)
Number of Passengers
Total Travel Time (s)
100 meters
150 meters
200 meters
250 meters
300 meters
(d)
Number of Passengers
Total Travel Time (s)
400 meters
450 meters
500 meters
550 meters
600 meters
650 meters
(e)
Number of Passengers
Different T2 (meters)
Different T1 (meters)
(f)
Time Cost (s)
Different T2 (meters)
Different T1 (meters)
(g)
Number of Delivered Passengers
Total Travel Time (s)
1.7 km
1.6 km
1.5 km
1.4 km
1.3 km
1.2 km
1.1 km
1.0 km
(h)
Average In/Out Degrees
Time Cost
Average In/Out Degrees
Time Cost (s)
Different δ Values (km)

Fig. 9.9 Results of parameter sensitivity study. (**a**) CDF results of cluster size under different T_1 ($T_2 = 500$ m). (**b**) CDF results of cluster size under different T_2 ($T_1 = 150$ m). (**c**) Skyline route results under different T_1 ($T_2 = 450$ m). (**d**) Skyline route results under different T_2 ($T_1 = 150$ m). (**e**) The maximum number of passengers under different T_1 and T_2 combinations. (**f**) Time cost under different T_1 and T_2 combinations. (**g**) Selected bus routes at different δ. (**h**) The route graph complexity and time cost under different δ

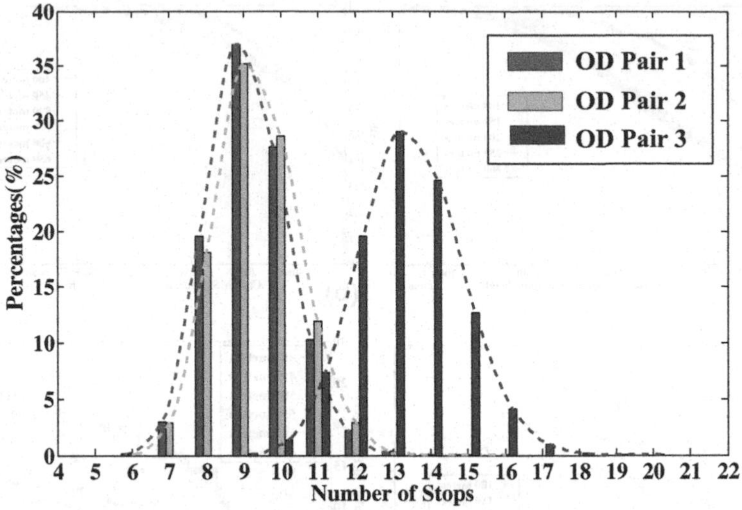
OD Pair 1
OD Pair 2
OD Pair 3
Percentages(%)
Number of Stops
0
5
10
15
20
25
30
35
40
4 5 6 7 8 9 10 11 12 13 14 15 16 17 18 19 20 21 22

Fig. 9.10 The number of stops of candidate route stops statistics for 3 OD pairs

3. *Candidate Routes Statistics:* Fig. 9.10 shows the statistical information about the number of stops of candidate routes. Several interesting observations can be obtained:

   (a) For OD pair 1, routes with 8 10 stops take up over 80% of the cases (both origin and destination are included). Few routes can reach the destination by traversing only 4 stops, or passing more than 11 stops.
   (b) For OD pair 2, over 60% of the routes contain 9 or 10 stops. Similar to the case of OD pair 1, some routes can reach the destination by passing 4 stops.
   (c) For OD pair 3, most of the routes contain 10 to 18 stops due to the longer OD distance, and almost half of the routes include 13 or 14 stops.
   (d) The statistical results comply with the intuition that the longer distance of a given OD pair, the more stops the route would contain.

We also provide the statistics of the total travel time of candidate routes having the same number of stops (mean and standard deviation), which is shown in Fig. 9.11. We can see that, for all three OD pairs, the average total travel time almost increases linearly with the number of stops, suggesting the total travel time constraint is related to the constraint of the total number of stops.

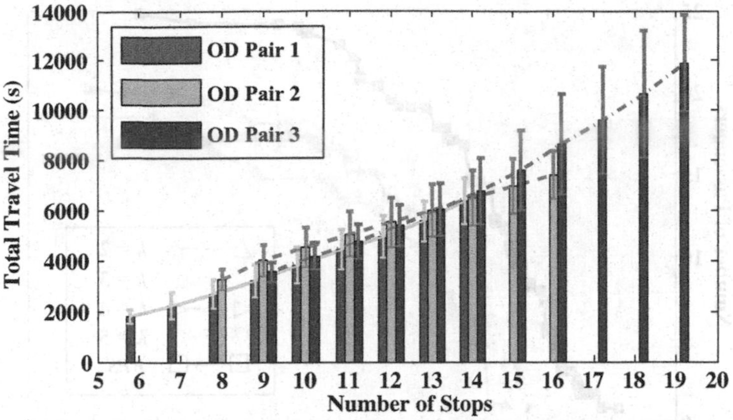
OD Pair 1
OD Pair 2
OD Pair 3
Total Travel Time (s)
Number of Stops
0
2000
4000
6000
8000
10000
12000
14000
5
6
7
8
9
10
11
12
13
14
15
16
17
18
19
20

Fig. 9.11 The relationship between the number of stops and total travel time statistics for 3 OD pairs

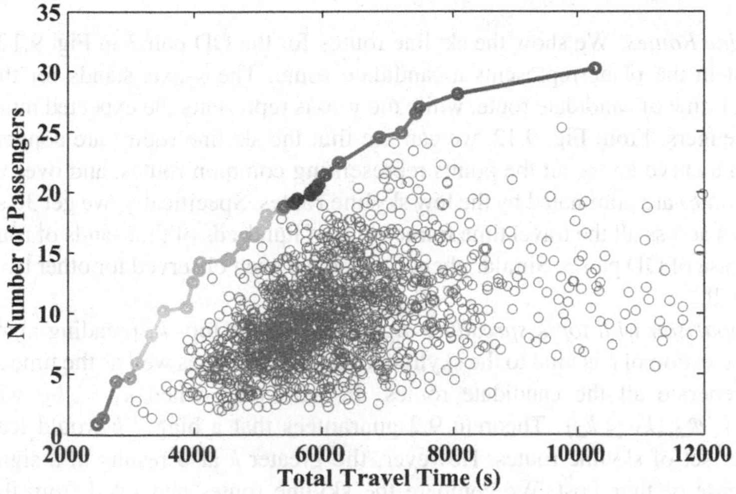
Number of Passengers
Total Travel Time (s)
0
5
10
15
20
25
30
35
2000
4000
6000
8000
10000
12000

Fig. 9.12 Detected *skyline routes* and other candidate routes

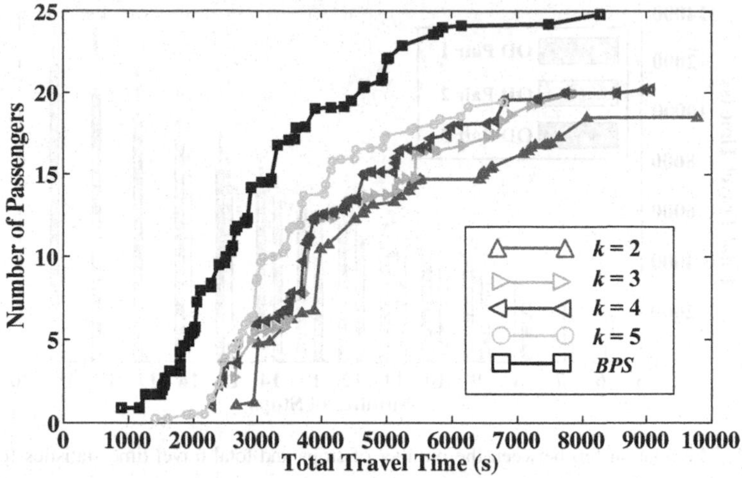
25
20
15
10
5
0
Number of Passengers
0 1000 2000 3000 4000 5000 6000 7000 8000 9000 10000
Total Travel Time (s)
k = 2
k = 3
k = 4
k = 5
BPS

Fig. 9.13 Comparison results with baseline under different k values

4. *Skyline Routes:* We show the skyline routes for the OD pair 3 in Fig. 9.12. Each point in the plane represents a candidate route. The x-axis stands for the total travel time of candidate route, while the y-axis represents the expected number of passengers. From Fig. 9.12, we can see that the skyline routes are connected to form a curve above all the points representing common routes, and over 99% of the routes are dominated by the few skyline routes. Specifically, we get 36 skyline routes across all the travel time frames, out of hundreds of thousands of routes for the case of OD pair 3. Similar phenomena have been observed for other two cases as well.
5. *Comparison with top-k spreading algorithm:* In the top- k spreading algorithm, the selection of k is vital to the skyline routes generated as well as the time needed to generate all the candidate routes. In particular, when $k_1 < k_2$, we have $\mathcal{R}_{k_1} \subseteq \mathcal{R}_{k_2}(k_1 \leq k_2)$. Theorem 9.2 guarantees that a bigger k would lead to a better set of skyline routes. However, the greater k also results in a significant increase of time cost. We compare the skyline routes generated from the *BPS* method with that from the *top- k spreading* method with different k values for the case of OD pair 1, which is shown in Fig. 9.13. We can see that the *BPS* approach outperforms the *top- k algorithms* even when k is set to 5. Again, a similar conclusion can be also drawn for the other two OD pairs.

9.5.3 Bidirectional vs. Single Directional Bus Route

In real life, bus route got by Algorithm 9.2 may (1) be the *skyline route* in both directions; (2) be the *skyline route* in only one direction; (3) not be the *skyline route* in any direction. It is noteworthy to compare the overall best bidirectional bus route obtained by Algorithm 9.3 to the best routes in single direction. We have drawn all

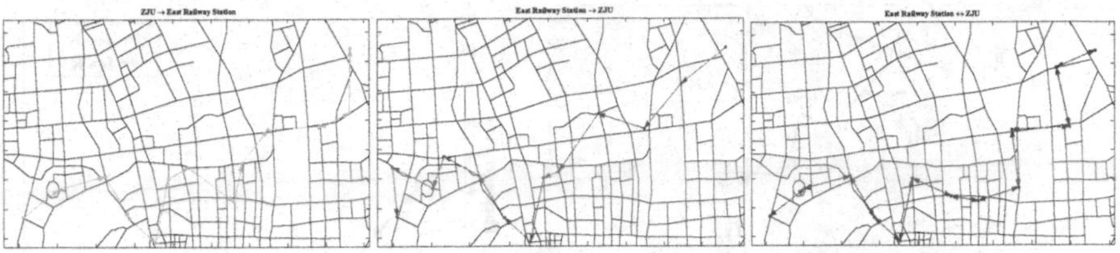
ZJU → East Railway Station
East Railway Station → ZJU
East Railway Station ↔ ZJU

Fig. 9.14 Comparison results of the selected bus routes in two directions to that in one direction. $R_{O \to D}$ (left); $R_{D \to O}$ (middle); $R_{O \to D}$ (right)

Table 9.2 Two metrics of the selected bus routes

	Direction	Average Travel Time (in second)	Number of Passengers
$R_{O \to D}$	ZJU → East Railway	5406.7	17.25
$R_{D \to O}$	East Railway → ZJU	5352.2	20.31
$R_{O \leftrightarrow D}$	ZJUEast ↔ Railway	5320.2	18.73

the selected bus routes on the city digital map in Fig. 9.14 for the OD pair 3. They are different routes, which means the bidirectional bus route is neither the skyline route in the ZJU → East Railway Station direction, nor in the East Railway Station → ZJU direction. A reasonable explanation is that the passenger flow and the travel time among stops is often asymmetrical, and thus the bus route which carries the maximum number of passengers under the give time constraints in one direction would probably fail to deliver the same performance in the opposite direction. However, they all have 13 stops in total and share several common stops near the ZJU stop, especially for the route $R_{O \to D}$ (left figure in Fig. 9.14) and $R_{O \leftrightarrow D}$ (right figure in Fig. 9.14). By further checking, we find that these common stops are popular night life centers.

We show the average travel time and the number of expected delivered passengers of these three bus routes in Table 9.2, and note that heavier passenger flow can be found from East Railway Station to ZJU direction ($R_{D \to O}$). While $R_{D \to O}$ takes slightly less time and delivers a larger number of passengers than $R_{O \to D}$, it carries about 48 more passengers on average per night. $R_{O \leftrightarrow D}$, however, takes the least time, and the average number of delivered passengers lies between $R_{O \to D}$ and $R_{D \to O}$.

9.5.4 Comparison with Real Routes and Impacts on Taxi Services

As the taxi GPS dataset we have was collected from April 2009 to March 2010, we are very interested in knowing if there was any new night-bus route created during this year and how the planned bus route generated with our approach compares with the manually created route. Fortunately, we were told that a night-bus route was

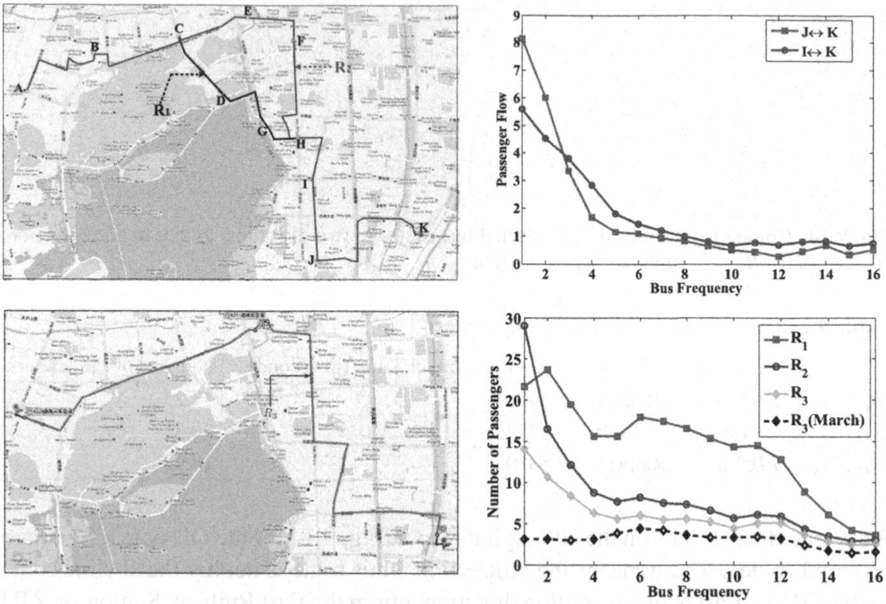
A
B
C
D
E
F
G
H
I
J
K
R1
R2
R3
Passenger Flow
Bus Frequency
J↔K
I↔K
Number of Passengers
Bus Frequency
R1
R2
R3
R3(March)

Fig. 9.15 Results comparison. Planned routes (top left); Passenger flow comparison of two segments at different frequencies (top right); Opened night bus route (bottom left); Number of delivered passengers at different frequencies (R_1, R_2, and R_3, bottom right)

created in February 2010. We could access all the taxi passenger flows before and after the route started date. It is noted that the route is designed by local experts and the user demands are obtained from the expensive human survey. We first draw the newly started night-bus route R_3 on Google map as shown in Fig. 9.15 (left bottom), then we draw our proposed night-bus route R_1 in Fig. 9.15 (left top). Through comparison, we see that they are quite different. With the newly started route, we decide to take a similar route in our selected candidate bus routes (not the best one), and we find R_2 as shown in Fig. 9.15 (left top). It is noted that the main difference between R_2 and the newly started route R_3 is that R_2 includes an additional Stop J in the route. By comparing the passenger flow in the segment $I \leftrightarrow K$ with that in segment $J \leftrightarrow K$ at different time slots, it is found that the passenger flow in path $J\$K$ is even greater than $I \leftrightarrow K$ in the first two time slots, as shown in Fig. 9.15 (right top). Considering further the accumulation effects, including Stop J in the bus route would significantly increase the expected number of passengers along the route. This is evidenced by Fig. 9.15 (bottom right). The accumulated effect is more remarkable at the first three frequencies. Thus, our candidate bus route R_2 would outperform the newly added bus route R_3, at the cost of adding one more bus stop and more travel time.

We also compare our proposed best route R_1 with the candidate route R_2. The difference between R_1 and R_2 lies in two different paths taken from C to H. While R_2 passes the famous shopping street (Yan'an Road) in Hangzhou ($C \leftrightarrow E \leftrightarrow F \leftrightarrow H$),

Table 9.3 Total travel time of the bus routes

Bus route	Total travel time (in second)
R_1	3583.8
R_2	4664.9
R_3	3624.0

R_1 traverses the famous night-club areas along the West Lake. If we compare the number of passengers in R_1 and R_2, it can be seen from Fig. 9.15 (right bottom) that the passenger flow of R_2 is heavier than that of R_1 only around 22:00, and it is much lighter soon after 23:00. With the rest of the stops being the same for both R_1 and R_2, there is no doubt about why R_1 has been selected as the best night-bus route. If we take a closer look at R_1, R_2, and the newly started route R_3, as R_1 takes a much shorter route than R_2 and needs similar travel time as the newly started route R_3 does (shown in Table 9.3), but R_1 expects much more passengers than R_2 and the newly started route R_3, thus it is reasonable to conclude that the selected night-bus route with our proposed approach is better than the current route-in-service in terms of travel time as well as the expected number of passengers.

It is understood that introducing new public services (i.e. new Metro/bus lines) would affect taxi services in the city [2]. It is interesting to compare the taxi passenger flow change along the new bus route before/after it was opened. We choose the new night bus route (R_3) opened in February, 2010 for this study. We prepare taxi GPS data collected in January and March, 2010, and calculate the corresponding taxi passenger flow along the new bus route across all bus frequencies, which is shown in the right bottom subfigure of Fig. 9.15. We can see that the number of passengers who travel by taxi along the bus route in March is much smaller but quite stable across all the bus frequencies. This may be interpreted by the fact that while some passengers might switch to public services, a certain number of passengers still prefer to take taxis at night.

9.5.5 Bus Capacity Analysis

After selecting the best bus route for operation, the next important thing is to determine the proper bus capacity to save operation costs. The essence of bus capacity estimation is to determine the maximum number of passengers on the bus across all the frequencies. For the bus route R_1 of OD pair 1, Fig. 9.16 shows the number of passengers on the bus across all the frequencies for both directions. As can be seen from the results, choosing buses with 20 seats could well meet the requirements. Besides, we also have the following three observations:

1. More passengers are often expected in both directions for the first operation frequency, except for the 11th and 12th frequencies when the bus runs from C to D.

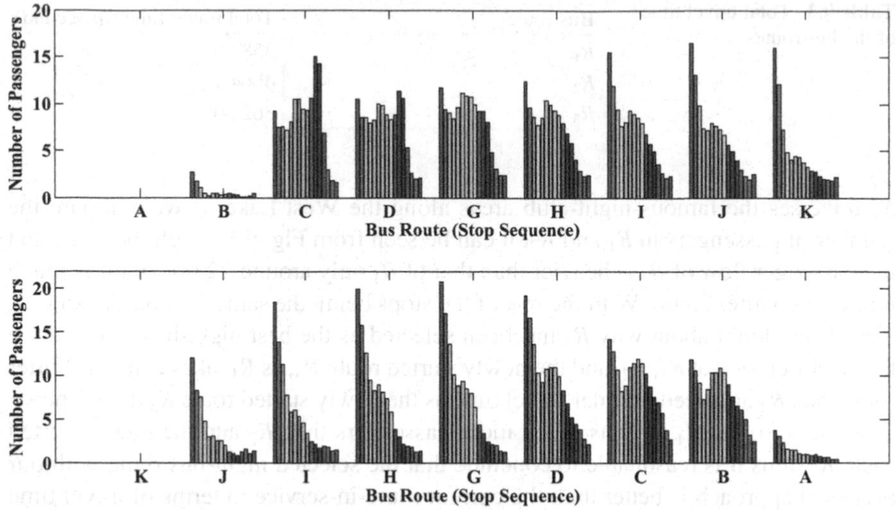

Number of Passengers
0
5
10
15
20
A
B
C
D
G
H
I
J
K
Bus Route (Stop Sequence)
Number of Passengers
0
5
10
15
20
K
J
I
H
G
D
C
B
A
Bus Route (Stop Sequence)

Fig. 9.16 The number of passengers on the bus before reaching the stop for OD pair 1

2. Buses running close to the capacity only last for 3 stops (from A to K) or 4 stops (from K to A).
3. Night buses heading towards different directions have a quite different passenger flow patterns.

9.6 Conclusions and Future Work

In this chapter, we propose a two-phase approach for night-bus route planning by leveraging the taxi GPS traces. In the first phase, we develop a process to cluster "hot" areas with dense passenger pick-up/drop-off and then propose effective methods to split big "hot" areas into clusters and identify a location in the cluster as the candidate bus stop. In the second phase, given the bus route origin, destination, candidate bus stops as well as bus operation frequency and maximum total travel time, we derive several criteria from building bus route graph and pruning the invalid stops and edges iteratively. Based on the graph, we further develop two heuristic algorithms to generate candidate bus routes in both directions automatically. Finally, we select the best route, which expects the maximum number of passengers under the given conditions. We compare our proposed candidate bus stop identification method with the popular k-means clustering method on a real-world dataset and show that our method can generate more reasonable and meaningful results. We further extensively evaluate our proposed BPS algorithm for automatic bus route generation and validate its effectiveness as well as its superior performance over the heuristic top-k spreading algorithm. Furthermore, we show the selected night-bus route with our proposed approach is better than a newly started night-bus route-in-service in Hangzhou, China.

In the future, we plan to broaden and deepen this work in the following directions:

1. We consider modeling the passenger flow patterns directly at the street-level instead of dividing the city into small grids in advance. In addition, the grid has some inherent constraints (such as physical barriers), and it is not straightforward enough to understand human mobility based on grids.
2. Bus route planning is essentially an optimal spatial search problem. We will employ some advanced algorithms of this problem, such as genetic algorithms, which can obtain satisfactory solutions under complex situations.
3. Bus route planning can be further extended to the design of the city-wide bus network. Towards this end, we plan to select several related bus routes for unified route planning. The main intuition behind this idea is, the passenger flow between bus routes passing through the same station is closely related.
4. The bus routes of a city are usually affected by the existing transportation infrastructure and human mobility. It makes sense for developing countries/cities to learn from well-designed bus routes when planning a new bus route. Therefore, during the bus route's design process, we can try to transfer knowledge from the bus systems of different cities. For example, we can learn which plan is the most effective in a similar road structure.
5. In the future, the design of bus routes can be more flexible. We plan to study how to design some customized, new bus routes to improve the public transport system's operating efficiency.

References

1. Aslam J, Lim S, Pan X, Rus D. City-scale traffic estimation from a roving sensor network. Toronto, Ontario, Canada: Proceedings of the 10th ACM Conference on Embedded Network Sensor Systems; 2012. p. 141–54.
2. Feng Z, Zhu Y. A survey on trajectory data mining: techniques and applications. IEEE Access. 2016;4:2056–67.
3. Li B, et al. Hunting or waiting? Discovering passenger-finding strategies from a large-scale real-world taxi dataset. New York, NY: 2011 IEEE International Conference on Pervasive Computing and Communications Workshops (PERCOM Workshops); 2011. p. 63–8.
4. Chen C, Liao C, Xie X, Wang Y, Zhao J. Trip2Vec: a deep embedding approach for clustering and profiling taxi trip purposes. Pers Ubiquit Comput. 2019;23(1):53–66.
5. Markou I, Kaiser K, Pereira FC. Predicting taxi demand hotspots using automated internet search queries. Transport Res Part C: Emerg Technol. 2019;102:73–86.
6. Liu Y, Wang F, Xiao Y, Gao S. Urban land uses and traffic 'source-sink areas': evidence from GPS-enabled taxi data in Shanghai. Landsc Urban Plan. 2012;106(1):73–87.
7. Djenouri Y, Belhadi A, Lin JC, Djenouri D, Cano A. A survey on urban traffic anomalies detection algorithms. IEEE Access. 2019;7:12192–205.
8. Zhao L, et al. T-GCN: a temporal graph convolutional network for traffic prediction. IEEE Trans Intell Transp Syst. 2020;21(9):3848–58.

9. Li Y, Fu K, Wang Z, Shahabi C, Ye J, Liu Y. Multi-task representation learning for travel time estimation. New York, NY, USA: Proceedings of the 24th ACM SIGKDD International Conference on Knowledge Discovery & Data Mining; 2018. p. 1695–704.
10. Yuan NJ, Zheng Y, Zhang L, Xie X. T-finder: a recommender system for finding passengers and vacant taxis. IEEE Trans Knowl Data Eng. 2013;25(10):2390–403.
11. Yuan J, Zheng Y, Xie X, Sun G. T-drive: enhancing driving directions with taxi drivers' intelligence. IEEE Trans Knowl Data Eng. 2013;25(1):220–32.
12. Bastani F, Huang Y, Xie X, Powell JW. A greener transportation mode: flexible routes discovery from GPS trajectory data. Chicago, Illinois: Proceedings of the 19th ACM SIGSPATIAL International Conference on Advances in Geographic Information Systems; 2011. p. 405–8.
13. Zhang W, Shemshadi A, Sheng QZ, Qin YL, Xu X, Yang J. A user-oriented taxi ridesharing system with large-scale urban gps sensor data. New York, NY: IEEE Transactions on Big Data; 2018. p. 1–1.
14. Chhabra R, Verma S, Krishna CR. A survey on driver behavior detection techniques for intelligent transportation systems. New York, NY: 2017 7th International Conference on Cloud Computing, Data Science Engineering—Confluence; 2017. p. 36–41.
15. Chen C, Zhang D, Zhou Z-H, Li N, Atmaca T, Li S. B-Planner: Night bus route planning using large-scale taxi GPS traces. New York, NY: 2013 IEEE International Conference on Pervasive Computing and Communications (PerCom); 2013. p. 225–33.
16. Newell GF. Some issues relating to the optimal design of bus routes. Transp Sci. 1979;13 (1):20–35.
17. Kim S, Shekhar S, Min M. Contraflow transportation network reconfiguration for evacuation route planning. IEEE Trans Knowl Data Eng. 2008;20(8):1115–29.
18. Guihaire V, Hao J-K. Transit network design and scheduling: a global review. Transp Res A Policy Pract. 2008;42(10):1251–73.
19. Zhang D, Guo B, Yu Z. The Emergence of Social and Community Intelligence. Computer. 2011;44(7):21–8.
20. Borzsony S, Kossmann D, Stocker K. The skyline operator. New York: Proceedings 17th International Conference on Data Engineering; 2001. p. 421–30.

Part V
Enabling Smart Urban Services: Travellers

Chapter 10
TripPlanner: Personalized Trip Planning Leveraging Heterogeneous Trajectory Data

10.1 Introduction

It is essential to plan an itinerary before paying a visit to a city. However, the preparation activities are always time-consuming. Specifically, to obtain feasible routes in a popular tourist city, people need to choose the point of interests (POIs) from hundreds of POIs according to their preferences, figure out the order of POIs in the route. To meet their time budget, the time of visiting each POI and transiting from one POI to the next need to be considered. In general, three main factors need to be considered for a trip planning system: (1) *the venue constraints*, which include the starting location and ending location of the trip, the POIs expected to be covered, the POI visiting order and the POI categories that might be added if time permits. (2) *the time constraints*, which include the operation time of each POI to visit, the visiting of each POI which can be estimated and controlled by users, the time-dependent diving time between POIs that depends on the traffic condition of the time and the given time budget of trip; (3) *and user's preference scores* about a specific POI and an itinerary at certain time of the day that are assumed to be computable. The trip planning system aims to interact with users to determine whether the user-specified POIs can all be covered under the time budget constraints.

Since the transit time between POIs is always assumed to be constant in the previous research [1], the proposed approaches cannot be used to solve the above problem. The above issue is also not similar to route recommendation which suggests routes directly according to the similarity between user's visiting history in other contexts and other people's trip records in the targeted city. Others suggest venues based on a user's preference and recommend routes according to certain

Part of this chapter is based on a previous work: C. Chen, D. Zhang, B. Guo, X. Ma, G. Pan and Z. Wu, "TripPlanner: Personalized Trip Planning Leveraging Heterogeneous Crowdsourced Digital Footprints," in IEEE Transactions on Intelligent Transportation Systems, vol. 16, no. 3, pp. 1259-1273, June 2015, doi: https://doi.org/10.1109/TITS.2014.2357835.

C. Chen et al., *Enabling Smart Urban Services with GPS Trajectory Data*,
https://doi.org/10.1007/978-981-16-0178-1_10

criteria (e.g., with the highest route score) [2, 3]. However, a user may have some additional constraints in the actual trip such as "the total time budget should be $\leq$ 6 h.", "go to park before lunch," and "need to go some specific places." The previous work provided some tedious and time consuming methods, for instance, they asked people to manually select and configure travel routes based on the recommendation [3]. Even worse, based on the given constrains, there might be no candidate routes. Another group of route planning work aims to find the optimal (e.g., fastest or shortest) paths on road networks according to the time-varying assumption of each road segment. These studies always pay little attention on the attributes associated with the nodes (POIs) and only focus on the edge information on road networks. In other words, they ignore the opening hours and duration of visit of each POI and care only about the time on the road. In our work, we aim to build a *personalized, interactive, and traffic-aware* trip planning service in which we consider the characteristics of each POI e.g., its attractiveness, operation hours, and order of visit.

In order to find the personalized routes for users, we build a POI network model. Specifically, we first acquire the information about all POIs in a targeted city and links among them. So far, there are mainly the following three parts of data sources which have been exploited: (1) websites, Wikipedia, web blogs which can be used to reveal preferences and experiences with POIs based on the contained information of tourists' profiles, and comments; (2) social media sites such as Flickr, Facebook, and location-based social networks (LBSNs) (e.g., Foursquare and Gowalla), from which the popularity, functions, and operating hours of the POIs and individual user's travel history can be extracted [1, 4]; and (3) GPS trajectories of people and taxis, which can inform transit time between two places and duration in each place and [5]. To obtain a more complete picture of the POI network, we can integrate heterogeneous data sources based on the strength and weakness of each data in characterizing certain facets of the POI nodes and edges. In this section, we proposed the TripPlanner which is a novel trip planning framework. Specifically, it provides an interactive way for users to specify their venues of interests with varied constraints and construct the POI network model based on heterogeneous crowdsourced digital footprints. TripPlanner could suggest a personalized route with the highest trip score under the total travel time constraint by a two-phase query resolution process. In summary, there are four main contributions of this section.

- First, we propose a new trip planning problem in which users can customize the must-visit venues and optional venue categories if the time permits, according to a total travel time budget. We also make more realistic assumption about the time-dependent transit time between venues according to traffic condition.
- Second, we extract relevant information about nodes and links from *heterogeneous crowdsourced digital footprints* and construct a dynamic POI network model by utilizing the strengths of various data source.
- Third, we propose a *two-phase approach* for *personalized, interactive, and traffic-aware* trip planning. Considering both the popularity and individual preference of venues, we also propose a new way to score an itinerary.

- Finally, extensive empirical studies are performed on two real-world data sets from the city of San Francisco, which contain more than 391,900 passenger delivery trips generated by 536 taxis in a month and 110,214 check-ins left by 15 680 Foursquare users in 6 months. The results verify the effectiveness and efficiency of the proposed TripPlanner.

10.2 Related Work

The related work can be grouped into two sections in the following. We first review the methods of building the POI network model based on the information from different data sources in previous work and then we introduce how to plan a travel route according to certain assumptions.

10.2.1 Construction of POI Network

In previous studies, the relevant information about node and edge are obtained from various data sources to model a POI network. For example, in [1, 6], based on geotagged photos from photo sharing sites (e.g., Flickr), many researchers have extracted the information about POIs, such as location, popularity, characteristics, and proper visiting time and order. What's more, demographics and social relationship of visitors to these POIs can be also derived. However, the dynamic transit time between POIs from geotagged photo data is difficult to obtain. More recently, the user-generated LBSN digital traces attract increasing attention due to that such data contain rich information which can be used to *directly* characterize each POI in a tourist city and users' preferences to each POI [7, 8]. Similar to geotagged photo data, the dynamic transit time between POIs also cannot be obtained from the LBSN traces. Fortunately, GPS trajectory data can be used to predict the fastest route at certain time of the day in a city. Researchers have shown that GPS trajectory traces can provide an accurate forecast for the transit time between POIs, which is more accurate than the results provided by Google Maps. It should be noted that the average point-to-point transit time estimated by Google maps is 35% lower than the actual values.

Building on existing work, we construct a POI network model contained the processes of node modeling and edge modelling based on the information extracted from the taxi GPS trajectory and LBSN trace data. On this basis, the proposed trip planning system is able to suggest routes which can better meet user's demands.

10.2.2 Trip Planning

The previous work on trip planning [9–12] can be divided into three main categories. Specifically, the first category is *route search*. Based on a given POI network, it suggests routes according to the user's queries. As a representative, *Traveling salesman problem* (TSP) has been proposed for many years. However, in the real world, situations may be much more complicated. The starting point and destination may be different. Furthermore, compared to a specific POI location, users may simply have in mind a type of POIs. The problem is addressed by *Trip planning query* (TPQ) which aims to find the shortest path containing all user-specified node categories. Some variations of TSP and TPQ problems have been defined by adding some new constraints. However, the transit time between POIs is always constant. In this chapter, we suggest routes to users under the conditions, i.e., the transit time between POIs is assumed to be time-dependent according to the traffic condition and both POIs and POI categories (i.e., types) are given in advance.

The second category is *route recommendation*. The studies mainly focus on two parts, i.e, modelling the user preference and suggesting the POIs and routes. In this problem, the POIs or POI categories of interests will not be given by users. For instance, a probabilistic model is proposed to plan personalized routes in which the user preferences, location, and available time are considered [4]. Given a predefined budget (e.g., time and money), a *personalized trip recommendation* framework is developed by Lu et al. with the objective of suggesting personalized arrangement of visit to venues [8]. By utilizing the check-in patterns of user, Hsieh et al. suggest popular time-sensitive trips [13]. Unlike to these studies, with the information of POIs and/or POI categories at hand, we assign a venue score to each POI according to its popularity and users' preferences.

The third category is *route planning*. The objective is to finding the optimal time-dependent routes. To name a few, Yuan et al. [14] and Ziebart et al. [15] plan optimal routes to user by using the information extract from the historical taxi GPS traces in which the travel time on the road segment is assumed to be dynamic due to the traffic condition. Different from this category of studies, we take more factors into account in the route planning framework, such as the POI priority, the preferred order of visit, and the visiting time constraint of each POI.

10.3 Trip Planner System

Here, several key terminologies are first introduced. Then, we formally define the personalized trip planning problem. Finally, we give a detailed description of the framework of TripPlanner system, which is composed of three major parts, i.e., a dynamic POI network model, a route search component, and a route augmentation component (see Fig. 10.1).

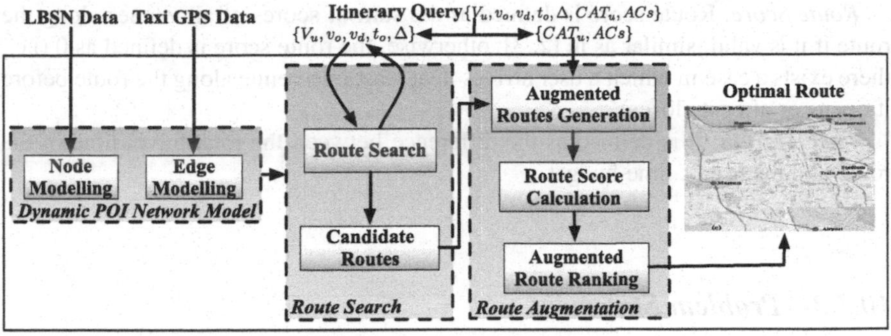
LBSN Data
Taxi GPS Data
Itinerary Query$\{V_u, v_o, v_d, t_o, \Delta, CAT_u, ACs\}$
$\{V_u, v_o, v_d, t_o, \Delta\}$
$\{CAT_u, ACs\}$
Node Modelling
Edge Modelling
Dynamic POI Network Model
Route Search
Candidate Routes
Route Search
Augmented Routes Generation
Route Score Calculation
Augmented Route Ranking
Route Augmentation
Optimal Route

Fig. 10.1 Framework of our proposed TripPlanner

10.3.1 Key Terminologies

Dynamic POI Network Model: The model can be represented as a *directed complete* graph G = (V, E). Each node in V denotes a venue (i.e., POI), which has five attributes, namely, *category, operation time, popularity, geographical location, and stay time (i.e., the duration of visit)*. Each directed edge (v_i,v_j) in *E* represents a link from node v_i to v_j, which carries the transit time between the two venues, denoted as $tt(v_i,v_j)$. The transit time is asymmetric and dynamically changing.

Lemma 10.1 *(Dynamic POI Network has FIFO Property)*: Given a dynamic network G = (V, E), where the transit time of each edge in G is time dependent. The network is first-input–first-output (FIFO) since, for any arc (i, j) in E, given user A leaves node v_i at time t_0 and user B leaves node v_i at time t_1 ($t_1 > t_0$), user B cannot arrive at node v_j before user A.

Itinerary Query: An itinerary query IQ consists of four parts: (1) a user-specified venue list V_u that the user intends to cover; (2) starting place v_o and starting time t_o, ending place v_d, and a travel time budget Δ; (3) a set of user-preferred venue categories CAT_u (optional venues to visit if time permits); and (4) additional constraints ACs such as constraints on the time and the order of venues are to be visited. For instance, a user may want to have lunch at noon and visit museums after that. In summary, the query IQ can be thus represented as $\{V_u, v_o, v_d, t_o, \Delta, CAT_u, ACs\}$. It should be noted that users may not impose ACs when planning their visit, and thus, the corresponding field is empty.

Valid Route: A route R = $<v_1,v_2 \ldots,v_n>$ is *valid* iif

$$aT(v_i) \geq oT(v_i), lT(v_i) \leq cT(v_i) \forall i \in \{1, 2, \ldots, n\}$$

This implies that the user should visit all venues while they are open. Here, $aT(\cdot)$, $lT(\cdot)$ are the users' arriving and leaving times for the given venue, whereas $oT(\cdot)$, $cT(\cdot)$ refer to the opening and closing times of the given venue, respectively.

Route Score: Route score is defined as the sum of scores of all venues along the route if it is valid, similar as in [2, 3]; otherwise, the route score is defined as 0 (i.e., there exists a case in which a user arrives in at least one venue along the route before it opens or after it closes).

Time Margin: It is defined as the difference between the total travel time of the route and the user's time budget.

10.3.2 Problem Statement

Personalized Trip Planning Problem: Given a *dynamic POI network G* in a targeted city and a user's *itinerary query IQ*, our objective is to find the *optimal valid* route with the maximum route score.

10.3.3 Framework

As shown in Fig. 10.1, the proposed framework contains three components, i.e., the dynamic POI network model, the route search component, and the route augmentation component. While the dynamic POI network model is prebuilt and maintained offline, the route search and route augmentation components collaboratively answer users' trip queries in real time.

1. *Dynamic POI Network Model*: The key problem of POI network model construction is to separately extract attributes of POI nodes from the Foursquare data set and information of the edges from the taxi GPS data set.
2. *Route Search*: Given user-specified venues to visit, the starting time, and the time budget, the route search component returns routes that traverse all the intended venues from the starting location to the destination. In particular, the returned routes with a time margin greater than a user-determined threshold become candidate input to the route augmentation component. However, users might list too many venues to cover within the time constraint, or the planned visiting time does not agree with the operating hours of certain venues. If the TripPlanner system detects any of those cases, it will interact with the user to manually modify the venue list.
3. *Route Augmentation*: This component aims to augment the candidate generated from the *route search* module with user-preferred venues inferred from the intended venue categories in the query, maximizing the route score under the given travel time budget. It first pulls together all of the venues that belong to user-preferred venue categories as candidate venues. Then for each candidate route, it tries to insert venues in the pool into it to *generate an augmented route* without breaking any constraint. In the end, TripPlanner presents the augmented

routes to the user, in an order sorted according to their scores in the *augmented route ranking* module.

In the following two sections, we elaborate on the offline construction of the dynamic POI network and the online route planning process, respectively.

10.4 Dynamic Network Modelling

10.4.1 Node Modeling

Each node in the model corresponds to a POI with five attributes, i.e., *operation time, category that the venue belongs to, popularity, geographical location, and stay time*. For each venue, users provide their expected stay time, whereas we extract information related to the first four attributes from Foursquare data (see Fig. 10.2).

Operation time of a venue may vary according to the day of the week and even time of the year.

A venue can be associated with two or more category labels with different granularities. Take the Nick's Crispy Tacos venue shown in Fig. 10.2 as an example.

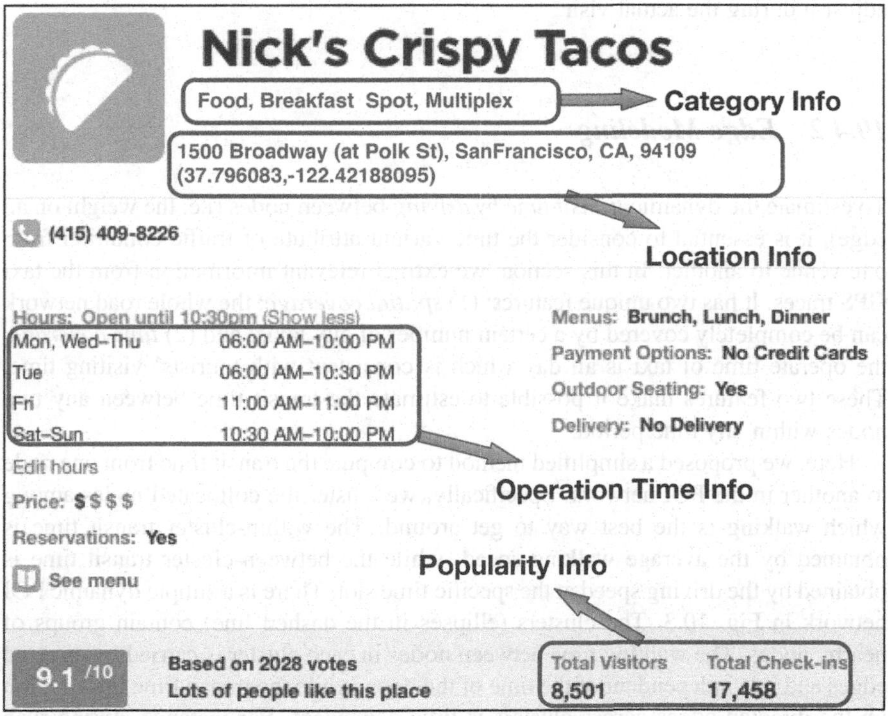

Nick's Crispy Tacos
Food, Breakfast Spot, Multiplex
Category Info
1500 Broadway (at Polk St), SanFrancisco, CA, 94109
(37.796083,-122.42188095)
(415) 409-8226
Location Info
Hours: Open until 10:30pm (Show less)
Mon, Wed–Thu 06:00 AM–10:00 PM
Tue 06:00 AM–10:30 PM
Fri 11:00 AM–11:00 PM
Sat–Sun 10:30 AM–10:00 PM
Menus: Brunch, Lunch, Dinner
Payment Options: No Credit Cards
Outdoor Seating: Yes
Delivery: No Delivery
Edit hours
Operation Time Info
Price: $ $ $ $
Reservations: Yes
See menu
Popularity Info
9.1 /10
Based on 2028 votes
Lots of people like this place
Total Visitors 8,501
Total Check-ins 17,458

Fig. 10.2 Relevant information of the node provided by Foursquare

It has three category labels, among which "Food" is a level-1 label, "Breakfast Spot" is a level-2 label, and "Multiplex" is a level-3 label.

To compute the popularity of a given venue, we use two indicators, i.e., the total number of visitors (*tvs*) and the total number of check-ins *tcs* (Eq. 10.1). The total visitor number is usually *smaller* than the total check-in number of the same venue since some user checked in repeatedly during a single visit.

$$\mathrm{Pop}(v_i) = \frac{2 \times \frac{tvs(v_i)}{c_1} \times \frac{tcs(v_i)}{c_2}}{\frac{tvs(v_i)}{c_1} + \frac{tcs(v_i)}{c_2}} \tag{10.1}$$

where c_1 is the maximum visitor number of all venues in the targeted city, and likewise, c_2 is the maximum check-in number of all venues. Note that the most visited venue may be different from the one with the most check-in records. The venue score is fused by the harmonic mean as we want both values to be relatively higher.

Regarding the *geographical location* of a given venue, Foursquare provides the longitude/latitude information together with its address.

Although the exact value of the *stay time* at a given venue cannot be precisely derived from the check-in data, it could be approximated using the average stay time of tourists. Note that users might estimate the stay time when planning the trip and adjust it during the actual visit.

10.4.2 Edge Modeling

To estimate the dynamic *transit time by driving* between nodes (i.e. the weight on an edge), it is essential to consider the time-variant attribute of traffic condition from one venue to another. In this section, we extract relevant information from the taxi GPS traces. It has two unique features: (1) *spatial coverage*: the whole road network can be completely covered by a certain number of city taxis; and (2) *time coverage*: the operate time of taxi is all day which is consistent with tourists' visiting time. These two features make it possible to estimate the transit time between any two nodes within any time period.

Here, we proposed a simplified method to compute the transit time from one node to another in the POI network. Specifically, we cluster the collocated nodes among which walking is the best way to get around. The within-cluster transit time is obtained by the average walking speed, while the between-cluster transit time is obtained by the driving speed at the specific time slot. There is a simple dynamic POI network in Fig. 10.3. The clusters (ellipses in the dashed line) contain groups of nearby nodes. The walking time between nodes in each cluster is carried by directed edges and it is independent of the time of the days, while the transit time information on the directed edges across clusters is time-dependent. For instance, during rush hours in the morning, the transit time from the upper right cluster to the bottom

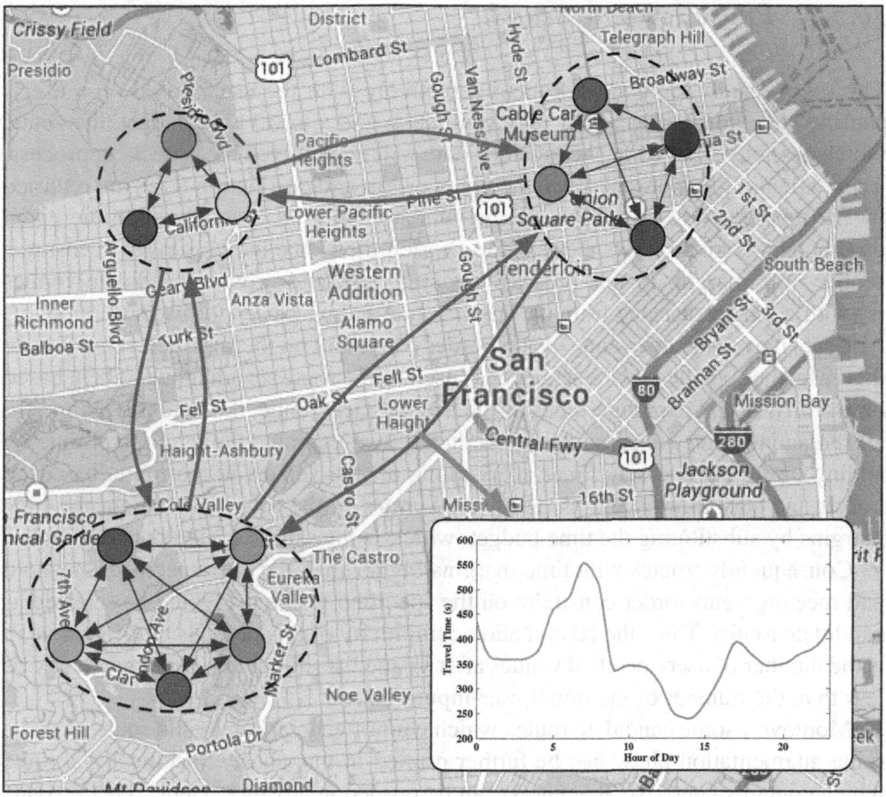
Crissy Field
Presidio
District
Lombard St
Telegraph Hill
Broadway St
Hyde St
Van Ness Ave
Gough St
Cable Car Museum
Pacific Heights
Lower Pacific Heights
Pine St
Union Square Park
Tenderloin
1st St
2nd St
South Beach
Western Addition
Anza Vista
Inner Richmond
Balboa St
Arguello Blvd
Geary Blvd
Turk St
Alamo Square
Fell St
Oak St
Lower Haight
San Francisco
Bryant St
Brannan St
3rd St
Mission Bay
Central Fwy
Haight-Ashbury
Jackson Playground
16th St
Cole Valley
The Castro
Eureka Valley
7th Ave
Market St
Noe Valley
Forest Hill
Portola Dr
Diamond
Travel Time (s)
Hour of Day

Fig. 10.3 Illustration of the dynamic network built with Foursquare and taxi GPS data sets

cluster is more than twice of the least travel time of the day (refer to the green curve in the bottom right of Fig. 10.3 for a whole-day view of dynamic transit time).

10.5 The Two-Phase Approach

We take a two-phase approach, i.e., *route search* and *route augmentation*, to perform trip planning. Route search retrieves candidate routes traversing all user-specified venues within the time budget. Route augmentation further enriches the candidate routes with user-preferred venues as long as time permits, and recommends to users the optimal routes with the highest scores. The set of *user-preferred* venues is a subset of venues in the targeted city, which are obtained based on the user-preferred venue categories CAT_u in the itinerary query (IQ).

10.5.1 Phase I: Route Search

The route search component works interactively with the user. Based on a user's starting and ending places, specified venue list, and a travel time budget, the venue which cannot be visited on the intended date will be removed by a checking process. Then, all possible routes between the given origin and destination (OD) are returned by the module. When a long venue list which cannot be covered within the given time budget is given, users are asked to shorten the venue list *iteratively* to ensure a proper time margin. In this process, the system would provide candidate venues (s) with a longer distance from the starting and ending places for users to remove.

Users may also have venue visiting order constraint on the specified venue list. In this case, the specified venue sequence is used as the route search inputs. Then, based on the arrival and departure times of the user at each venue, the total travel time for all remaining routes is computed. Specifically, we compute the total travel time by adding up transit and stay times spent along all the venues in the route, based on transit and stay times spent along all the venues in the route. We obtain the time margins by substituting the time budget, which is the same for all remaining routes.

Consequently, routes with time margins bigger than the user-specified threshold and meeting venue order constraint on the specified venue list would be selected as candidate routes. Thus, the computation complexity of this phase is $O(m!)$, where m is the number of user-specified venues after venue list shortening. Note that it may be less than the number of the initial user-input venues.

Moreover, some candidate routes which cannot generate any valid route after the route augmentation phase can be further pruned in this phase. Specifically, a route which contains later-arrival venue(s) can be pruned in advance. Here, "later arrival" refers to arriving at a venue after its closing time; "earlier arrival", on the contrary, means arriving at a venue before it opens. The rationale behind is *inserting a new venue before a "later-arrival" venue will further push back the arrival time at this venue, whereas the "later-arrival" venue would be still late when inserting a venue after it*. In other words, the outputs of the route search phase are all candidate routes that have enough time margins, meet venue visiting order constraint, and do not contain any "later-arrival" venues.

Remark: As this section targets the travel route planning issue for tourists visiting a city for a day trip by renting a car, we thus assume that the number of user-specified venues is not big. Therefore, the route search problem is just finding all the routes that meet the visiting order and time constraints, and the time spent on route search in a city scale is relatively short.

For the case that users specify no venue, the route search phase will generate one candidate route from the starting place to the ending place with no other POIs in between. The route augmentation phase (discussed in the following section) would insert possible POIs as time permits to generate a near-optimal route with maximal score.

10.5.2 Phase II: Route Augmentation

The *route augmentation* component tries to insert optional *user-preferred* venues into the candidate routes returned from the previous phase. The route augmentation problem is NP-hard and suffers from combination explosion. On one hand, some candidate routes may allow the user to visit only one more optional *user-preferred venue*, whereas some of them allow to visit more than one, depending on their time margins and the locations of the user-preferred venues. On the other hand, there are many possible orders to visit the optional user-preferred venues along each candidate route, and different visiting orders would result in different total travel times. The aim for optimization is to maximize the route score without exceeding the time budget. It is very challenging as it tries to satisfy two competing requirements: (1) the route should contain as many *user-preferred* venues as possible; and (2) the route should meet the travel time budget and the venue visiting time constraints. We have to consider the following two factors when selecting new valid venues to optimize the route score.

Arrival Time Delay by Adding New Venues: Apparently, inserting new venues into a given route would increase its total visiting time, adding additional transit time and stay time. The arrival time to some of the existing venues may be delayed. Furthermore, the transit time needed between existing venues might be also different due to the time shift. Taking the diagram in Fig. 10.4 as an example, after inserting venue v_c in the route, the arrival time to v_4,v_5,v_6,v_7 would be delayed, and the transit time between v_4 and v_7 might also change as the traffic conditions might be different later in the day.

Total Route Score Increased by Adding New Venues: Generally, adding more user-preferred venues would increase the score of a route but may violate the given constraints if not done properly. We designed a method for route augmentation, which consists of two steps, i.e., *venue inserting* and *score maximization*. The former aims to find a suitable position in the candidate route to insert a selected venue, whereas the latter is responsible for maximizing the score of the updated route.

1. *Venue Inserting Algorithm*: There are two principles that we should follow when inserting a new venue: The augmented route should be valid, and we should minimize the extra cost in time. For a candidate route with n venues and a new venue v_c to insert, if the candidate route does not contain any "earlier-arrival" venue, we need to check $n - 1$ positions to determine the final augmented route. It should be noted that, for the case that users have order constraint on the *user-preferred venues*, the positions in which a new venue can be inserted is limited ($<n-1$). Only the positions that will not violate the venue order constraint are

Fig. 10.4 Illustrative example of inserting a venue into a candidate route

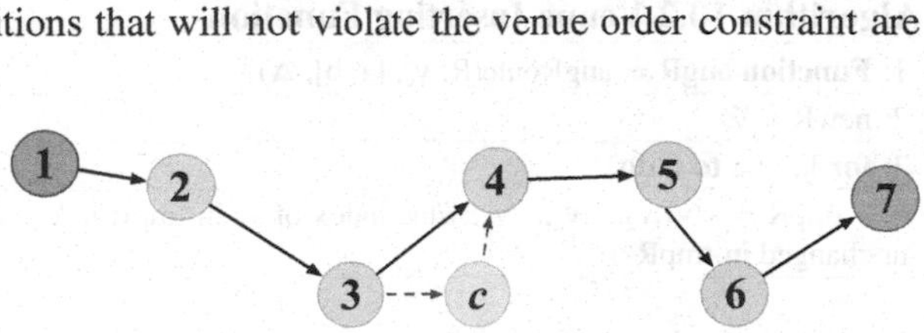
1
2
3
c
4
5
6
7

considered. However, if the candidate route does contain “earlier-arrival” venues, we only need to check $k - 1$ ($<n-1$) positions, where k is the position of the first “earlier-arrival” venue in the candidate route according to Theorem 10.1. Again, the possible positions can be further narrowed down for the case that users have venue visiting order constraint.

Theorem 10.1 For a candidate route that contains “earlier-arrival” venues, inserting a candidate user-preferred venue be-hind the first “earlier-arrival” venue could not lead to a valid route.

The pseudocode of the venue inserting algorithm is shown in Algorithm 10.1. For clarity, we assume that users do not have venue visiting order constraint on *user-preferred venues* in the scope of this section. We first check whether the candidate route contains any “earlier-arrival” venue (line 1). If it does, the possible positions where the new venue can be inserted are in [2,k]; otherwise, the range is [2,n] (lines 2–5). Note that the “wait” for a venue to open is not considered in this section the total travel time is a hard constraint in our case. The core function of Algorithm 10.1 is the *augRoute* function shown in Algorithm 10.2. In this function, the candidate venue is inserted into the given route at each possible position (lines 3–8). Note that not every position where the candidate venue is inserted can lead to a valid route (lines 5–7). If no augmented routes are valid or the total travel time cost of all the generated augmented routes exceeds the time budget, the function returns the original input route (lines 9–11); otherwise, it returns the augmented route with the minimum total travel time (lines 12–13).

Algorithm 10.1 Venue Inserting Algorithm

Input: A candidate route $R = <v_1,v_2,...,v_n>$;
A candidate venue v_c
A user-specified total travel time budget Δ;
Output: An augmented route augR
1: **if** R has “earlier-arrival” venues **then**
2: k = pos(R)//pos(R) gets the index of the first “earlier- arrival” venue in R
3: augR = augRoute(R, v_c, [2,k], Δ)
4: **else**
5: augR = augRoute(R, v_c, [2,n], Δ)
6: **end if**

Algorithm 10.2 Venue Inserting Function

1: **Function** augR = augRoute(R, v_c, [a, b], Δ)
2: newR = $\varnothing$
3: **for** k: = a **to** b **do**
4: tmpR = $<v_1,v_2,...,v_c,...,v_n>$//the index of v_c in tmpR is k, and the venue orders in R kept unchanged in tmpR

(continued)

if tmpR is valid **then**
  newR = newR ∪ tmpR
end if
end for
if newR is empty or min[TC(newR)] > Δ **then**
  augR = R
else
  augR =argminTC(newR)//select the newly augmented route with the minimal total travel time cost
end if

The algorithms above illustrate how to insert one venue to a candidate route. If there are multiple venues to add, this process will iterate through the list, again following the pro-posed principles. In the rest of this section, we use the expression R + {v_{c1},v_{c2}, . . .,v_{cn}} to denote the operation of inserting the venue list {v_{c1}, v_{c2}, . . .,v_{cn}} to the candidate route R sequentially. Note that, for the same set of candidate venues, different inserting orders may result in different augmented routes (e.g., $R + \{v_{c1}, v_{c2}\} \neq R + \{v_{c2}, v_{c1}\}$).

2. *Route Score Maximization Algorithms*: We first present mathematical formulation of our route score maximization algorithms and then introduce how to compute the route score according to the user's preferences. In the end, we propose three heuristic algorithms to maximize the route score.

   Mathematical Formulation: For a given user u_i, a set of candidate venues $\{v_{ci}\}_{i=1}^{N}$, and a candidate route R, the route score maximization problem is

$$max\ \mathcal{RS}\left(u_i, R + \{x_i v_{ci}\}_{i=1}^{N}\right) \tag{10.2}$$

   Subject to:

$$x_i \in \{0, 1\} \tag{10.3}$$

$$x_1 v_{c1} \cdot \text{cat} \cup x_2 v_{c2} \cdot cat \cup \ldots \cup x_n v_{cN} \cdot cat \subseteq CAT_u \tag{10.4}$$

$$TC\left(R + \{x_i v_{ci}\}_{i=1}^{N}\right) \leq \Delta \tag{10.5}$$

   where Eq. (10.2) refers to the objective function (i.e., the route score) for maximization. It is subjected to three constraints, as shown in Eqs. (10.3)–(10.5). Eq. (10.4) defines the constraint for the augmented venue selection, i.e., only the user-preferred venues can be selected for route augmentation but not necessarily covering all venue categories due to the total travel time constraint. Eq. (10.5) emphasizes that the total time cost of the newly augmented route should be within the predefined travel time budget Δ.

   Route Score Calculation: The route score calculation algorithm is the core of the route augmentation component, which estimates the attractiveness of a candidate route to a particular user. The *route score is defined as the sum of all its venue scores*, and thus, the venue scoring method is vital.

Venue Scoring: On one hand, the score of a venue is determined by its popularity [*Pop*, as shown in (Eq. 10.1)], which is objective (denoted as VS_{obj}). On the other hand, the venue score is also related to individual user's personal interests revealed in his/her check-in history, which is subjective. For instance, the scores of "Art & Museum" venues should be higher for a user if he/she visits venues in this category more often than the others, as shown in the Foursquare check-in records. The normalized check-in preference value VS_{sub} of the venue v_i for user u_j is calculated by (Eq. 10.6). For simplicity, only the level-1 category labels (i.e., the nine category labels defined by Foursquare) are used in the scope of this section.

$$\mathcal{VS}_{\text{sub}}\left(u_j, v_i\right) = \frac{\text{tcs}\left(u_j, \{v_i \cdot \text{cat}\}\right)}{\text{tcs}\left(u_j\right)} \tag{10.6}$$

where $tcs(u_j)$ represents the total number of check-ins that the user u_j conducted in Foursquare, whereas $tcs(u_j,\{v_i.cat\})$ stands for the total number of check-ins at venues belonging to the same category v_i.

Finally, the venue score can be computed according to (Eq. 10.7), considering both the venue popularity and the user preferences.

$$\mathcal{VS}\left(u_j, v_i\right) = \mathcal{VS}_{\text{obj}}(v_i) + \mathcal{VS}_{\text{sub}}\left(u_j, v_i\right) \tag{10.7}$$

Three Heuristic Algorithms: As we mentioned previously, the operator of route augmentation includes two parts, i.e., *selecting* new venues and *inserting* them into the candidate routes sequentially. It is an important thing that we should consider the individual venue scores and the total number of venues that can be added at the same time. For instance, the time budget will be exhausted if a faraway venue which has a high venue score is added. In this case, the more new venues would be forbidden in the subsequent process. In contrast, if we insert a close-by venue with an average venue score first, it would allow including more new venues. It is hard to say which method would return a route with a higher score in the end. Therefore, three heuristic algorithms which aim to maximize the route score are proposed in the route augmentation phase. Note that added venues are all *user-preferred* venues.

Travel time minimizer: The basic idea of this algorithm is to insert as many new venues as possible, given the fact that the route score would be higher as the number of venues increases in general. Thus, at each venue inserting iteration, our proposed heuristic is that the venue closest to the candidate route (measured by the additional travel time) would be selected first for insertion, *regardless of its venue score*.

Venue score maximizer: The basic idea of this algorithm is to prioritize high-scored venues. Thus, in each iteration, the venue with the highest venue score that can lead to a valid route would be inserted first *no matter how far away it is from the candidate route*.

The above two algorithms are used as baseline methods. The first heuristic algorithm only considers the number of the venues added, whereas the second one emphasizes merely on the scores of the inserted venues. As a result, the routes of the first algorithm would be generally longer (i.e., containing more venues), compared with the second algorithm. It is because the second heuristic algorithm, given the same time budget constraint, favors having one venue with a high venue score over two nearby average venues, although the latter case might lead to a higher route score. To overcome the limitations of these two baseline methods, we propose our gravity maximizer.

Gravity maximizer: Inspired by *Newton's law of universal gravitation*, which is capable of modeling human mobility patterns (the travel behaviors to places, travel patterns, etc.), we introduce a *gravity model* that uses the venue scores and the venue distances to the candidate route collectively for route augmentation. In our gravity model, the *spherical distance* between the candidate route and the new venue is analogous with the distance defined in *Newton's gravity model*, where the location of the candidate route is obtained by averaging the locations of all venues that it contains. Likewise, the average venue score of the candidate route and the score of new venue correspond to the *mass*. Finally, the gravity can be computed using

$$G(v_{ci}, R) = \frac{\mathcal{VS}(u_j, v_{ci}) \times \frac{1}{n} \sum_{i=1}^{n} \mathcal{VS}(u_j, v_i)}{\text{dist}(v_{ci}, R)^{\lambda}} \tag{10.8}$$

In the proposed gravity maximizer, the new venues are sorted in the descending order of their gravity values computed via (8), instead of the venue scores. The rest of the procedure is exactly the same as that of the *venue score maximizer*. Thus, the two methods are similar in the computation complexity, with an extra cost of the venue's gravity computation in the *gravity maximizer*.

In fact, the ranking based on gravity values would be degraded to that of venue scores if we set $\lambda = 0$, as gravity values would be determined by venue scores only. In other words, the gravity maximizer and the venue score maximizer algorithms would reach the same result when $\lambda = 0$. On the contrary, as can be inferred from (Eq. 10.8), if we set λ to be extremely high (e.g., $\lambda > 5$), the gravity values would be dominantly influenced by the distance to the candidate route, introducing a bias toward the "closest" venue (i.e., with the smallest distance to the candidate route). This agrees with the basic idea of the travel time minimizer algorithm. Furthermore, with a large negative λ (e.g., $\lambda < -5$), "distant" venues would be ranked higher, which should be avoided.

3. *Augmented Route Ranking*: The algorithms discussed in Sect. 10.5 aim to optimally augment the set of candidate routes returned from Phase I (i.e., route search). *Augmented route ranking* operation then picks out the augmented route with the highest route score to answer the user's itinerary query (IQ). Note that if multiple "optimally" augmented routes possess the same route score as they may

contain the same venues but in different order, the route with a smaller "total travel time" would be ranked higher.

10.6 Evaluations

Here, we present the evaluation results that aim to validate the efficiency and effectiveness of the trip planning algorithms. We first describe the experimental setup, results of the parameter sensitivity study, and evaluation on algorithm efficiency and effectiveness.

10.6.1 Experimental Setup

Data Preparation: We used Foursquare check-in data of San Francisco from April 2010 to October 2010 and the taxi GPS traces of the same city from the CabSpotting project (http://cabspotting.org) to construct the POI network of San Francisco. The Foursquare data contain 110,214 check-ins generated by 15,680 users. The taxi GPS data contain 391,938 passenger-delivery trips generated by 536 taxis in June 2008.

Evaluation Environment: All the evaluations in this section are run in MATLAB on an Intel Xeon W3500 PC with 12-GB RAM and running Windows 7 operation system.

10.6.2 Parameter Sensitivity Study

We have only one internal parameter λ in the proposed *gravity maximizer* algorithm [1] and no internal parameter in the other two baselines. We are thus interested in how it affects the optimal route score. We do not set λ to extreme values as discussed; instead, we vary λ in the range of $[-3, 3]$ with an interval of 0.1. The optimal scores under different λ values, in comparison with the two baseline algorithms, are shown in Fig. 10.5a. As the figure suggests, the optimal route score generated by the travel time minimizer algorithm is always the lowest since it does not take the individual score of candidate venues into consideration. As expected, the optimal route score computed by the gravity maximizer algorithm and the venue score maximizer algorithm are the same when λ is around 0. We also find that the gravity maximizer algorithm yields higher optimal route score than the venue score maximizer algorithm when λ is within the range of [0.5 2.3].

We also show the change in computation time of the gravity maximizer algorithm under different λ values in Fig. 10.5b. More specifically, the computation time fluctuates with the increase in λ. However, the maximum time cost is no longer

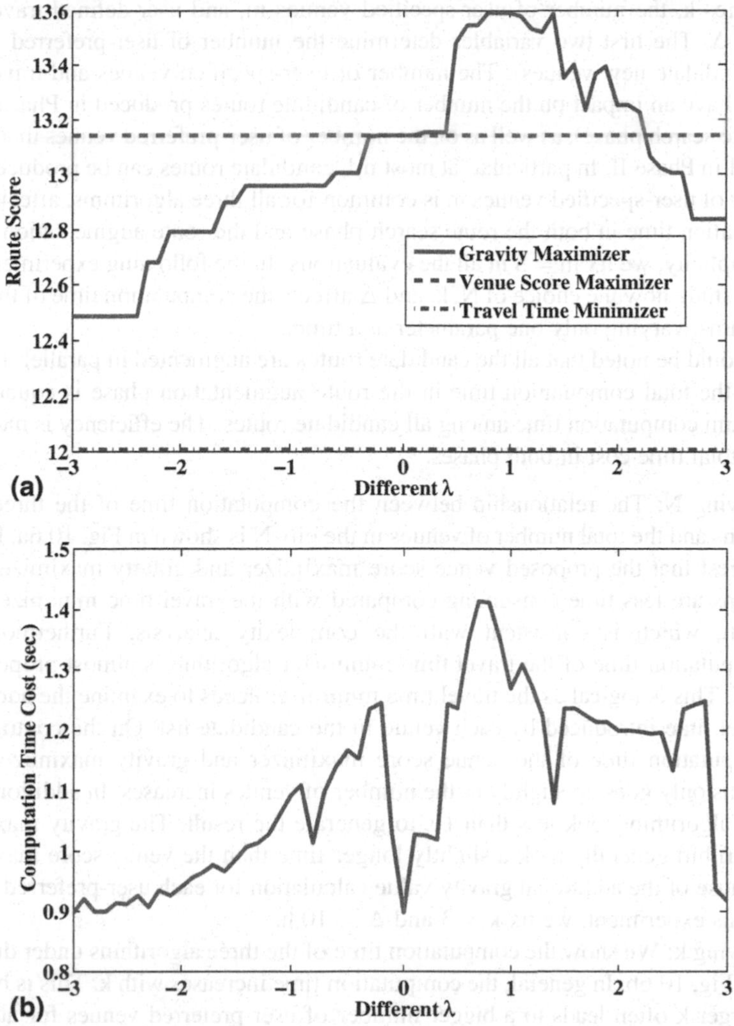

13.6
13.4
13.2
13
12.8
12.6
12.4
12.2
12
Route Score
Gravity Maximizer
Venue Score Maximizer
Travel Time Minimizer
−3
−2
−1
0
1
2
3
(a)
Different λ
1.5
1.4
1.3
1.2
1.1
1
0.9
0.8
Computation Time Cost (sec.)
(b)
Different λ

Fig. 10.5 Results of parameter sensitivity study. **(a)** Optimal route scores under different λ. **(b)** Computation time cost under different λ

than 1.45 s, which is acceptable. Considering the tradeoff between route score and computation time, we choose $\lambda = 1.5$ for the rest of the evaluations.

10.6.3 Efficiency Evaluation

The efficiency of the three algorithms depends on several parameters such as the total number of venues N in the targeted city, the number of user-preferred venue categories k, the number of user-specified venues m, and user-defined travel time budget Δ. The first two variables determine the number of user-preferred venues (i.e., candidate new venues). The number of user-specified venues and travel time budget have an impact on the number of candidate routes produced in Phase I (i.e., the route search phase), as well as on the number of user-preferred venues that can be inserted in Phase II. In particular, at most m!, candidate routes can be produced. The number of user-specified venues m is common for all three algorithms, affecting the computation time in both the route search phase and the route augmentation phase. For simplicity, we fix $m = 5$ in all the evaluations. In the following experiments, we mainly study how the choice of N, k, and Δ affects the computation time of the three algorithms, varying only one parameter at a time.

It should be noted that all the candidate routes are augmented in parallel. In other words, the total computation time in the route augmentation phase is equal to the maximum computation time among all candidate routes. The efficiency is measured by the total time cost in both phases.

1. Varying N: The relationship between the computation time of the three algorithms and the total number of venues in the city N is shown in Fig. 10.6a. Results suggest that the proposed venue score maximizer and gravity maximizer algorithms are less time consuming compared with the travel time minimizer algorithm, which is consistent with the complexity analysis. Furthermore, the computation time of the travel time minimizer algorithm is almost proportional to N. This is logical as the travel time minimizer needs to examine the additional travel time introduced by each venue in the candidate list. On the contrary, the computation time of the venue score maximizer and gravity maximizer algorithms only goes up slightly as the number of venues increases. In addition, these two algorithms took less than 1 s to generate the result. The gravity maximizer algorithm generally took a slightly longer time than the venue score maximizer because of the additional gravity value calculation for each user-preferred venue. In this experiment, we fix $k = 3$ and $\Delta = 10$ h.
2. Varying k: We show the computation time of the three algorithms under different k in Fig. 10.6b. In general, the computation time increases with k. This is because a larger k often leads to a bigger number of user-preferred venues for augmentation. Again, the computation time of the travel time minimizer algorithm is much longer than that of the other two algorithms under the same setting, due to the same reason as when N varies. For the venue score maximizer and gravity maximizer algorithms, their computation time increases more significantly as *k* becomes bigger, as compared with that under different N. This is indeed caused by the increase in the number of user-preferred venues. As N increases, both the number of user-preferred and non-user-preferred venues would increase. However, all non-user-preferred venues can be excluded from the route augmentation process and thus have no impact on the computation time. In contrast, any change in k would be completely and directly reflected on the change in the number of user-preferred venues. In this experiment, we fix $N = 300$ and $\Delta = 8.5$ h.

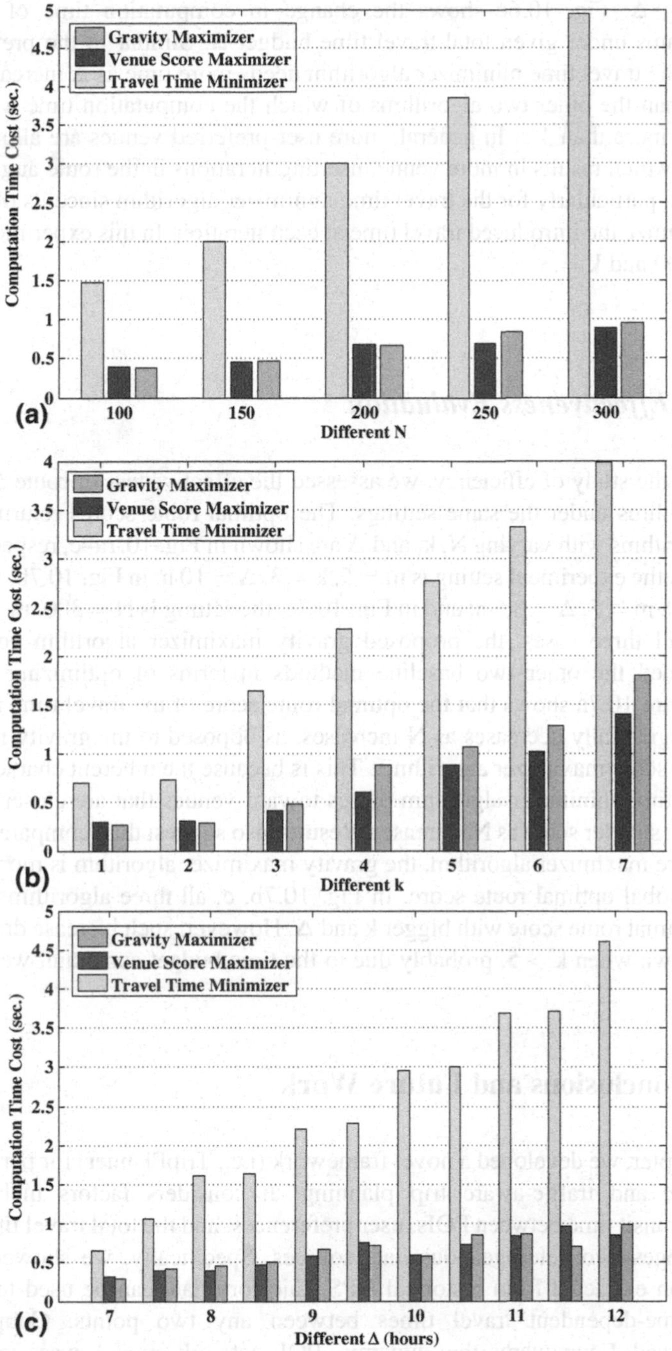
Gravity Maximizer
Venue Score Maximizer
Travel Time Minimizer
Computation Time Cost (sec.)
(a)
100
150
200
250
300
Different N
(b)
1
2
3
4
5
6
7
Different k
(c)
7
8
9
10
11
12
Different Δ (hours)

Fig. 10.6 Results of efficiency evaluation. **(a)** Computation time cost by varying N. **(b)** Computation time cost by varying k. **(c)** Computation time cost by varying Δ

3. Varying Δ: Fig. 10.6c shows the change in computation time of the three algorithms under given total travel time budget Δ. Similar to the previous two cases, the travel time minimizer algorithm needs more time as Δ increases, much more than the other two algorithms of which the computation time was similar and no more than 1 s. In general, more user-preferred venues are allowed to be added, which results in more venue inserting iterations in the route augmentation process, particularly for the travel time minimizer algorithm since its objective is to minimize the introduced travel time at each iteration. In this experiment, we fix $\mathrm{N} = 300$ and $\mathrm{k} = 3$.

10.6.4 Effectiveness Evaluation

Similar to the study of efficiency, we assessed the effectiveness of route augmentation algorithms under the same settings. The optimal route scores returned by the three algorithms with varying N, k, and Δ are shown in Fig. 10.7a–c, respectively. In Fig. 10.7a, the experiment setting is $\mathrm{m} = 5$, $\mathrm{k} = 3$, $\Delta = 10$ h; in Fig. 10.7b, the setting is $\mathrm{N} = 300$, $\mathrm{m} = 5$, $\Delta = 8.5$ h; and in Fig. 10.7c, the setting is $\mathrm{N} = 300$, $\mathrm{m} = 5$, and $\mathrm{k} = 3$. In all three cases, the proposed gravity maximizer algorithm consistently outperformed the other two baseline methods in terms of optimizing the route score. Figure 10.7a shows that the optimal route score of the travel time minimizer algorithm gradually decreases as N increases, as opposed to the gravity maximizer and venue score maximizer algorithms. This is because the inherent characteristic of the travel time minimizer algorithm biases toward venues that are closer but probably with a smaller score as N increases. Results also suggest that, compared with the venue score maximizer algorithm, the gravity maximizer algorithm is more likely to find the global optimal route score. In Fig. 10.7b, c, all three algorithms achieved higher optimal route score with bigger k and Δ. However, such increase dramatically slowed down when $\mathrm{k} > 5$, probably due to the time budget constraint we impose.

10.7 Conclusions and Future Work

In this chapter, we developed a novel framework (i.e., TripPlanner) for personalized, interactive, and traffic-aware trip planning. It considers factors including the dynamic transit time between POIs, user preferences, and the total travel time budget and leverages two heterogeneous data sources. Specifically, we showed that the information extracted from historical GPS trajectory data can be used to infer the typical time-dependent travel times between any two points. Coupled with crowdsourced Foursquare, the dynamic POI network model was constructed. Then, a two-phase approach was proposed for personalized trip planning which had a comprehensive route scoring method and a novel route search-augmentation-

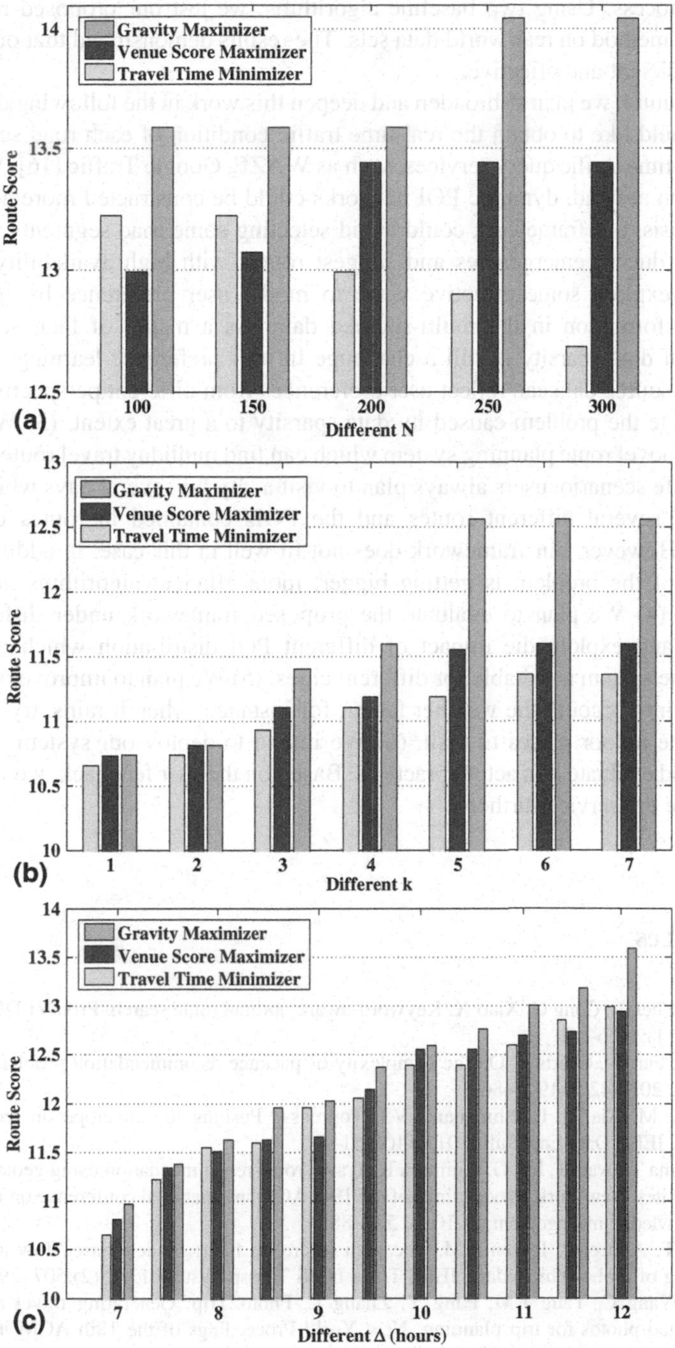
Gravity Maximizer
Venue Score Maximizer
Travel Time Minimizer
Route Score
14
13.5
13
12.5
100
150
200
250
300
(a)
Different N
Gravity Maximizer
Venue Score Maximizer
Travel Time Minimizer
Route Score
13
12.5
12
11.5
11
10.5
10
1
2
3
4
5
6
7
(b)
Different k
Gravity Maximizer
Venue Score Maximizer
Travel Time Minimizer
Route Score
14
13.5
13
12.5
12
11.5
11
10.5
10
7
8
9
10
11
12
(c)
Different Δ (hours)

Fig. 10.7 Results of effectiveness evaluation. (**a**) Optimal route score by varying N. (**b**) Optimal route score by varying k. (**c**) Optimal route score by varying Δ ranking process. Using two baseline algorithms, we test our proposed route augmentation method on real-world data sets. The results demonstrated that our method is more efficient and effective.

In the future, we plan to broaden and deepen this work in the following directions: (1) we would like to obtain the real-time traffic condition of each road segment by some real-time traffic query services, such as WAZE, Google Traffic [16]. With such information at hand, dynamic POI networks could be constructed more accurately. On this basis, our framework could avoid selecting some road segments which are congested due to emergencies and suggest routes with high availability. (2) We intend to explore some effective ways to model user preference by integrating relevant information in the multi-sourced data. As a matter of fact, solving the problem of data sparsity is still a challenge in user preference learning. We think that multi-source data can reflect user preferences from different perspectives which can alleviate the problem caused by data sparsity to a great extent. (3) We plan to propose a novel route planning system which can find multiday travel routes for user. In a real-life scenario, users always plan to visit a city for several days which means they need several different routes and the POIs contained in routes cannot be repeated. However, our framework does not fit well in this case. In addition, since the scale of the problem is getting bigger, more efficient algorithms need to be proposed. (4) We plan to evaluate the proposed framework under different POI networks and explore the impact of different POI distribution which can verify whether the system is suitable for different cities. (5) We plan to improve our system by taking into account the weather factor, for instance, when it rains, try to recommend some indoor places to visit. (6) We intend to deploy our system on mobile devices and evaluate it in actual practices. Based on the user feedback, we would like to improve the service further.

References

1. Cao X, Chen L, Cong G, Xiao X. Keyword-aware optimal route search. Proc VLDB Endowm. 2012;5(11):1136–47.
2. Deng T, Fan W, Geerts F. On the complexity of package recommendation problems. SIAM J Comput. 2013;42(5):1940–86.
3. Khabbaz M, Xie M, Lakshmanan LVS. Toprecs+: Pushing the envelope on recommender systems. IEEE Data Eng Bull. 2011;34(2):61–8.
4. Kurashima T, Iwata T, Irie G, Fujimura K. Travel route recommendation using geotags in photo sharing sites. New York: Proceedings of the 19th ACM international conference on Information and knowledge management; 2010. p. 579–88.
5. Hunter T, Abbeel P, Bayen AM. The path inference filter: model- based low-latency map matching of probe vehicle data. IEEE Trans Intell Transp Syst. 2013;15(2):507–29.
6. Lu X, Wang C, Yang J-M, Pang Y, Zhang L. Photo2Trip: Generating travel routes from geo-tagged photos for trip planning. New York: Proceedings of the 18th ACM international conference on Multimedia; 2010. p. 143–52.

7. Liu B, Fu Y, Yao Z, Xiong H. Learning geographical preferences for point-of-interest recommendation. New York: Proceedings of the 19th ACM SIGKDD international Conference on Knowledge Discovery and Data Mining; 2013. p. 1043–51.
8. Lu EH-C, Chen C-Y, Tseng VS. Personalized trip recommendation with multiple constraints by mining user check-in behaviors. New York: Proceedings of the 20th International Conference on Advances in Geographic Information Systems; 2012. p. 209–18.
9. Souffriau W, Vansteenwegen P. Tourist trip planning functionalities: State-of-the-art and future. Proc Current Trends Web Eng. 2010;2010:474–85.
10. Abdelrahman A, El-Wakeel AS, Noureldin A, et al. Crowdsensing-based personalized dynamic route planning for smart vehicles. IEEE Netw. 2020;34(3):216–23.
11. Bock JD, Verstockt S. SmarterROUTES—A data-driven context-aware solution for personalized dynamic routing and navigation. ACM Trans Spatial Algorithm Syst. 2020;7(1):1–25.
12. Jeong MG, Lee EB, Lee M, et al. Multi-criteria route planning with risk contour map for smart navigation. Ocean Eng. 2019;172:72–85.
13. Hsieh H-P, Li C-T, Lin S-D. Exploiting large-scale check-in data to recommend time-sensitive routes. New York: Proceedings of the ACM SIGKDD International Workshop on Urban Computing; 2012. p. 55–62.
14. Yuan J, Zheng Y, Xie X, Sun G. T-drive: Enhancing driving directions with taxi drivers' intelligence. IEEE Trans Knowl Data Eng. 2011;25(1):220–32.
15. Ziebart BD, Maas AL, Dey AK, Bagnell JA. Navigate like a cabbie: probabilistic reasoning from observed context-aware behavior. New York: Proceedings of the 10th international conference on Ubiquitous computing; 2008. p. 322–31.
16. Li Y, Yiu ML. Route-saver: leveraging route apis for accurate and efficient query processing at location-based services. IEEE Trans Knowl Data Eng. 2014;27(1):235–49.

Chapter 11
ScenicPlanner: Recommending the Most Beautiful Driving Routes

11.1 Introduction

Most current automatic trip planning systems recommend either a single POI or a sequence of POIs, neglecting the detailed travel route planning issue between two suggested consecutive POIs. Although the travel routes with the shortest travel distance or time can be obtained from available online map services or commercial GPS navigators, it is still difficult to meet the diverse user requirements in many cases [1]. For instance, the scenic view along the travel routes would be important when users intend to travel by driving for leisure purposes. In this case, the sceneries along the route should be in a high priority [2]. Thus, in addition to the shortest distance (or time), the quality of scenic view along the suggested routes is taken into account in this chapter. We aim to find the travel routes which minimize distance and offer high quality of sceneries along the way as well.

Intuitively, with the objective of planning an optimal travel route from one POI to another one considering multiple factors, i.e., the travel distance (or time) and quality of the scenery, we can assign a score for each factor and aggregate them into a combined score by a weighting score function based on the user preference. Then, we can find the travel route composed of edges with the highest total combined score satisfying the travel distance budget, which can be viewed as an arc orienteering problems (AOP) [3, 4]. Unfortunately, the solution is problematic and may be invalid in practice. This is because: (1) the weight of each factor should vary depending on users which has been proved to be nontrivial [5, 6]. (2) A budget limitation has to be imposed by users, and only the optimal result within the defined

---

Part of this chapter is based on a previous work: C. Chen, X. Chen, L. Wang, X. Ma, Z. Wang, K. Liu, B. Guo and Z. Zhou, "MA-SSR: A Memetic Algorithm for Skyline Scenic Routes Planning Leveraging Heterogeneous User-Generated Digital Footprints," in IEEE Transactions on Vehicular Technology, vol. 66, no. 7, pp. 5723–5736, July 2017, doi: https://doi.org/10.1109/TVT.2016.2639550.

C. Chen et al., *Enabling Smart Urban Services with GPS Trajectory Data*,
https://doi.org/10.1007/978-981-16-0178-1_11

budget can be returned. In this process, if the restriction is not appropriately set or the user intends to set another new limit, the algorithm needs to be invoked repeatedly which degrades the system usability. In order to solve the problem of factor weighting, the skyline operator is used to rank all the possible travel routes instead, which is able to find all routes being optimal with respect to an arbitrary linear weighting of the underlying criteria. In addition, we can obtain a set of optimal travel routes with diverse travel distances by a single-system run. The one that satisfies the user most can be chosen. By this way, the route planning system would become more usable and efficient. To ensure the feasibility of the proposed idea, there are two major research challenges that need to be addressed in the following.

1. *How to model the scenic road network effectively?* The objective of the scenic road network modeling is to assign the beautifulness score for each road segment (i.e., edge enrichment) accurately, based on its nearby scenic environment [2]. However, this is a challenging task because of the following two aspects: (1) the term of "nearby" is based on personal cognition, which is, thus, a qualitative term and difficult to be mathematically measurable. For instance, due to a good visibility, POIs to a road segment with a distance value of 50 m can be still near sometimes, while POIs with a distance value of 30 m cannot be near sometimes due to occlusions; (2) the term of "scenic environment" is subjective, which might relate to many factors such as the edge's popularity, the user's preference and the user's visiting time.
2. *How to identify and discover the skyline scenic routes (SSRs) efficiently?* This problem focuses on the methods of selecting road segments and determining the order of traveling. In addition, the total travel distance for the same set of selected road segments would be different under different traveling orders. There is a more challenging problem, i.e., traversing a given road segment in two directions would result in different total travel distances as well. Last but not least, it is nontrivial to identify an SSR due to that it is defined according to a set of travel routes; we need to generate a set of travel routes and evaluate them. Furthermore, it is hard to find an efficient way to discover a set of SSRs with different travel distances simultaneously.

Considering the above-mentioned research objective and challenges, there are three main contributions of the chapter as shown in the following.

1. We propose a novel two-phase framework, i.e., a memetic algorithm for skyline scenic routes planning (MA-SSR). It has two functional components (i.e., the scenic road network modeling and the scenic route planning) to suggest travel routes for the given origin and destination in a city. The objective is finding a set of SSRs, which can make a tradeoff between the travel distance and the quality of the scenic view for user selection.
2. We first obtain the road network from the crowd-sourcing platform, i.e., OpenStreetMap (OSM). For each edge on it, we assign a scenic score in a comprehensive way by using the complementary information which is provided

by a combination of geotagged images and check-ins from two LBSNs platforms (i.e., Flickr and Foursquare), respectively.

3. A memetic algorithm (MA) is used in our route planning to find SSRs. We conduct extensive experiments to validate its effectiveness and efficiency bases on the road network of the Bay Area of the city of San Francisco (SFC), CA, USA, which contains more than 31,000 geotagged images generated by 1571 Flickr users in a year, and 110,214 check-ins left by 15,680 Foursquare users in 6 months. Compared with two baseline approaches, the MA can obtain high-quality solutions which are reasonably close to the optima but within desirable computation time. Moreover, the results are prominently better than the solutions obtained by genetic algorithms (GAs).

11.2 Related Work

11.2.1 Geotagged Image and Check-in Data Mining

In recent years, geotagged images and check-in data have been mined to support a variety of applications, which have attracted extensive attention of researchers. To give a few examples, how to manage them effectively has become a key issue as geotagged image data are rapidly accumulating; thus, for a better accessible, a framework of automatically select a summary set of photos is proposed in [7]. Due to that geotagged images taken by different users at different locations may involve a same landmark, a lot of researches have been carried out to identify and classify landmarks by data mining algorithm [8]. These geotagged image data are the result of group sharing photos on social media sites, and the wisdom of the crowed is behind it [9]; therefore, knowledge and patterns of our human society can be discovered, such as frequent associated POI sequences suggesting, (personalized) landmark recommendation/search, the heatmap of landmark popularity at different time under-standing [8]. Along the same lines, researchers extracted the knowledge from check-in data which is used to support similar applications [10]. Compared to geotagged image, there is rich and explicit attribute information about the POIs in check-in data. To the best of our knowledge, we first propose to quantify the beautifulness for each road segment comprehensively based on the complementary information in two heterogeneous user-generated digital footprints.

11.2.2 Travel Route Planning

A considerable amount of work has been done in the area of planning travel routes from point to point with nonstop in a city. Many commercial navigators and online map services focus on planning the travel routes with the shortest distance or time which are the most common objective. Moreover, the time-dependent shortest time travel route is also one of the common objectives since the travel time on a road segment is affected by the traffic condition which is highly dynamic at different times of the day or under different weather conditions. Considering the real-time traffic information, researchers have also done some work, such as avoiding roads with high congestion. In addition to the objectives of the shortest distance or time, there are some new objectives in route planning have been recently appeared in the literature [11–13]. To number a few, the healthy routes with the objective of maximizing the physical activity are recommended for travelers in [14]. Quercia et al. [15] considered the recommendation of "happy" and "pleasant" travel routes to pedestrians. Based on the sentiment of geotagged image data, Kim et al. [16] suggested the friendlier, more enjoyable, and potentially safer routes. Similarly, Galbrun et al. [17] made use of crime data which can be openly available to recommend travel routes according to multiple criteria, such as the distance and the safety. In addition, some constraints are often considered, such as the maximum travel time and bypassing some given POIs (with some specific attributes). Similar to our work, Zheng et al. [18] presented a GPSViewer system to suggest a driving route with scenery and sightseeing qualities, in which the scenic score is assigned according the information in geotagged image data. The weight of edges in the road network can be negative and it is a weighted summary of the travel distance and scenic view. The shortest path-finding algorithm [18] which can deal with the negative weights in network may fail to provide a valid travel route to users due to the problem of negative loops, i.e., all weights in the loop are negative. In addition, the method proposed in [18] cannot provide a controllable driving distance to users for the travel route selection. Compared to GPSView system, our proposed MA-SSR have two main improvements in the following: (1) we proposed a more reasonable and accurate way to model the scenic score for each edge on the road network, specifically, the information contained in Flickr geotagged image data and Foursquare check-in data was fully extracted and utilized and (2) we found a set of SSRs by the MA; then, according the distance and the scenic view along, user can select a route that satisfies them most.

11.3 Preliminary

11.3.1 Basic Concepts

Definition 11.1 (A Travel Route): A travel route TR is a sequence of nodes (n_1,n_2, . . .,n_i, . . .,n_k), in which between any two consecutive nodes <n_i, n_{i+1}>, there exists a road segment (i.e., an edge) e (n_i, n_{i+1}) in the road network. Alternatively, a travel route can be also defined as a sequence of road segments, in which between any two consecutive road segments, they share a unique node.

Definition 11.2 (A Geotagged Image): A geotagged image is defined as a quadruple gim = (u_{id},x_i,y_i,t_i), showing a user with id u_{id} took an image at location (x_i,y_i) at time t_i using Flickr.

Definition 11.3 (A Check-in): A check-in is defined as a triple ck = (u_{id},v_{id},t_i), showing a user with id u_{id} checked-in a venue (i.e., POI) with id vid at time t_i using Foursquare. In addition, Foursquare provides the physical coordinates, tags, and the category information about an any given venue.

Definition 11.4 (Scenic Score of a Road Segment): Given a road segment e(i, j), function S(e(i, j)) → R+ assigns a score to reflect its quality of scenic view (i.e., beautifulness). The higher the score, a more beautiful scene it is. A road segment with score greater than 0 is also called scenic road segment.

Definition 11.5 (Scenic Score of a Travel Route): The scenic score of a travel route TR can be defined as follows:

$$S(\text{TR}) = \frac{\sum_{i=1}^{m} S(e_i)}{1+\theta}; \theta = \sum_{i=1}^{m} (\text{freq}_i - 1) \times S(e_i)$$

where m is the number of edges in the travel route, θ is a penalty factor for the purpose of reducing the total scenic score if including repeated road segments in the travel route, and freq_i is the times of appearance of the edge e_i. Obviously, θ = 0if the travel route does not have repeated edge or the scenic score of the repeated edge is equal to 0.

Definition 11.6 (SSR): Given a set of travel routes, in which each travel route has two metrics, i.e., the travel distance (dist) and the scenic score (S), respectively, we say a travel route (TR_i) dominates another travel route (TR_j) iif 1) dist(TR_i) ≤ dist (TR_j) and 2) S(TR_i) > S(TR_j). The travel route that is not dominated by others in the set is defined as an SSR.

Definition 11.7 (A Travel Route Query): A travel route query (TRQ) consists of three parts: (1) a user-specified starting point, p_o = (x_o,y_o); (2) a user-specified ending point, p_d = (x_d,y_d); and (3) a travel distance budget for the targeted travel route $dist_{bgt}$. It should be noted that the third part is optional. The system would return only one SSR satisfying the budget constraint if the user imposed an additional budget limitation; otherwise, all SSRs with different travel distances will be returned.

11.3.2 Problem Formulation

The problem of planning SSRs can be formulated as follows:
Given:

1. a user's TRQ;
2. a collection of geoimages and check-ins from the targeted city;
3. a road network G(N, E) of the targeted city;

model the scenic road network by designing a proper function to score each road segment leveraging the collection of geotagged images and check-ins and, then, find the SSRs, while satisfying the user's TRQ.

11.4 The Two-Phase Approach

11.4.1 System Overview

As shown in Fig. 11.1, the proposed MA-SSR framework contains two phases, i.e., scenic road network modeling and MA for SSRs planning, respectively. The input of Phase I is the geo-tagged image and the check-in data and raw road network data. On top of the raw road network data, besides the property of distance, each edge is enriched by assigning a scenic score to it, leveraging the information hidden behind the two heterogeneous user-generated digital footprints. Phase I is prebuilt and maintained offline. Phase II mainly consists of four modulars, i.e., region of interest (RoI) determination, chromosome encoding and decoding, chromosome partition and crossover, and local improvement-based mutation. Phase II is invoked to work online once receiving the TRQ from the user. Before returning the SSRs to the user, Phase II performs the last two procedures iteratively until meeting the stopping criteria, as shown in the rightmost part of Fig. 11.1. In the next two sections, we elaborate on the offline modeling of the scenic road network and the online SSR planning process sequentially.

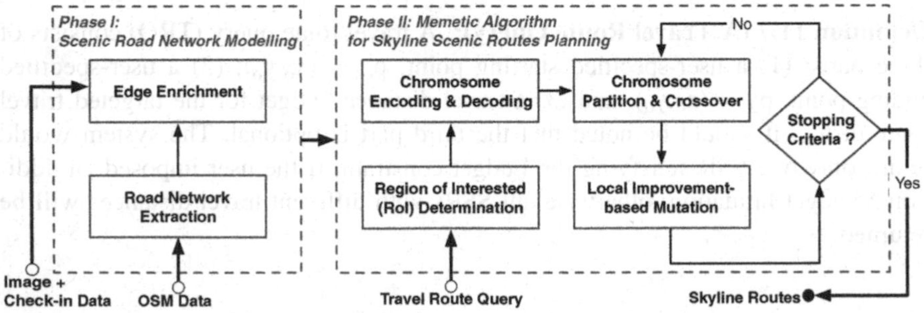
Phase I:
Scenic Road Network Modelling
Edge Enrichment
Road Network Extraction
Image + Check-in Data
OSM Data
Phase II: Memetic Algorithm for Skyline Scenic Routes Planning
Chromosome Encoding & Decoding
Chromosome Partition & Crossover
Stopping Criteria ?
No
Yes
Region of Interested (RoI) Determination
Local Improvement-based Mutation
Travel Route Query
Skyline Routes

Fig. 11.1 Overview of the MA-SSR framework

11.4.2 Phase I: Scenic Road Network Modeling

As discussed, the key issue of modeling the scenic road net-work is to score the scenic view of each road segment (i.e., the degree of beautifulness) according to its landscape quality. Previous studies on quantifying factors contributing to the scenic beauty of routes concluded that scenic routes usually have a higher density of surrounding geotagged photos and some specific landscape features (i.e., a good visibility to POIs with some specific categories, such as the foreground river and garden) [2]. Bringing this idea in and going a step further, we leverage the complementary information provided by geotagged image and check-in data to score the scenic view for each road segment, detailed as follows.

1. *Geotagged Image Data on Scoring*: one of the important salient indicators of quality of scenic view for a road segment is a high density of surrounding geotagged image. However, a better scenic view does not entirely depend on a higher value of density. There is an important factor for reference, i.e., the dominate direction of surrounding image distribution. Specifically, when the dominated direction is consistent with the road direction, we can obtain a better sightseeing from the road. This is because that if the overall view of a road attracts people, they would probably take photos along the route, while people would take photos from different standpoints around a center location if they are attracted by a nearby landmark. As the two distributions shown in Fig. 11.2, with the same density (i.e., number of images), the road segment in the left case should have a higher score. Thus, we assign the scenic score for each road segment based on both the density and dominate direction of the geoimage data distribution, which can be computed based on the following:

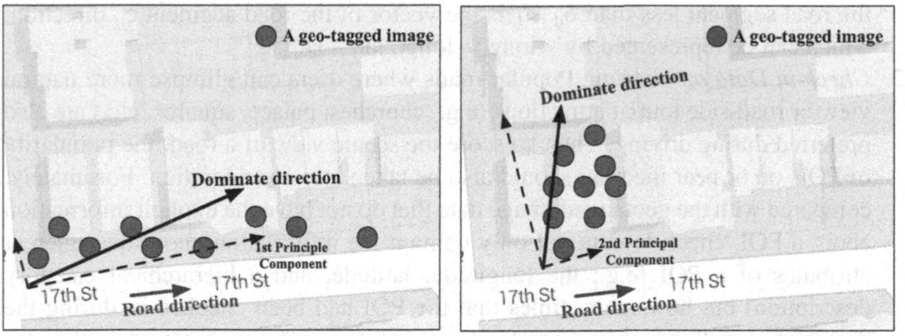
A geo-tagged image
Dominate direction
1st Principle Component
17th St
17th St
Road direction
A geo-tagged image
Dominate direction
2nd Principal Component
17th St
17th St
Road direction

Fig. 11.2 Illustrative example of two geotagged image distributions with the same density but different dominate directions

$$S_{\text{inage}}\left(e_{ij},\{\text{gim}\}\right) = w\left(e_{ij},\{\text{gim}\}\right) \times \log\left[\text{size of}\left(\{\text{gim}\delta_d\}\right)\right] \tag{11.1}$$

where $dist((x_i,y_i), e_{ij})$ computes the geodistance from point (x_i,y_i) to the road segment e_{ij}; δ_d is a predefined distance threshold when counting the number of images near the road segment. Only the geotagged images with the distance less than δ_d are counted when calculating the density to ensure the visibility. Note that the threshold should be set differently according to the road types. On one hand, the width of the highways is generally bigger than that of the residential streets. On the other hand, the visibility is varied while driving on different roads. For instance, users can still enjoy the distant sceneries while driving on highways, on the contrary, only a limited and narrow vision can be obtained due to building/tree occlusions while driving on residential streets. Hence, we set the distance threshold value to 20 m for roads with tags of "residential," "tertiary," and "secondary," and 50 m for roads with tags of "primary," "trunk," and "motorway," respectively. w is a weighting factor, which considers the road direction and the dominate direction of the image distribution, and it is calculated as

$$\left(e_{ij},\{\text{gim}\}\right) = \left|\vec{v}_1 + \vec{v}_2\right| \times \cos\frac{\left(\vec{v}_1 + \vec{v}_2\right)\cdot\vec{d}_r}{\left|\vec{v}_1 + \vec{v}_2\right|\cdot\left|\vec{d}_r\right|}$$
$$\{v_1, v_2\} = \text{PCA}\left(\{\text{gim} \mid \text{dist}\left(\text{gim}.(x_i, y_i), e_{ij} < \delta_d\right\}\right) \tag{11.2}$$

where $\vec{v_1}$ and $\vec{v_2}$ are the eigenvectors corresponding to the first and second principal components, respectively, as shown in Fig. 11.2, when applying the principal component analysis (PCA) algorithm to the image set with distance to the road segment less than δ_d. $\vec{d_r}$ is the vector of the road segment e_{ij} direction, which can be represented by $< long_j - long_i, lat_j - lat_i >$.

2. *Check-in Data on Scoring*: Popular roads where users can glimpse more natural view or road-side tourist attractions (e.g., churches, palace, squares, etc.) are also preferred during driving. Thus, to score the scenic view of a road, the popularity of POIs on or near the road should also be taken into consideration. Fortunately, compared with the geotagged image data that do not have the explicit information about a POI, check-in data not only contain the information about the inherent attributes of a POI (e.g., the longitude, latitude, and a hierarchical category description) but how many times that the POI had been checked-in during the past time as well, which is a good indicator of its popularity. Inspired by the idea that POIs with some specific categories would contribute relatively more on its scenic view, we, thus, intentionally divide the POIs into three groups according to their category labels, as shown in Table 11.1. Then, we apply the widely used term frequency–inverse document frequency method, which not only considers how frequently that the venue had been checked but also how commonly that the

Table 11.1 Divided three groups POIs

Group name	Category labels
G_1: Natural scenery	park, garden, lake, forest, mountain, beach, sea, river, bridge, harbor, scenic, hiking.
G_2: Tourist attraction	museum, palace, church, gallery, memorial, monument, square, zoo, university, historic site, square.
G_3: Others	restaurant, cafe, hotel, and etc.

group of the venue belongs to appears in the city, to compute the popularity of the venue based on check-in data, as shown in the following equation:

$$\text{popularity}(v_c) = \frac{\text{size of}(\{ckv_{id} = v_c\})}{\text{size of}(\{ckv_{id}) = G(v_c)\})} \times \log \frac{\text{size of}(\{v_{\text{id}}v_{\text{vid}}\})}{\text{size of}(\{v_{id}v_{id}) = G(v_c)\})} \tag{11.3}$$

where size of($\{ck|ck.v_{id} = v_c\}$) refers to the number of times of the given venue vc that had been checked before the time of data collection, size of($\{ck|G(ck.v_{id}) = G(v_c)\}$ refers to the number of times of venues belonging to the same group as the given venue vc that had been checked, size of($\{v_{id} | ck.v_{id}\}$) refers to the number of all unique venues in the city, and size of($\{v_{id}|G(ck.v_{id}) = G(v_c)\}$) refers to the number of venues belonging to the same group to the given venue v_c.

The scenic view of a road segment e_{ij} based on the check-in data can be computed as follows:

$$\begin{aligned} S_{\text{checkin}}(e_{ij}, \{ck\}) = \sum w(G(ck \cdot v_{id})) \\ \times \text{popularity}(\{v_{id}v_{id}, e_{ij}) < \delta_d\}) \end{aligned} \tag{11.4}$$

where dist(v_{id},e_{ij}) measures the geodistance from the venue to the road segment e_{ij}, and only venues having a distance to the road segment less than δd are used to quantify the score, which is the same to the case of geotagged image data. Popularity(v_{id}) is the popularity of the venue v_{id}, which is computed according to (Eq. 11.3). Moreover, those popularities are weighted differently based on the venue group. The venue group is roughly following the idea in [2], in which the scenic view and the surrounding POI categories of a travel route are investigated quantitatively. Results suggest that POIs belonging to the natural scenery and tourist attraction generally contribute more on the scenic view of the travel route than the remaining one. Therefore, check-ins at venues belonging to the first two groups are higher weighted. Specifically, we empirically set $w(G_1) = 0.65$,w $(G_2) = 0.3$, and $w(G_3) = 0.05$.

3. *Integration of Heterogeneous Data on Scoring*: Given the heterogeneous data of geotagged images and check-ins, the scenic view score of a road segment is integrated based on

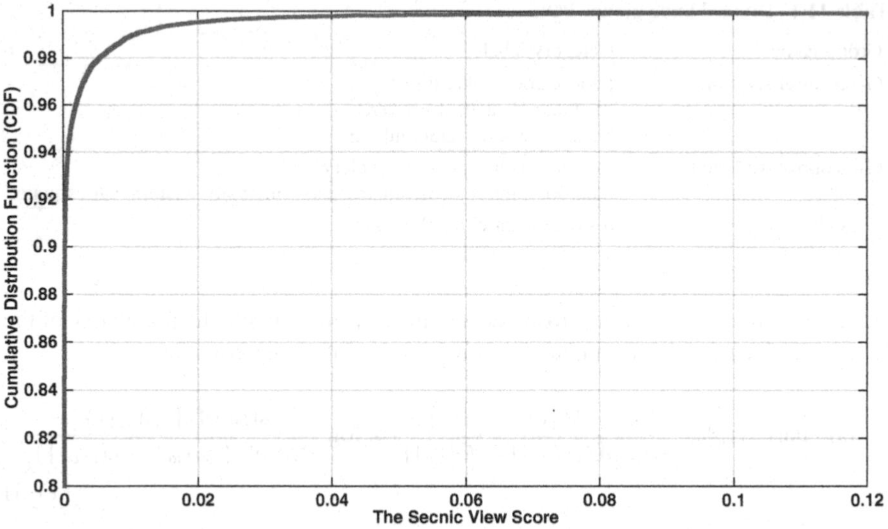
Cumulative Distribution Function (CDF)
The Secnic View Score
1
0.98
0.96
0.94
0.92
0.9
0.88
0.86
0.84
0.82
0.8
0
0.02
0.04
0.06
0.08
0.1
0.12

Fig. 11.3 CDF results on the scenic view score of road segments in SFC

$$S(e_{ij}, \{\text{gim}\}, \{ck\}) = \sqrt{S_{\text{image}}(e_{ij}, \{\text{gim}\}) \times S_{\text{checkin}}(e_{ij}, \{ck\})} \tag{11.5}$$

Figure 11.3 shows cumulative distribute function (CDF) results on the scenic view score of all road segments in the Bay Area of SFC. We can observe that most of the roads are normal ones without much beautiful landscape, and only a very small percentage of them have score bigger than 0. Note that the two parts of scores are normalized separately before the integration, in which the min–max normalization is utilized.

Actually, there are many other different weighting algorithms to integrate the contributions of geotagged images and check-ins on scoring a road segment, as well as different weighting mechanisms would result in different values of score. Furthermore, one kind of data might contribute more than the other one. At the current stage, we are more focused on validating the effectiveness of the introduction of the new kind of user-generated data (i.e., check-ins) on quantifying the score of the scenic view of the road segment. We also have tried two more means of the two parts, i.e., harmonic and quadratic mean, and find that the chosen weighting algorithm quantifies the road scenic view accurately most. Due to the space limitation, we do not show the results of the other two weighting algorithms. In the near future, we plan to explore different weighting algorithms more comprehensively and investigate the related influence on the performance.

11.4.3 Phase II: The MA for SSR Planning

With the results returned from Phase I (i.e., scenic road net-work modeling), the objective of Phase II (i.e., SSR planning) is to find a set of SSRs from the starting point to the ending point for users to pick from, considering both travel distance and beautifulness of the travel route. The problem is a well-known NP-hard one and suffers from combination explosion, which is proved to be difficult or even impossible to get the exact solution particularly if the road network is large and complex [4]. To find high-quality solutions within an acceptable computation time, approximate algorithms are mainly employed in practice [19], among which, the MA is widely used due to its superior performance. Thus, in the section, we propose an effective MA-SSR to deal with the above-mentioned challenges, the overall process of which consists of four modulars. We elaborate the details for each modular in the following.

1. *RoI Determination*: The goal of RoI determination is to obtain the set of candidate road segments for traveling, according to the TRQ. Only road segments locating within the RoI are qualified to be included in the travel routes, while those outside the RoI are impossible to be traveled theoretically (see Theorem 11.1 for the details). In this simple way, the number of road segments is reduced significantly, and thus, the efficiency can be improved greatly.

Theorem 11.1 Given two points (e.g., s and d) in a road network, and the maximal travel distance ($dist_{max}$), if any node of a road segment that is outside the ellipse Elip (s, d, $dist_{max}$) with foci are s and d and the major axis is $dist_{max}$, then it cannot be traveled and can be safely pruned in advance.

Proof: Suppose a given road segment, one node of which locates outside the ellipse (e.g., p_j as shown in Fig. 11.4). According to the ellipse property, we have the following in equation:

$$\text{dist}(s, p_j) + \text{dist}(p_j, d) > \text{dist}_{max}$$

where dist(a, b) computes the pointwise shortest distance. For the given road segment, it can be traversed in two directions. If the node pi is first traversed, then the total distance of the path is

$$\text{path.dist} = \text{dist}(s, p_i) + \text{dist}(p_i, p_j) + \text{dist}(p_j, d) \tag{11.6}$$

According to the property of the triangle inequality, we have

$$\text{dist}(s, p_i) + \text{dist}(p_i, p_j) > \text{dist}(s, p_j) \tag{11.7}$$

By substituting (Eq. 11.7) into (Eq. 11.6), we can derive the following formula:

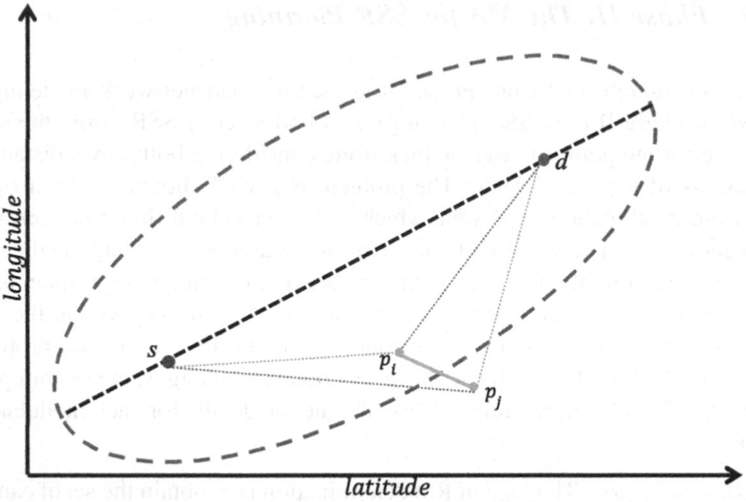
longitude
latitude
s
d
p_i
p_j

Fig. 11.4 Proof of ellipse pruning

$$\text{path.dist} > \text{dist}(s, p_i) + \text{dist}(p_j, d) > \text{dist}_{max} \tag{11.8}$$

Therefore, the total travel distance would be greater than dist_{max} if traveled, indicating that it cannot be included in the travel routes and can be pruned in advance. Same conclusion can be also drawn if the node p_j is first traversed.

2. *Chromosome Encoding and Decoding*: In our MA, each chromosome is generated by a permutation of the set of scenic road segments which are selected from the RoI, as we care more about the scenic road segments, expect for the starting and the ending genes. Specifically, the starting gene is the index of the road segment where it locates while the ending gene is the index of the road segment where it locates. Since users may specify the starting and ending points freely in the road network, since users may specify the starting and ending points freely in the road network, these two road segments may not necessarily be scenic ones. We also add some constraint in the *chromosome* encoding process. Specifically, a gene nonrepetition constraint, which means that a gene cannot contain two identical genes, due to that people seldom travel the same road segments more than once. As a result, in addition to the starting and ending genes, each chromosome is encoded by a sequence of scenic road segment Ids. Noted that each chromosome has gaps between any two consecutive genes, which need to be further augmented and determined in the operation of chromosome decoding.

   Chromosome decoding aims to convert it into an operational driving route. The decoded chromosome is measures in terms of the distance and the score of the scenic view. Due to that the detailed driving paths between any two consecutive genes is unknown, the actual driving distance and the beautifulness score directly cannot be obtained for a given encoded chromosome. Therefore, we need

Table 11.2 Four detailed shortest paths and distances for a pair of genes

<s,d>	Shortest path	Distance (km)
<23(f), 3(f)>	23(f), 4(f), 4(r), 7(r), 7(f), 3(r), 3(f)	1.3
<23(f), 3(r)>	23(f), 4(f), 4(r), 7(r), 7(f), 3(r)	**1.1**
<23(r), 3(f)>	23(r), 2(f), 2(r), 8(r), 8(f), 3(r), 3(f)	1.5
<23(r), 3(r)>	23(r), 2(f), 2(r), 8(r), 8(f), 3(r)	1.6

to determine the detailed driving routes for any two consecutive genes. Here, we adopt the shortest paths to augment their gap for a pair of two consecutive genes. Specifically, there are two nodes in each gene; we, thus, have four shortest candidate paths for a pair of genes. To distinguish different nodes in a gene, we use symbols f and r, respectively, where f is short for front and r is short for rear. For an instance, 23(f) denotes the front node of the road segment with the Id of 23, and 23(r) denotes its rear node. For a pair of genes (e.g., 23 3),), we would have four shortest paths and corresponding distances, as shown in Table 11.2.

A general procedure of the decoding is shown in Fig. 11.5. To make it clear, we take the chromosome in Fig. 11.5 as an example. The chromosome has five genes. For the first pair of consecutive genes, comparison of four shortest paths and distances (i.e., ①–④) needs to be done before determining the detailed driving route between them. The one that leads to the shortest distance would be chosen. Meanwhile, the driving direction of the second gene in this pair can be also fixed (e.g., from 3(r) to 3(f)). Thus, for the rest pairs of genes in the chromosome (the number of pairs of genes is (N − 1) for a chromosome with N genes), only comparison of two shortest paths is required to determine the detailed operational driving routes, since the driving direction of the first gene in each rest pair (i.e., the second gene in the former pair) has been known. Therefore, there need to be (2 N + 2) shortest distance computation and comparison in total, which can be inefficient especially when N is extremely large. Fortunately, on one hand, the number of scenic road segments in a city is small and limited, even for a megacity with millions of road segments, such as NYC, LA, as evidenced by the results shown in Fig. 11.3. Moreover, the number of scenic road segments in the ROI is even smaller. On the other hand, the pairwise shortest paths and distances between any two scenic road segments can be precomputed and stored a prior offline. As a consequence, the efficiency can be guaranteed.

Next, we continue to compute the travel distance and the score of scenic view for a chromosome, respectively. The travel distance of the chromosome is just the total distance of the detailed driving route obtained via decoding, which can either be represented by a sequence of nodes or a sequence of edges in the road network. It can be computed as

Fig. 11.5 Demonstration on the general procedure of gene decoding operation

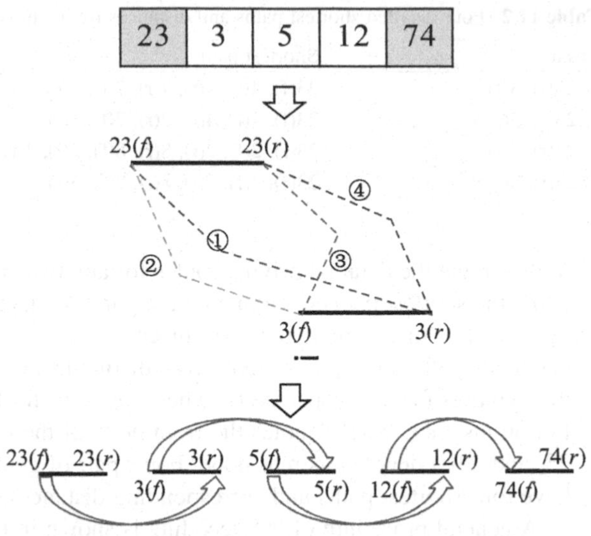
23
3
5
12
74
23(f)
23(r)
①
②
③
④
3(f)
3(r)
23(f)
23(r)
3(f)
3(r)
5(f)
5(r)
12(f)
12(r)
74(f)
74(r)

$$\text{path.dist} = \sum_{i=1}^{n-1} \text{dist}(n_i, n_{i+1}) \tag{11.9}$$

where n_i refers to the *i*th node in the planned travel route. It should be noted that a gene in a chromosome may also appear in the detailed shortest paths after decoding even though a gene nonrepetition constraint is imposed in the chromosome encoding, thus probably leading to the reappearance of some road segment in the planned scenic travel routes.

For a given chromosome, its scenic score is the total scenic score of all the road segments included in the decoded driving route and can be computed according to Definition 11.6.

3. *Chromosome Partition and Crossover*: A population of chromosomes is randomly generated in the initial phase via chromosome encoding; in traditional GAs, two of them would be randomly chosen as parent chromosomes to generate new off-spring. However, tradition GAs usually only get a near-optimal solution. Hence, they cannot be applied to solve the SSRs planning problem directly, since there exists a set of SSRs with different travel distances (i.e., a set of solutions) for an origin-destination pair. Here, to make traditional GAs work adaptively for our problem, we first partition the whole population5 into several sets; then, in each set, two chromosomes are randomly selected as the parents to generate new offspring, as illustrated in Fig. 11.6. More specifically, bound chromosomes are first identified. A chromosome is bounded if and only if no other chromosomes have a shorter travel distance and a higher score of scenic view than itself simultaneously (the red points in Fig. 11.6). Then, we conduct the set partition based on the locations of two consecutive bound chromosomes on the 2-D plane, in which the x-axis refers to the total travel distance and the y-axis refers to the

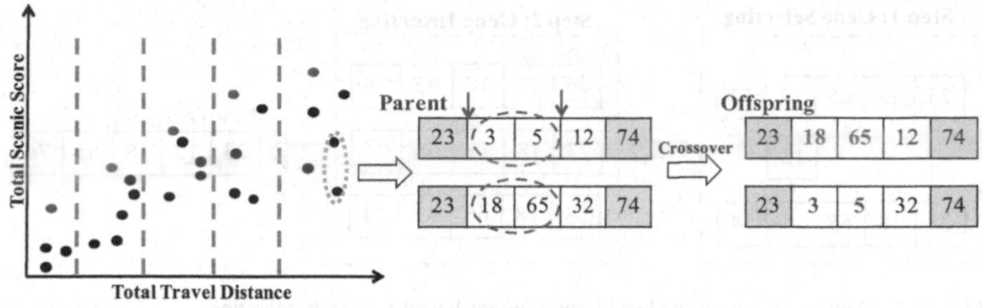
Total Scenic Score
Total Travel Distance
Parent
23 3 5 12 74
23 18 65 32 74
Crossover
Offspring
23 18 65 12 74
23 3 5 32 74

Fig. 11.6 Demonstration on the chromosome partition and crossover operation

total scenic score of the chromosomes, respectively. The borderline between the two sets can be determined as

$$x = \frac{x_i + x_{i+1}}{2}, i \in \{1, 2, \ldots, k-1\} \tag{11.10}$$

where x_i and x_{i+1} are the distances of two consecutive bound chromosomes, and k is the number of the identified bound chromosomes.

Afterward, two parent chromosomes are randomly picked from each partitioned population pool to generate two offspring chromosomes via the crossover operation, as shown in the middle and rightmost parts of Fig. 11.6. It should be noted that the crossover operation is controlled by two parameters, i.e., P_c and ρ_1, where P_c is a user-specified constant and ρ_1 is a randomly generated value in the range of [0,1]. The crossover operation would be activated when $P_c > \rho_1$, in which a big P_c implies a more frequent crossover between two parent chromosomes, producing more new offspring chromosomes.

4. *Local Improvement-Based Mutation*: In order to achieve better performance, we devise a local improvement-based mutation on the top of the offspring chromosome returned from the crossover operation. Only one of two offspring chromosomes would be input to the mutation operation. The essence of the local improvement is the gene reordering for the offspring chromosome, with the objective of finding the newly generated chromosome with the shortest driving distance. For a chromo-some with N genes, it has (N − 2)! different gene combinations (the first and last gene cannot be changed), making it very time consuming to get the ideally best one with the shortest driving distance. To produce a satisfactory chromosome efficiently, we take the following two simple steps to approximate, as illustrated in Fig. 11.7. The first step is the gene selecting, in which a gene is randomly picked from the given offspring chromosome (except for first and last genes). Meanwhile, the selected gene will be removed from the chromosome, resulting in a newly shortened chromosome with (N − 1) genes. The second step is the gene inserting, in which the selected gene is inserted into every possible positions of the new chromosome (e.g. 23 18 65 74, in Fig. 11.7). As expected, there will be (N − 2) new chromosomes generated after the gene

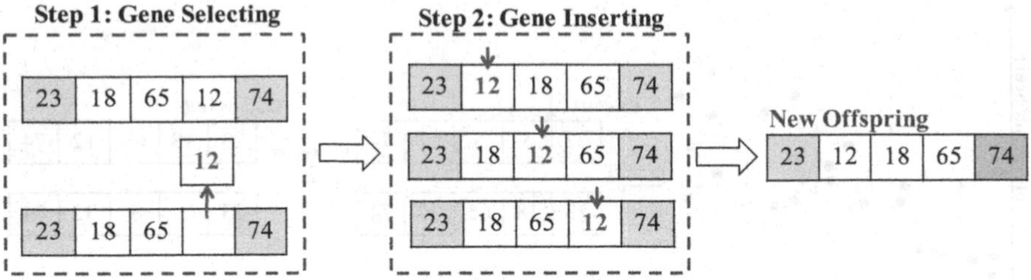
Step 1: Gene Selecting
23 18 65 12 74
12
23 18 65 74
Step 2: Gene Inserting
23 12 18 65 74
23 18 12 65 74
23 18 65 12 74
New Offspring
23 12 18 65 74

Fig. 11.7 Demonstration on the local improvement-based mutation operation

inserting, among which the one with the shortest distance will be finally survived. Similar to the crossover operation, the local improvement-based mutation is also controlled by two parameters, i.e., P_m and ρ_2, where P_m is a user-specified constant and ρ_2 is a randomly generated value in the range of [0,1]. The mutation operation would be trigged when $P_m > \rho_2$, and a bigger P_m indicates a higher probability of the mutation operation, resulting in a higher computation cost.

11.5 Evaluations

In this section, we first describe the experimental setup and, then, present the evaluation results on scenic network modeling and SSR planning extensively, respectively.

11.5.1 Experimental Setup

1. *Data Preparation*: Three datasets in the Bay Area of SFC are used, i.e., the road network, the geotagged image data, and the check-in data. Statistical information about the three datasets is shown in Table 11.3.
2. *Comparison Algorithms*: Two baseline algorithms are used for comparison. The first one is the adaptive generic algorithm (GA), which is similar to our proposed MA. We develop the adaptive GA by adding the novel chromosome partition operation to the traditional GA to work for our problem. In the rest of this section, we use "GA" to represent adaptive GA with chromosome partition, and use "traditional GA" to represent GA without this feature. The only difference between GA and MA is the mutation operation, in which no local improvement is applied for GA. The other algorithm is an exact one that is based on the mixed integer programming and embedded in the commercial solver CPLEX[6] 12.5 (64 bit). It obtains the optimal solution in an exhaustive manner and, thus, needs a highly intensive computation resource.
3. *Evaluation Environment*: The MA algorithm is coded using MATLAB language, and CPLEX is compiled based on the C++ environment. All programs are run on

Table 11.3 Statistics of three datasets

Dataset	Platform	Properties	Statistics
Geotagged image	Flickr	# of images # of users time duration	31,022 1571 12 months
Check-in	Foursquare	# of check-ins # of users time duration	110,214 15,680 6 months
Road network	OpenStreetMap	# of nodes # of edges	3771 5940

an Intel Core i5-4460 PC with 8-GB RAM and running Windows 7 operation system.

11.5.2 Evaluation on Scenic Road Network Modeling

The scenic road network modeled by leveraging two user-generated digital footprints is shown in the left figure in Fig. 11.8. As a comparison, we also provide a part of 49-Mile Scenic Drive for the selected rectangle region in the road network in Fig. 11.8. From the results, we can see that most of the road segments in the urban area are scored extremely low. Overall, the road segments that are scored higher are generally consistent with the designated scenic road tour recommended by the local government, demonstrating the effectiveness of our proposed scenic road network modeling approach.

To quantitatively assess the accuracy of scenic road network modeling, we first select 30 representative road segments and recruit ten volunteers (five females) to give a scenic score to each road segment based on its Google Street View. The score ranges from 1 to 10, with 10 referring to the best scenic view. To eliminate the individual cognitive error, for each road segment, we simply average all its scores contributed by the ten volunteers and use the mean value to represent its score. The rank based on the average score of the selected road segments is regarded as the ground truth rank. We can also obtain a scenic score using our proposed modeling approach for each select road segment and get a new rank. We use the difference between the new rank and the ground truth rank to assess the accuracy of our method. Here, we adopt the commonly used normalized discounted cumulative gain (nDCG) metric to measure the closeness between the two ranks, which is defined as follows:

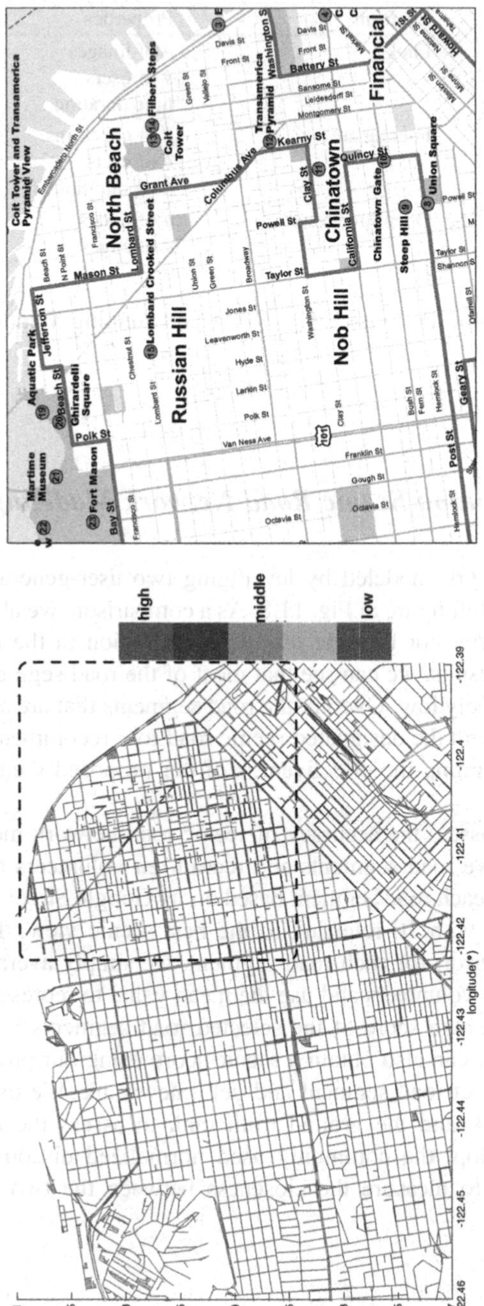
high
middle
low
longitude(°)
latitude(°)
North Beach
Russian Hill
Nob Hill
Chinatown
Financial
Colt Tower and Transamerica Pyramid View
Lombard Crooked Street
Union Square

Fig. 11.8 Result of scenic road network modeling (left) and a part of the 49-Mile Scenic Drive (right). (Best viewed in an enlarged digital version)

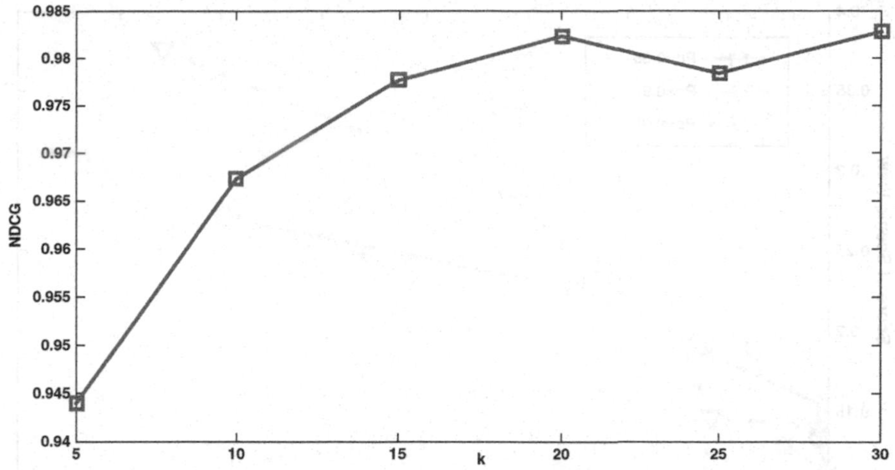
0.985
0.98
0.975
0.97
0.965
0.96
0.955
0.95
0.945
0.94
NDCG
5
10
15
20
25
30
k

Fig. 11.9 Accuracy result of scenic road network modeling

$$\mathrm{nDCG}(k) = \frac{\mathrm{rel}_1' + \sum_{i=2}^{k} \frac{\mathrm{rel}_1'}{\log(i)}}{\mathrm{rel}_1 + \sum_{i=2}^{k} \frac{\mathrm{rel}_1}{\log(i)}} \tag{11.11}$$

where rel_i is the graded relevance of the *i*th road segment, which can be computed by M + 1 − rank(i) (M is the number of selected road segments), while rel'_i is the new graded relevance of the *i*th road segment, which can be computed by M + 1 − rank'(i). For the *i*th road segment, rank(i) gets its position in the ground truth rank; rank'(i) gets its position in the new rank. k refers to the number of road segments considered when calculating nDCG. More specifically, only road segments that are among the top-k in the ground truth rank are counted.

Figure 11.9 shows the accuracy result of scenic road network modeling in terms of nDCG under different choices of k. As one can observe, nDCG generally increases as k becomes bigger. It is high for all k values, i.e., over 94%, demonstrating the reliability of the proposed modeling approach.

11.5.3 Evaluation on Skyline Routes Planning Algorithm

1. *Varying Parameter*: There are four major parameters in our proposed MA-SSR. As discussed, P_c and P_m control the frequency of the crossover operation and the local improvement-based mutation operation, respectively. Usually, a relatively big value of P_c (e.g., > 0.8) would be chosen to ensure the diversity of the new generation of the population, while a much smaller value of P_m (e.g., < 0.2)

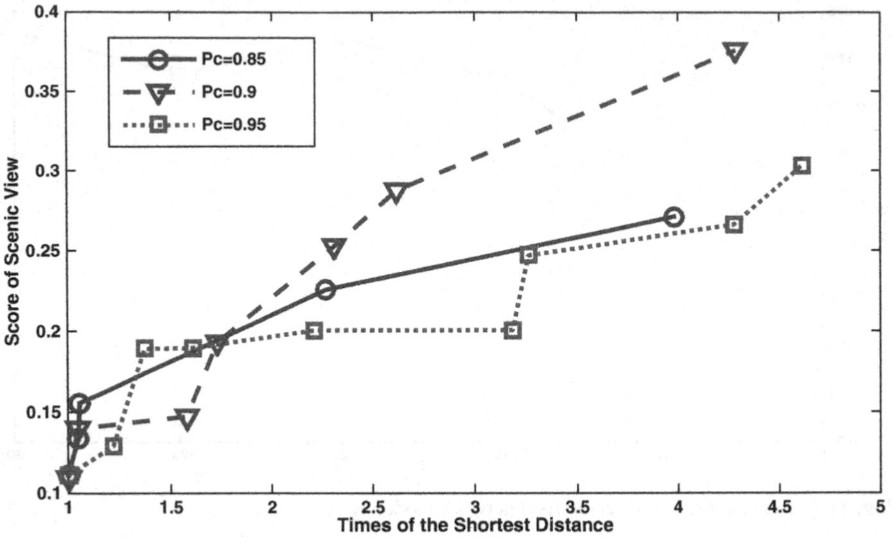
Pc=0.85
Pc=0.9
Pc=0.95
Score of Scenic View
Times of the Shortest Distance
0.4
0.35
0.3
0.25
0.2
0.15
0.1
1
1.5
2
2.5
3
3.5
4
4.5
5

Fig. 11.10 Comparison results on the SSRs by varying P_c. The shortest distance from the starting point to the ending point is 2.447 km

would be chosen to avoid the intense consumption of computer resources. pop_size is the size of the population. max_iter is related to the stopping criteria and the algorithm would terminate when the number of iterations reaches the input max_iter. In the following experiments, we mainly study how the choice of P_c, P_m, pop_size, and max_iter affects the performance in terms of the SSRs and the computation time, varying only one parameter at a time. To ensure that the best SSRs have been obtained under a given parameter setting, we mainly take the following two steps: (1) Repeat MA for ten rounds; in each round, MA terminates when it reaches the user-specified maximum iterations and output a set of SSRs; and (2) the final SSRs are obtained by applying the skyline operator based on all the sets of SSRs returned at each round. It should be noted that the computation time needed under a given parameter setting is the most time-consuming one of all rounds, rather than the accumulative times, since MA at each repeated round runs independently and thus can easily be implemented in a parallel manner.

(a) *Varying P_c:* Fig. 11.10 shows the comparison results of different SSRs obtained by MA-SSR under different choices of P_c, in which the *x*-axis refers to the times of the shortest distance, the *y*-axis refers to the scenic score of the travel route, and a point refers to an SSR. Note that all SSRs with the same choice of P_c are connected to form a line for better performance comparison. As observed, for each line, the score of the scenic view along the travel route increases monotonously, as the travel distance becomes larger. This is because a larger travel distance allows users to travel more distant and high-quality scenic road segments, offering a better driving experience. Moreover, for most of cases, SSRs for the choices of $P_c = 0.85$ and

Table 11.4 Comparison result on the computation time by varying P_c

	Varying P_c		
P_c	0.85	**0.9**	0.95
Computation time (in seconds)	5.2811	**6.8477**	9.2171

$P_c = 0.95$ are dominated by SSRs for the choice of $P_c = 0.9$. We also show the computation time of the proposed MA-SSR by varying P_c in Table 11.4. We can see that the computation time when choosing $P_c = 0.9$ (i.e., 6.8477 s) is in-between compared with the other two choices, which is quite efficient and desirable. As concluded, we set $P_c = 0.9$ as the optimal through the experiment. Here, we fix $P_m = 0.05$, pop_size = 100, and max_iter = 3000.

(b) *Varying P_m*: Fig. 11.11 presents the comparison results on the SSRs by varying P_m. In terms of the SSRs, we can see that MA-SSR achieves similar performance under different choices of P_m when the travel distance is less than twice of the shortest distance value. Gaps between the line connecting by SSRs obtained when choosing $P_m = 0.15$ and the other two lines become wider when the travel distance is larger than the twice of the shortest distance value. In terms of the computation time, it increases monotonously as Pm becomes larger, which is due to the fact that more frequent mutation operation is generated. We can also observe that the computation time is within 20 s for all choices, as shown in Table 11.5. As a result, we set $P_m = 0.15$ as the optimal though it needs a bit more computation time. In this experiment, we fix $P_c = 0.9$, pop_size = 100, and max_iter = 5000.

(c) *Varying pop size*: We show the SSRs results in Fig. 11.12 under different sizes of the population. We can see that SSRs obtained by MA-SSR when pop_size = 500 dominate those under the other two settings, except when the travel distance is around 2.7 times of the shortest distance value. In this case, SSR obtained by MA-SSR when pop_size = 300 achieves the best performance, but it fails to discover more skyline routes as the travel distance increases. The computation time under different choices of pop size is shown in Table 11.6. As expected, the computation time increases when pop_size becomes greater. Consequently, we set pop_size = 500 as the optimal as it achieves the best performance in terms of the SSRs, though with the highest computation time. In this experiment, we fix $P_c = 0.95$, $P_m = 0.1$, and max_iter = 5000.

(d) *Varying max_iter*: Comparison results on SSRs by varying max_iter are shown in Fig. 11.13. In terms of the SSRs, MA-SSR under the choices of max_iter = 3000 and max_iter = 7000 achieves almost equally good performance. Though MA-SSR when setting max_iter = 5000 can discover more SSRs as the travel distance increases, it fails to improve the score of scenic view of the travel route any more. In terms of the computation time, the computation time increases gradually as max_iter becomes larger, as shown in Table 11.7. Thus, we choose max_iter = 3000 as the optimal setting. In this experiment, we fix $P_c = 0.9$, $P_m = 0.1$, and pop_size = 100.

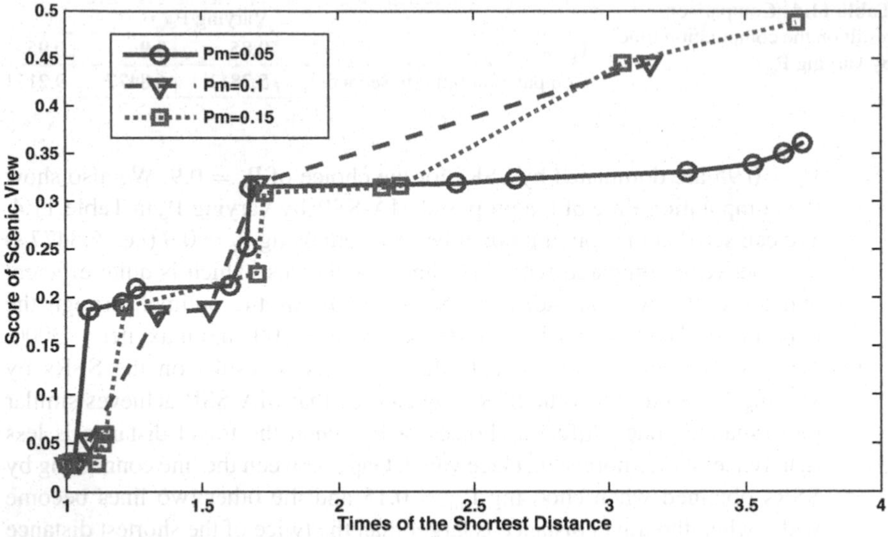

Pm=0.05
Pm=0.1
Pm=0.15
Score of Scenic View
Times of the Shortest Distance
0
0.05
0.1
0.15
0.2
0.25
0.3
0.35
0.4
0.45
0.5
1
1.5
2
2.5
3
3.5
4

Fig. 11.11 Comparison results on the SSRs by varying Pm. The shortest distance from the starting point to the ending point is 3.050 km

Table 11.5 Comparison result on the computation time by varying P_m

	Varying P_m		
P_m	0.05	0.1	**0.15**
Computation time (in seconds)	11.1135	15.0663	**17.718**

To sum up, we determine $P_c = 0.9$, $P_m = 0.15$, pop_size = 500, and max_iter = 3000 as the optimal combination of parameter setting, which will be fixed for the rest experiments.

2. *Comparison to Baseline Approaches*: To demonstrate the superior performance of the proposed MA-SSR, we compare it with two baseline approaches and show the corresponding results in Fig. 11.14. Through the comparison results between the MA and the GA, we can conclude that SSRs obtained by the MA consistently dominate ones that obtained by the GA, i.e., for any SSR obtained by the GA, we can always find at *least* one SSR that with a higher score of scenic view and shorter travel distance in the set of SSRs obtained by the MA, demonstrating the effectiveness of the local improvement-based mutation. In terms of the computation time, the MA needs 22.7858 s, while GA needs 16.4057 s. The time difference is exactly the extra time cost induced by the local improvement.

We further compare the results of SSRs obtained by the MA with that obtained by CPLEX, to gain an in-depth understanding on the performance gap between our proposed algorithm and the truly optimal one. It should be noted that CPLEX can only work to get one optimal travel route, given the constraint of the total

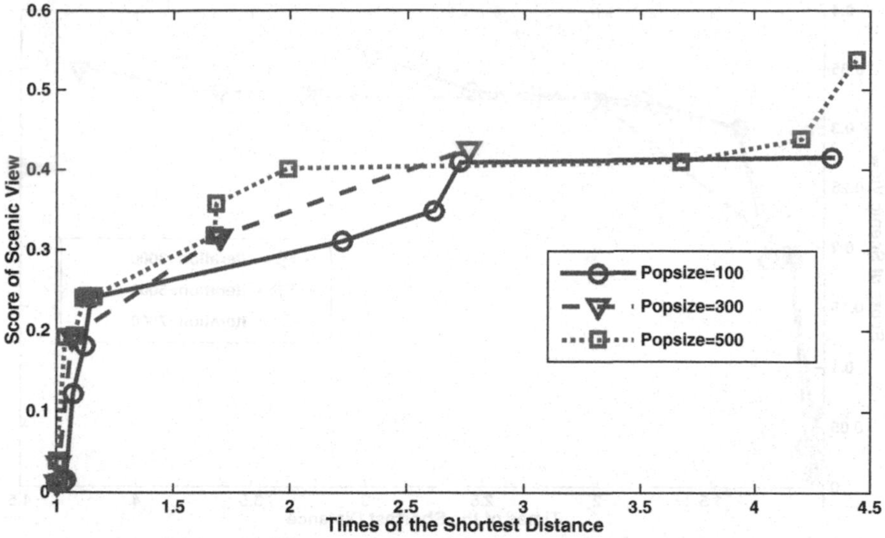

Score of Scenic View
Times of the Shortest Distance
Popsize=100
Popsize=300
Popsize=500
0.6
0.5
0.4
0.3
0.2
0.1
0
1
1.5
2
2.5
3
3.5
4
4.5

Fig. 11.12 Comparison results on the SSRs by varying pop size. The shortest distance from the starting point to the ending point is 2.014 km

Table 11.6 Comparison result on the computation time by varying pop_size

	Varying P_m		
pop_size	100	300	500
Computation time (in seconds)	18.609	22.2321	28.1086

travel distance. For better comparison, we, thus, get the corresponding optimal travel route separately, given the total travel distance of each SSRs obtained by the MA, as shown in Fig. 11.14, i.e., points in blue line and red line are with the same x-axis values. As expected, CPLEX performs consistently better than the MA, but the gap between them is quite narrow. In terms of the computation time, the result of CPLEX is shown in Fig. 11.15. The computation time goes up exponentially as the travel distance increases. CPLEX is highly time consuming, even if the constraint of the total travel distance is set small. As an instance, the computation time is around 244 s when the total travel distance is 1.25 times of the shortest distance value, which is ten times more than that of the MA. Given 4.45 times of the shortest distance value, CPLEX needs over 2.6 h to return an optimal travel route. What is more is that, as discussed, MA-SSR needs to run only once to return a set of SSRs, while CPELX has to be executed repeatedly given different travel distance budgets; thus, the computation time should be accumulated. Hence, our proposed MA-SSR is rather efficient while guaranteeing high quality of SSRs.

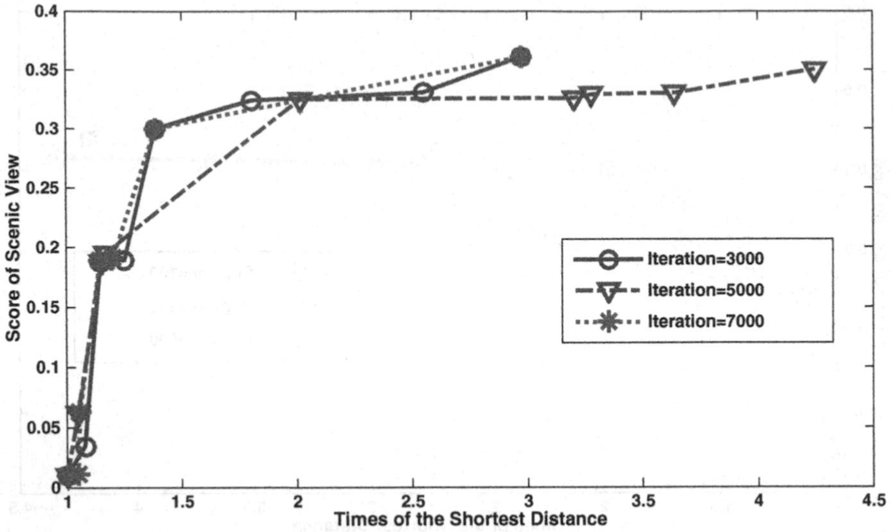

Score of Scenic View
Times of the Shortest Distance
0
0.05
0.1
0.15
0.2
0.25
0.3
0.35
0.4
1
1.5
2
2.5
3
3.5
4
4.5
Iteration=3000
Iteration=5000
Iteration=7000

Fig. 11.13 Comparison results on the SSRs by varying max iter. The shortest distance from the starting point to the ending point is 2.651 km

Table 11.7 Comparison result on the computation time by varying max_iter

	Varying P_m		
max_iter	3000	5000	7000
Computation time (in seconds)	10.0253	14.1719	19.1983

11.6 Conclusions and Future Work

In this chapter, we proposed a novel framework called MA-SSR to recommend a set of SSRs to users given users' trip queries. There are two main phases in the framework, i.e., scenic road network modeling and scenic route planning. To be more specific, we extracted the relevant information from two datasets (i.e., the geo-tagged image and the check-in data) to model the scenic score for each road segment. Then, the MA is developed for SSR planning with a novel and comprehensive process in which the RoI determination, chromosome encoding and decoding, chromosome partition and crossover, and local improvement-based mutation are conducted. Based on the real-world road network with 3771 nodes and 5940 edges obtained from OSM platform and a large-scale geotagged image data left by 1571 Flickr users in a year, and check-in data generated by 15,680 Foursquare users in 6 months in the Bay Area in SFC, we evaluated our framework in terms of the effectiveness and efficiency.

In the future, we plan to broaden and deepen this work in the following directions: (1) We intend to explore the time information hiding in the data and construct dynamic road networks in which the travel time and scenic score on each road

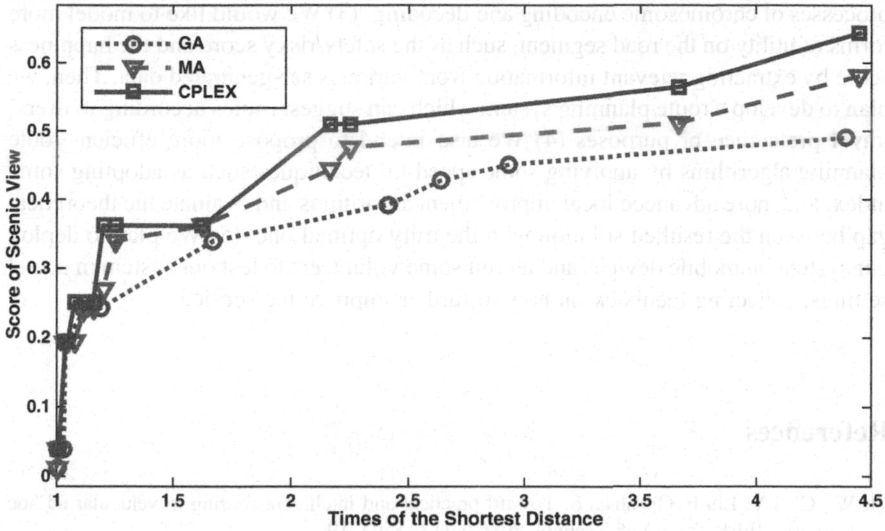
GA
MA
CPLEX
Score of Scenic View
Times of the Shortest Distance

Fig. 11.14 Comparison results on the SSRs of three algorithms, i.e., MA, GA, and CPLEX, respectively. The shortest distance from the starting point to the ending point is 2.014 km

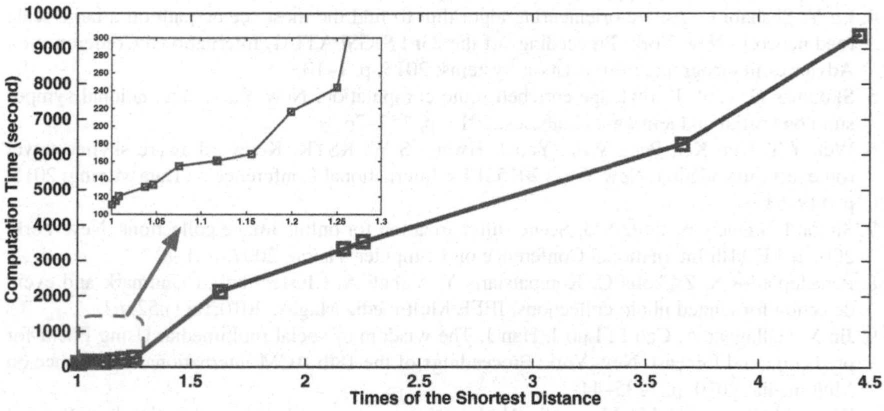
Computation Time (second)
Times of the Shortest Distance

Fig. 11.15 Result on the computation time of CPLEX

segment are time-dependent. Specifically, in addition to the travel time which is affected by dynamic traffic conditions, the scenic score on the road segment is also different at different times of the day and different seasons of the year. On this basis, we are able to obtain the optimal route at different departure time. (2) We plan to infer the driving direction for the road segments based on the given location of source and the destination by mining the vehicular GPS trajectory data (e.g., taxi). The rationale behind is that, we can obtain the frequent bypassing direction from the source to the destination for a road segment according to the historical GPS trajectory data. Then, the direction information can be used to accelerate the processes of chromosome encoding and decoding. (3) We would like to model more forms of utility on the road segment, such as the safety/risky score and the happiness score by extracting relevant information from various user-generated data. Then, we plan to develop a route planning system which can suggest routes according to users' travel preference or purposes (4) We also intend to propose more efficient route planning algorithms by applying some speed-up techniques such as adopting some index, and more advanced local improvement algorithms and evaluate the theoretical gap between the resulted solution with the truly optimal one. (5) We plan to deploy our system on mobile devices and recruit some volunteers to test our system in actual settings, collecting feedback on how to further improve the service.

References

1. Wu C, Ji Y, Liu F, Ohzahata S. Toward practical and intelligent routing in vehicular ad hoc networks. IEEE Trans Veh Technol. 2016;64(12):5503–19.
2. Alivand M, Hochmair H, Srinivasan S. Analyzing how travelers choose scenic routes using route choice models. Comput Environ Urban Syst. 2015;50:41–52.
3. Gunawan A, Lau HC, Vansteenwegen P. Orienteering problem: a survey of recent variants, solution approaches and applications. Eur J Oper Res. 2016;255:315–32.
4. Lu Y, Shahabi C. An arc orienteering algorithm to find the most scenic path on a large-scale road network. New York: Proceedings of the 23rd SIGSPATIAL International Conference on Advances in Geographic Information Systems; 2015. p. 1–10.
5. Skoumas G, et al. Knowledge-enriched route computation. New York: International symposium on spatial and temporal databases; 2015. p. 157–76.
6. Wen Y-T, Cho K-J, Peng W-C, Yeo J, Hwang S-W. KSTR: Keyword-aware skyline travel route recommendation. New York: 2015 IEEE International Conference on Data Mining; 2015. p. 449–58.
7. Simon I, Snavely N, Seitz SM. Scene summarization for online image collections. New York: 2007 IEEE 11th International Conference on Computer Vision; 2007. p. 1–8.
8. Papadopoulos S, Zigkolis C, Kompatsiaris Y, Vakali A. Cluster- based landmark and event detection for tagged photo collections. IEEE Multimedia Magaz. 2010;18(1):52–63.
9. Jin X, Gallagher A, Cao L, Luo J, Han J. The wisdom of social multimedia: Using Flickr for prediction and forecast. New York: Proceedings of the 18th ACM international conference on Multimedia; 2010. p. 1235–44.
10. Wang H, Terrovitis M, Mamoulis N. Location recommendation in location-based social networks using user check-in data. New York: Proceedings of the 21st ACM SIGSPATIAL International Conference on Advances in Geographic Information Systems; 2013. p. 374–83.
11. Weng D, Chen R, Zhang J, et al. Pareto-optimal transit route planning with multi-objective monte-carlo tree search. New York: IEEE Transactions on Intelligent Transportation Systems; 2020.
12. Choo MJ, Lee HJ, Park YH. Proposal of personalized path recommendation algorithm considering time and space. Korea: Proceedings of the Korea Information Processing Society Conference. Korea Information Processing Society; 2020. p. 424–6.
13. Chen C, Gao L, Xie X, et al. Enjoy the most beautiful scene now: a memetic algorithm to solve two-fold time-dependent arc orienteering problem. Front Comput Sci. 2020;14(2):364–77.
14. Sharker MH, Karimi HA, Zgibor JC. Health-optimal routing in pedestrian navigation services. New York: Proceedings of the First ACM SIGSPATIAL International Workshop on Use of GIS in Public Health; 2012. p. 1–10.

15. Quercia D, Schifanella R, Aiello LM. The shortest path to happiness: Recommending beautiful, quiet, and happy routes in the city. New York: Proc. 25th ACM Conf. Hypertext Social Media; 2014. p. 116–25.
16. Kim J, Cha M, Sandholm T. SocRoutes: Safe routes based on tweet sentiments. New York: Proceedings of the 25th ACM conference on Hypertext and Social Media; 2014. p. 179–82.
17. Galbrun E, Pelechrinis K, Terzi E. Urban navigation beyond shortest route: The case of safe paths. Inf Syst. 2016;57:160–71.
18. Zheng Y-T, et al. GPSView: A scenic driving route planner. ACM Trans Multimedia Comput Commun Appl (TOMM). 2013;9(1):1–18.
19. Gavalas D, Konstantopoulos C, Mastakas K, Pantziou G, Vathis N. Approximation algorithms for the arc orienteering problem. Inf Process Lett. 2015;115(2):313–5.

Part VI
Enabling Smart Urban Services: Beyond People Transportation

Chapter 12
CrowdDeliver: Making Citywide Packages Arrive as soon as Possible

12.1 Introduction

The logistics industry has been developing sustainably as the result of the exponentially increased popularity of round-the-clock online shipping. For example, the number of package deliveries generated by online ordered products grows from over one billion in 2013 to over 1.3 billion in 2018. It is reported by the online retailing research community that most people shop online mainly for convenience, however, they cannot get their goods immediately [1]. Thus, for online shoppers, how fast that the buying products can arrive at their hands has become a major concern and also is one of the competitors among different logistics companies. Therefore, to make them stand out, an increasing number of online retailers and logistics service providers aim to offer *the same-day delivery service* to attract more shoppers.

It is easy to understand that more delivery resources (e.g., people and vehicles) should be devoted if requiring the higher shipping speed. Unfortunately, the incurred increased express shipping fee may not be covered by the amount that shoppers are willing to pay [2]. To make matter worse, there has not yet been any cost-effective solution to cutting down the operational cost and increasing the profit margins. As a result, the same-day delivery service is still in its infancy and its road to the success in the long run remains unclear and controversial. In this chapter, with the idea of making use of the existing human mobility on the street as the resource of package hitchhiking. In more detail, we propose to have packages take hitchhiking rides with already passenger-occupied taxi rides. In this manner, packages and passengers can be sent with the common transporting vehicle, thus lowering the cost potentially. To

Part of this chapter is based on a previous work: C. Chen et al., "Crowddeliver: Planning City-Wide Package Delivery Paths Leveraging the Crowd of Taxis," in IEEE Transactions on Intelligent Transportation Systems, vol. 18, no. 6, pp. 1478–1496, June 2017, doi: https://doi.org/10.1109/TITS.2016.2607458.

C. Chen et al., *Enabling Smart Urban Services with GPS Trajectory Data*,
https://doi.org/10.1007/978-981-16-0178-1_12

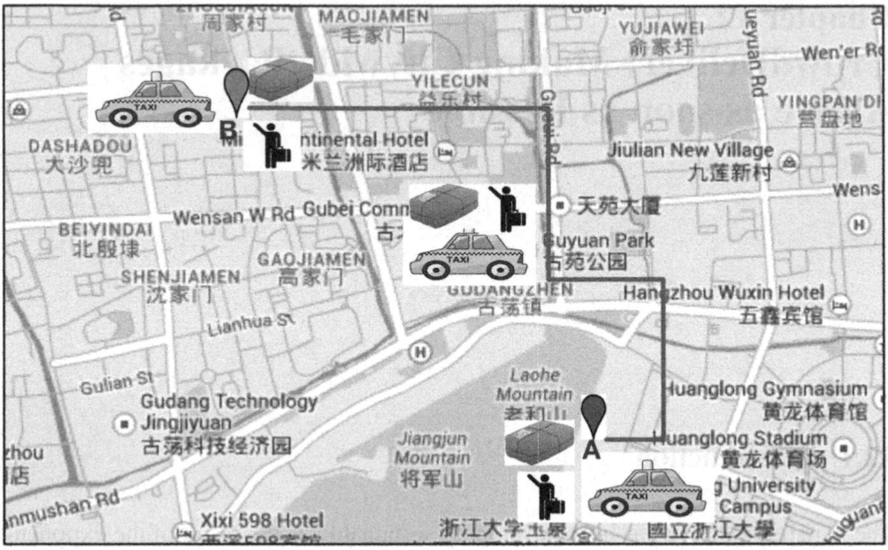
MAOJIAMEN
毛家门
周家村
YUJIAWEI
俞家圩
YILECUN
益乐村
营盘地
TAXI
B
DASHADOU
大沙兜
米兰洲际酒店
Jiulian New Village
九莲新村
Wensan W Rd
天苑大厦
BEIYINDAI
北殷埭
GAOJIAMEN
高家门
SHENJIAMEN
沈家门
古苑公园
古荡镇
五鑫宾馆
Lianhua St
Gulian St
Gudang Technology
Jingjiyuan
古荡科技经济园
Laohe
Mountain
老和山
Jiangjun
Mountain
将军山
A
黄龙体育馆
黄龙体育场
Campus
Xixi 598 Hotel

Fig. 12.1 An example case when a package takes a hitchhiking ride from its origin to the destination

make our idea clear, we illustrate the basic procedure using the following simple example, as shown in Fig. 12.1.

There is a user request that one package needs to be sent from *A* to *B*. To simplify the case, we assume that there are facilities that can store and manage packages *temporally* both at places *A* and *B*. By chance, there is a passenger at *A* intends to go to *B* and makes a taxi ride request using her smartphone. Once submitting the real-time request by the passenger, the platform (e.g., Uber and DiDi), a suitable taxi can be assigned to meet the require. As well, the package can be also assigned to the taxi before the taxi driver go to pick up the passenger. After picking up package and passenger, the assigned taxi is responsible to send them to the destination together, i.e., by dropping off the passenger first, then leaving the package at a specifically appointed equipment (e.g., smart postbox) at *B*. In this way, the quality of service to the passenger would not degrade significantly.

As illustrated in the example, it only needs very little effort and time from the recruited taxi. It has the following merits when comparing to the traditional solutions that maintaining a dedicated shipping resource (e.g., crew and vehicles). (1) *Abundant hitchhiking rides and opportunities*. As one of the most important transportation means, travelling by taking taxis becomes common and it can provide abundant hitchhiking rides for package deliveries within the city. (2) *24/7 operation*. Taxis are operating 24/7. Hence, package deliveries by taking hitchhiking taxi rides can be completed round-the-clock, and the working hours can be extended to include late night, early morning, weekends and holiday. More promisingly, the delivery can be even more efficient during holidays and weekends since the number of taxi passengers shall increase during that time. (3) *Green and Economic*. Since the proposed solution works based on the already existing mobility from taxis for the package deliveries, it does not induce extra air and noise pollutions.

However, it is usually more complicated in real life than the case illustrated in the example when exploiting passenger-occupied taxis to ship packages. In addition to the issues including package tracing and station placement for easy access of taxis, due to financial considerations, few passengers would travel by taxi if the destination is too far. The most frequent OD distance is around 3 km. It is thereby needing interchanges and more hitchhiking rides from a number of taxis (i.e., taxi relays) if intending to send a distant package. Here, in this chapter, we hence formulate the problem of taxi-based package delivery as a route planning one (i.e., finding the station combination first, then scheduling taxis according to the real-time taxi ride requests) with the objective of minimizing the total delivery time (i.e., maximizing the delivery speed). To make our idea feasible, we need to address the following two main research challenges:

1. **How to accurately estimate the non-stop package delivery time between two interchange stations?** It is easy to understand that the delivery time cost consists of two components, i.e., the waiting time each the incoming hitchhiking ride and the driving time when sending with the passenger in the taxi. Even for a fixed OD pair, it is non-trivial to model and estimate the waiting time for the hitchhiking ride since the event of passenger taking the ride for the specific pair is quite random. On the other hand, the estimation of the second component (i.e., the driving time) is also not easy due to the dynamic of road traffic condition. Even more challenging, both two components vary at different time of the day.
2. **How to adaptively schedule taxis according to the real-time incoming taxi ride requests in order to achieve the near-optimal package delivery paths?** The route planning problem seems to become the well-known time-dependent shortest path-finding problem [3] since we are able to estimate the delivery time between any two interchange stations. However, the paths discovered via such method only expect to require the least delivery time only in the statistical sense. In real dynamic situations, a rider corresponds to a longer path statistically can be available much earlier, resulting in a more efficient path actually than the expected shortest path.

In this chapter, we mainly make the following contributions to address the above-mentioned research challenges:

1. To strike a nice trade-off between the delivery speed and cost, we propose a cost-effective solution that leverages the unintentional cooperation among a crowd of passenger-occupied taxis to deal with the problem of citywide package express. In more detail, we use the existing 24 h convenience shop that distribute across the entire city as package interchange stations, reducing the cost of building dedicated storage facilities as well.
2. To obtain the fastest package delivery path for each OD request, we propose a two-phase path-planning framework called **CrowdDeliver**. In the first phase, based on the historical taxi trajectory data, we estimate the non-stop travel time

from one interchange station to another if there exists sufficient passenger flow in history at different time of the day [4], and discover the shortest delivery time with the estimated total travel time for any give OD pair. In the second phase, referring to the results obtained in the first phase offline, we propose an online algorithm named **AdaPlan** that schedules taxis adaptively and dynamically to re-plan the package delivery path for any incoming delivery request, according to the taxi ride requests generated in real time.

3. Using several data sets collected from the real world including the POI data, road network data and the large-scale taxi trajectory data generated by over 7600 taxis in 1 month in the city of Hangzhou, China, we conduct extensive experiments to verify the effectiveness and efficiency of the proposed **CrowdDeliver** system. We further showcase a case to demonstrate the feasibility of our proposed cost-effective solution.

12.2 Related Work

Our study is actually a specific application of the idea of Crowdsourcing in the area of logistics, which is also called as a special application scenario of spatial crowdsourcing. There are several terms appeared in the reference, including Crowdshipping, crowdsourced logistics and so on [5, 6]. There are two representative studies on Crowdshipping which utilizes the spatial and temporal overlaps between crowdsourcing carriers. In more detail, inspired by the idea that nearby Twitter users can work together to deliver packages since they may share and reveal their locations when posting notes. The package can be passed from one Twitter user to another if they have spatial and temporal overlaps [7]. Despite the novel idea, the solution suffers from the feasibility issue, more specifically, (1) it is not easy to trace and coordinate Twitter uses since they seldom share their location information continuously [8]. As a result, the number of users that can be recruited may be quite limited. Eventually, the delivery time may be long and uncontrollable. (2) It is not practical to require the one Twitter user to pass the package to another user, because, on one hand, the actual distance may be quite long even they share close Twitter geo-locations. On the other hand, such cooperation may interrupt his/her ongoing activities that cannot know or be inferred from the user's geo-twitters data. Similarly, the authors in [9] recruited mobile users based on the spatial and temporal overlaps revealed by the cell towers. They also share the similar limitations: (1) It is difficult or even not feasible for two users located in the same cell tower to handover packages since the coverage of cell towers in rural areas is quite wide. (2) The number of mobile users that can be recruited to participate tasks is also limited since only opportunities that when people making calls can be utilized.

In order to make a more realistic and practical form of Crowdshiping, there are several studies that intend to make use of the abundant existing passenger-sending trips to facilitate citywide package deliveries published during the last 3 years [10, 11]. To name a few, in their vision paper, Chen et al. [12] discuss the unique features (4F), key research issues and potential solutions when comparing to traditional logistics systems, to attract more scholars to enter this novel transport paradigm. By addressing a wide range of issues from social concerns (e.g., evaluating the public acceptance of this new delivery form) to operation optimizations (e.g., vehicle routing, path planning, vehicle scheduling), there are a bunch of Crowdshipping systems based on the shared mobility generated when delivering passengers (e.g., public buses, taxis) [10, 13]. Most current studies fail to take the distinct spatial and temporal patterns between people and package in transportation into consideration, although they combine both passenger flow and package flow to be sent mixed in their proof-of-concept systems. More specifically, they formulate the path planning problem as the share-a-ride problem and insert the package delivery requests into the passenger-sending trips. However, as we argued previously, the distance of package trip may be much longer than that of passenger trip, resulting in the failure in real cases. One package-sending trip may need a number of relays from passenger-sending trips. Worse still, the taxis are asked to make several dedicated detours and stops during sending passengers to participate the required tasks, which severely degrades the quality of service to passengers. Almost all of current studies focus on the package routing issues in Crowdshipping and ignore the issues of design and optimization of package transport network completely [10, 13, 14]. We believe the design and optimization of package transport network including the determinations of number and location of package is vital and should separately investigated [15]. In our current study, we simply cluster locations with frequent pick-ups and drop-offs to obtain package interchange stations without any optimization techniques [14]. The interchange station is responsible for storing packages temporary, thus no time overlap and pairwise cooperation between the recruited taxi drivers are needed. As mentioned, our proposed solution can transport packages over a long distance but requires only little effort from the participants. What is more, we aim to lower the impact at the quality and experience of passenger service.

12.3 Basic Concepts, Assumptions and Problem Statement

In this section, we provide definitions of some basic concepts, elicit assumptions we have made, and give a formal problem statement.

12.3.1 Basic Concepts

We define the basic concepts used in this chapter as follows:

Definition 12.1 (Taxi Trajectory): A taxi trajectory Tr is a sequence of GPS points corresponding to a single passenger-delivery trip. Each GPS point $p_i = (t_i, x_i, y_i)$ consists of a time-stamp t_i, a longitude x_i, and a latitude y_i. Thus Tr can be

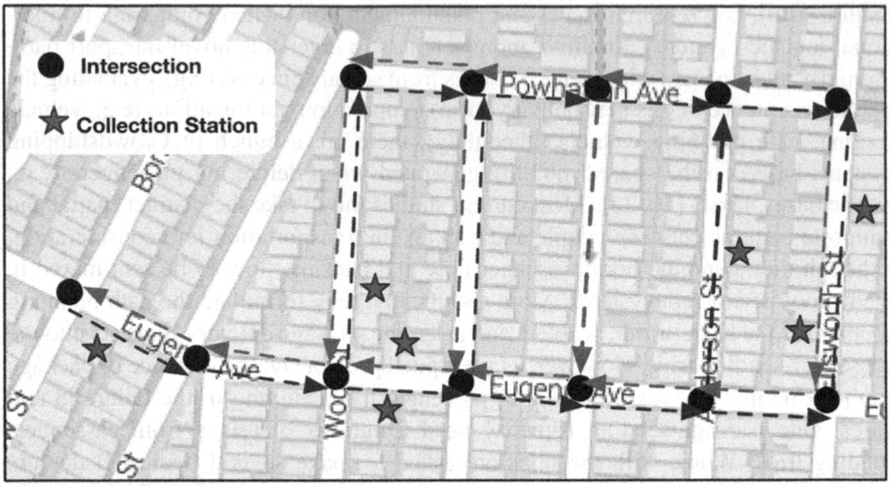
Intersection
Collection Station
Powhatan Ave
Eugen Ave
Eugenia Ave
Ellsworth St

Fig. 12.2 An example of the road network. Each edge is labeled with its driving directions

represented as $p_1 \rightarrow p_2 \rightarrow \cdots \rightarrow p_n$, where p_1 and p_n indicate when and where the passenger was picked up and dropped off respectively.

Definition 12.2 (Package Interchange Station): A package interchange station cs is a Point of Interest (POI) in a city, which is used to store-and-forward packages temporarily during the relay shipping process. The set of package interchange stations is denoted as CS.

Here, we select the 24-h convenience stores near the road sides as interchange stations, marked by red stars in Fig. 12.2. We exploit the convenience stores mainly because: (1) There are usually hundreds of such POIs[2] distributing around the city, not only guaranteeing a good coverage, but possible parking facilities available nearby; (2) Most convenience stores locate in or near the residential districts, which can help users upload and offload packages conveniently and safely; and (3) No need to introduce any dedicated hardware such as lockers and safe boxes, incurring no extra cost in terms of installation and maintenance.

Definition 12.3 (Package Delivery Request): A package delivery request pr is defined as a triplet $\langle o_p, d_p, t_p \rangle$., where o_p and d_p refer to the origin and destination of the package delivery respectively; t_p refers to the time when the user submits the request (i.e. the birth time). Note that here $o_p \in \mathrm{CS}$, $d_p \in \mathrm{CS}$, indicating that packages originate and end at interchange stations.

Definition 12.4 (Real-Time Taxi Ordering Request): A taxi ordering request tor is defined as a triplet $\langle o_t, d_t, t_t \rangle$, where o_t and d_t refer to the passenger's origin and intended destination respectively. t_t refers to the time that the passenger submits the request.

Definition 12.5 (Package Delivery Path): A package delivery path is a sequence of interchange stations that reflects the package routing order $cs_1 \rightarrow cs_2 \rightarrow \cdots \rightarrow cs_n$. It is easy to know the first interchange station in the path is the origin of the package $cs_1 = pr.\ o_p$; the last one is the destination of the package $cs_n = pr.\ d_p$.

Definition 12.6 (Package Operational Path): A package operational path refers to the package delivery path with the detailed taxi ride information added. A different taxi is scheduled to transport a package between consecutive interchange stations along a package delivery path. Therefore, a package operational path can be represented by $cs_1 \overset{taxi_i}{\rightarrow} cs_2 \overset{taxi_2}{\rightarrow} \ldots \overset{taxi_{n-1}}{\rightarrow} cs_n$.

Definition 12.7 (Package Delivery Time): The package delivery time of a package operational path is defined as the total time spent on the trip. It mainly consists of three parts: the total **waiting time**, the total package **dealing time**, and the total **driving time**.

For each segment of package operational path (e.g. $cs_i \overset{taxi_i}{\rightarrow} cs_{i+1}$), the **waiting time** is the time difference between when the package and the scheduled $taxi_i$ arrive at the interchange station cs_i. The package **dealing time** is the time that the driver of $taxi_i$ needs to handle the package, such as getting off the taxis, walking to the interchange station, and uploading or offloading the package. The **driving time** is the time that takes for $taxi_i$ to drive from cs_i to cs_{i+1}.

12.3.2 Assumptions

In this chapter, we make the following assumptions.

Assumptions 1: Taxis cannot be recruited to take part in the package delivery tasks neither during rush hours, nor in the course of sending passengers. We further assume the recruited taxis have enough room for the package storage.

We exclude rush hours since the demands of passengers are high and meeting their demands is the top priority during that period. Moreover, to minimize potential impact on passengers' experiences, the taxis are asked to collect packages before picking up passengers, and offload packages after dropping off passengers. This assumption tries to guarantee the quality of taxi services for passengers.

Assumptions 2: The taxi drivers are willing to accept the assigned package delivery tasks.

We believe that this assumption can be realistic given proper incentive mechanisms. In the design of incentive mechanisms, a prime principle is to ensure that the reward matches the amount of efforts put in by the drivers. For example, some places are harder to get to or park the taxi, then the incentive may be higher. However, designing a proper incentive mechanism is beyond the scope of this chapter.

Assumptions 3: The packages are traceable.

In the delivery process, a package is either stored at the interchange station or forwarded by a scheduled taxi. Each interchange station is authorized and has a unique Id; each taxi is registered in taxi management department and also has a unique Id. This assumption tries to address the package security issues. Any package damage or loss will impair this novel package delivery service.

12.3.3 Problem Statement

Leveraging the crowd of taxis for package delivery can be viewed as an optimal route discovery problem, and thus can be formulated as follows:

Given:

1. A road network $G(N, E)$ and a set of package inter-change stations $\mathrm{CS} = \{\mathrm{cs}_1, \mathrm{cs}_2, \ldots, \mathrm{cs}_m\}$ in the designated city.
2. A historical set of taxi trajectory records {Tr}, such as from the past month in the designated city.
3. A set of real-time taxi ordering requests TOR from mobile phone apps, and a set of real-time package delivery requests PR. Note that these two requests come in stream.

Find the optimal package operational paths for each package request (pr), with the **objective** of minimizing its package delivery time.

Constraints:

1. Only taxis that accept the real-time taxi ordering requests after the package delivery request is posted can be scheduled, i.e. $\{\mathrm{tor}.t_t\} > \mathrm{pr}.t_p$.
2. A recruited taxi can be available to participate again only after completing the current task (i.e. dropping off the package at the predefined interchange station).

12.4 Overview of CrowdDeliver

Our proposed **CrowdDeliver** system consists of three components, namely, *Data Pre-processing*, *Offline Trajectory Data Mining*, and *Online Package Routing*. Its inputs include offline data sources such as POI data, road network data, and taxi trajectory data, and online requests (i.e. real-time taxi ordering requests *TOR* and package delivery requests PR). The outputs are the final optimal package operational paths, as shown in Fig. 12.3.

1. **Data Pre-processing**. *Data Pre-processing* in **CrowdDeliver** is a two-step procedure, as illustrated in the leftmost part of Fig. 12.3. *POIs Filtering* selects proper POIs to serve as package interchange stations based on POI map and road network data. More specifically, we first identify and select the POIs near the road sides according to their geographical locations. Then, for each driving direction

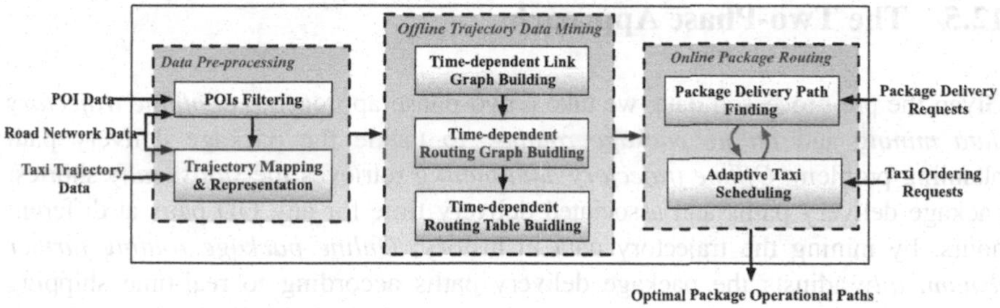
Offline Trajectory Data Mining
Data Pre-processing
Online Package Routing
POI Data
Road Network Data
Taxi Trajectory Data
POIs Filtering
Trajectory Mapping & Representation
Time-dependent Link Graph Building
Time-dependent Routing Graph Buidling
Time-dependent Routing Table Buidling
Package Delivery Path Finding
Adaptive Taxi Scheduling
Package Delivery Requests
Taxi Ordering Requests
Optimal Package Operational Paths

Fig. 12.3 The overview of CrowdDeliver

of an edge, we just keep at most one POI. There thus can be two POIs for a single edge, each for a driving direction. After filtering, we get 852 POIs in the designated city of Hangzhou. Note that POIs may locate densely in some small areas, such as near the Railway Station.

The *Trajectory Mapping and Representation* module maps time-stamped GPS points onto the road network based on our previous algorithm [16]. Thus one trajectory (a sequence of time-stamped GPS points) can be transferred to a sequence of edges (with the driving direction). We formally represent a trajectory as a quadruple $\langle e_+(u,v), e_-(m,n), t_s, t_e \rangle$, where $e_+(u,v)$, $e_-(m,n)$ are the original and destination edges of the trajectory respectively; t_s, t_e are the starting and ending time of the trajectory, thus $t_e - t_s$ is the *driving time* of the trajectory.

2. **Offline Trajectory Data Mining**. The main objective of the component is to discover the spatial-temporal patterns of taxi rides from trajectory data in history, mainly including the frequency of passenger taking taxis and the driving time needed between different POIs at different hours. We discuss this problem in details in *Time-dependent Link Graph Building* section (Step 1 in Sect. 12.5.1). Considering the fact that most passengers only take a taxi when their destinations are within a certain range of distance due to financial concerns, packages to be shipped to a distant place may need to take several rides with stops at intermediate stations. Thus a question arises naturally: *how to choose the intermediate stations to minimize package delivery time?* We address this question in *Time-dependent Routing Graph and Table Building* sections (Steps 2 and 3 in Sect. 12.5.1).
3. **Online Package Routing**. The objective of this component is to find the near optimal package operational paths. The shortest paths obtained through mining the historical trajectory data offline do not take the real-time taxi ordering requests into consideration, and thus they might not be the optimal in actual settings. The *Online Package Routing* component first retrieves the *statistically* shortest package delivery path and estimated delivery time, according to the package delivery request (i.e. *Package Delivery Path Finding*). Then, it decides whether to assign the package to a currently available taxi or not, by comparing its expected delivery time to the one suggested by the offline computation (i.e. *Adaptive Taxi Scheduling*). These two parts work *interactively* to find the final optimal package operational path. Its detailed procedure is elaborated in Sect. 12.4-B.

12.5 The Two-Phase Approach

Given the pre-processed data, we take a two-phase approach, i.e. *offline trajectory data mining* and *online package routing*, to tackle the package delivery path planning problem. *Offline trajectory data mining* retrieves the *statistically* shortest package delivery paths and associated delivery time for any OD pairs at different hours, by mining the trajectory data in history; *Online package routing further dynamically* adjusts the package delivery paths according to real-time shipping request and taxi availability. We present the technical details in the next two sections.

12.5.1 Phase I: Offline Trajectory Data Mining

1. ***Time-Dependent Link Graph Building:*** The link graph (LG) is a directional graph, in which the nodes are the interchange stations (CS); the weight of an edge from one interchange station (cs_i) to another one (cs_j) denotes the time needed to ship a package from cs_i to cs_j directly, which can be derived from the trajectory data. Obviously, the edge value varies at different time of a day, since the passengers' demands for taxis and the traffic on the road are time-dependent. For simplicity, we classify days into *work days* and *rest days*; divide a *work day* into three time slots (i.e. night-time hours, day-time hours, and rush hours) and a *rest day* into two time slots (night-time hours and day-time hours), as shown in Fig. 12.4. As mentioned before, packages are not delivered during rush hours. Thus, we build a link graph for each of the four time slots separately; two time slots for work days, and two for rest days. Building a link graph building takes two steps, detailed as follows.
Step 1: **From Trajectory Data to Passenger Flow**. It is straightforward to infer the passenger flow between any two interchange stations during a given time slot from the trajectory data. Specifically, we *first* group the trajectories according to their starting time (t_s). *Second,* to compute the passenger flow from cs_i to cs_j, we count the number of the trajectories satisfying Eqs. (12.1) and (12.2). It should be noted that there could be no passenger flow between some interchange station pairs. Prior research has observed regular spatial and temporal patterns in human

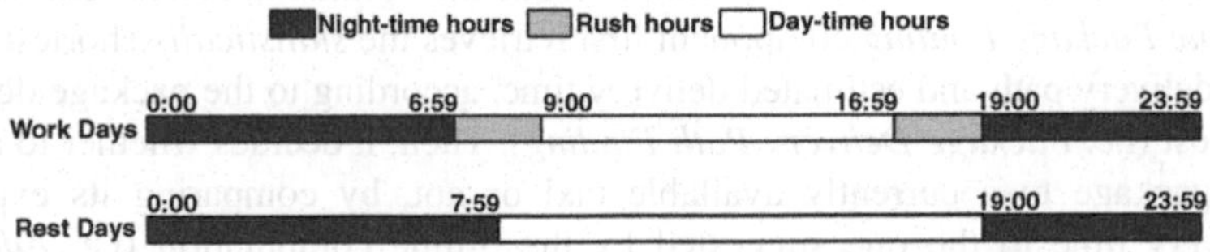
Night-time hours
Rush hours
Day-time hours
Work Days
0:00
6:59
9:00
16:59
19:00
23:59
Rest Days
0:00
7:59
19:00
23:59

Fig. 12.4 A work day is divided into night-time hours, rush hours, and day-time hours; a rest day is divided into night-time and day-time hours trajectory data (including taxi GPS traces), and consequently utilized these patterns in various applications [17].

$$\text{Ddist}\,(\text{Tr}_i.o, \text{loc}(\text{cs}_i)) < \delta \tag{12.1}$$

$$\text{Ddist}\,\big(\text{Tr}_i.d, \text{loc}\big(\text{cs}_j\big)\big) < \delta \tag{12.2}$$

where $\text{Tr}_i.\ o$ and $\text{Tr}_i.\ d$ are the original and destination points of Tr_i, respectively; loc(·) gets the latitude and longitude location of the given interchange station; Dist($a \cdot b$) calculates the *driving distance* from point a to point b; δ is a user-specified parameter. The physical meaning of δ is that any passenger-delivery ride which starts and ends near a pair of interchange stations (i.e., with driving distance less than δ) can be hitchhiked for the package delivery between this pair. Hence, for a given OD pair, a bigger δ would result in a bigger number of passenger flow. It is worth noting that, for a specific trajectory, there could be multiple interchange station pairs that satisfy Eqs. (12.1) and (12.2), in other words, can provide package hitchhiking ride between all these pairs. Therefore, a bigger δ also leads to a bigger number of interchange station pairs, suggesting that the corresponding trajectory can be more capable of providing hitchhiking rides. However, for passengers, a bigger δ may mean a longer waiting time for the reserved taxis, since the taxi driver might have to travel farther to collect the package before picking up passengers. To control for the additional waiting time, we set δ to 500 m.

Step 2: **From Passenger Flow to Edge Value.** To estimate an edge value, we need to estimate two parts, i.e. the waiting time and the driving time. The waiting time on the edge is defined as the time required to wait for a suitable hitchhiking ride that can transport a package from cs_i to cs_j directly. To address this problem, we employ the Non-Homogeneous Poisson Process (NHPP) to model the behavior of passenger taking taxis [4]. According to the statistical frequency of passenger taking taxis from cs_i to cs_j in history (i.e., passenger flow), we can estimate the waiting time of packages at different time slots at the interchange stations. Under the Poisson hypothesis within a time slot, we could derive the probability distribution of the waiting time for the next suitable hitchhiking ride event (i.e. t_{next}, the event of a passenger taking taxi from cs_i to cs_j), which can be expressed in Eq. (12.3):

$$\begin{aligned} Pt_{\text{next}} \leq t \quad &= 1 - Pt_{\text{next}} > t \\ &= 1 - P\{N(t) = 0\} \\ &= 1 - e^{-\lambda \cdot t} \end{aligned} \tag{12.3}$$

Here $N(t)$ represents the number of events occurring within t, and $P\{N(t) = k\} = e^{-\lambda \cdot t \cdot \left((\lambda \cdot t)^k / k!\right)}$. Then the probability density function (pdf) of t_{next} is just the derived function of Pt_{next}, as can be seen in Eq. (12.4).

$$p(t) = \lambda \cdot e^{-\lambda \cdot t} \tag{12.4}$$

Thus, we can deduce the expected t_{next} (i.e. the waiting time for the hitchhiking ride event occurring):

$$\mathrm{E}[t_{\text{next}}] = \int_0^\infty t \cdot \lambda \cdot e^{-\lambda \cdot t} \cdot \mathrm{dt} = \frac{1}{\lambda} \tag{12.5}$$

Note that λ in the model is the *frequency* of passenger taking taxis from cs_i to cs_j, which can be easily estimated by the Eq. (12.6)

$$\widehat{\lambda} = \frac{\bar{N}}{\Delta T} \tag{12.6}$$

where $\bar{N}$ is the average number of passengers taking taxis from cs_i to cs_j during the given time slots; ΔT is the length of the that time slot. For example, $\Delta T = 9$ h for the day-time time slot of work days.

Therefore, the *waiting time* on the edge from cs_i to cs_j is:

$$\text{waiting time} = \frac{1}{\widehat{\lambda}} = \frac{\Delta T}{\bar{N}} \tag{12.7}$$

For each passenger-delivery ride from cs_i to cs_j, the *driving time* is simply the average over all such rides, as shown in Eq. (12.8)

$$\text{driving time} = \frac{\sum_{\mathrm{i=1}}^{\mathrm{N}} \mathrm{Tr}_i \cdot (te - ts)}{N} \tag{12.8}$$

where N is the number of passenger-delivery rides during the target time slot in the observed days. $(te - ts)$ is the time cost of the corresponding taxi ride.

Finally, the edge value is the sum of waiting time and driving time on the respected edge, as shown in Eq. (12.9)

$$\begin{aligned} tc &= \text{waiting time} + \text{driving time} \\ &= \frac{\Delta T}{\bar{N}} + \frac{\sum\limits_{i=1}^{N} \mathrm{Tr}_i \cdot (te - ts)}{N} \end{aligned} \tag{12.9}$$

Note that the edge value would be $+\infty$ if there was no passenger flow on the respected edge in history.

Edge Value Statistics. The histogram of all edge values at a given time slot (e.g. day time in work days) in shown in Fig. 12.5a. Note that edges with $+\infty$ value are excluded. The total number of edges is 40,600, and the number of edges with value less than 1 h is around 2000—a very small percentage. To further

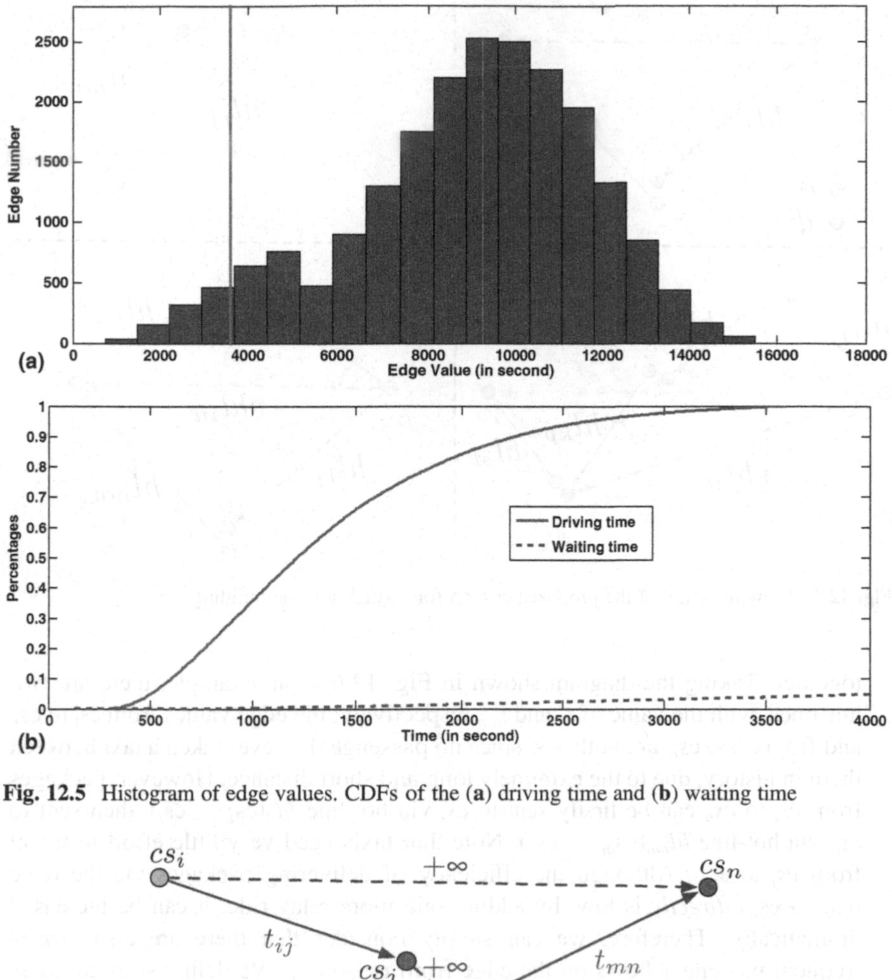
Edge Number
2500
2000
1500
1000
500
0
(a)
0
2000
4000
6000
8000
10000
12000
14000
16000
18000
Edge Value (in second)
Percentages
Driving time
Waiting time
(b)
Time (in second)

Fig. 12.5 Histogram of edge values. CDFs of the (a) driving time and (b) waiting time

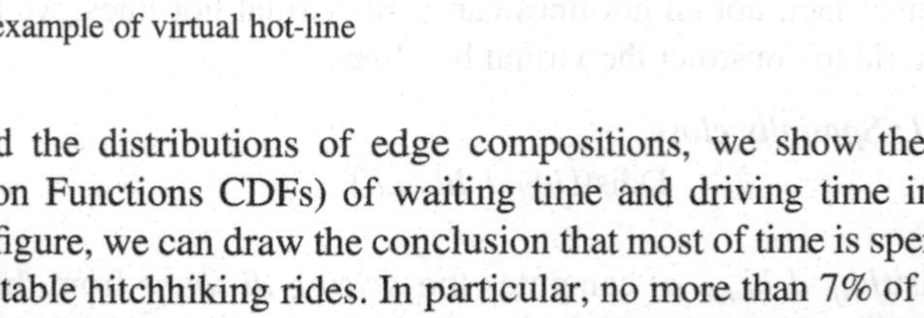
cs_i
$+\infty$
cs_n
t_{ij}
cs_j
$+\infty$
t_{mn}
cs_m

Fig. 12.6 An example of virtual hot-line

understand the distributions of edge compositions, we show the Cumulative Distribution Functions CDFs) of waiting time and driving time in Fig. 12.5b. From the figure, we can draw the conclusion that most of time is spent on waiting for the suitable hitchhiking rides. In particular, no more than 7% of edges have a waiting time less than an hour; while for the driving time, over 70% of edges are less than half an hour.

2. ***Time-Dependent Routing Graph Building:*** Intuitively, delivering packages via edges with smaller values is more efficient as their time costs are relatively low. We thus define such edges as hot-lines. Some hot-lines can be further connected

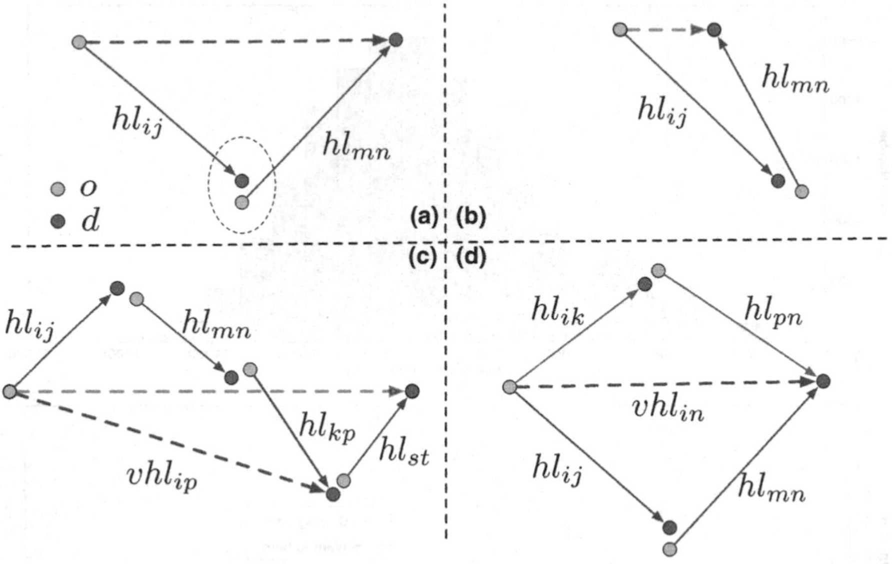

hl_{ij}
hl_{mn}
o
d
(a)
(b)
(c)
(d)
hl_{kp}
hl_{st}
vhl_{ip}
hl_{ik}
hl_{pn}
vhl_{in}

Fig. 12.7 Demonstration of the proposed criteria for virtual hot-line building

together. Taking the diagram shown in Fig. 12.6 as an example, there are two hot-lines, with the value of t_{ij} and t_{mn} respectively; the edge values from cs_i to cs_n and from cs_j to cs_m are both $+\infty$ since no passenger has ever taken a taxi between them in history, due to the extremely long and short distance. However, packages from cs_i to cs_n can be firstly sent to cs_j via hot-line $hl_{ij}(cs_i \rightarrow cs_j)$, then sent to cs_n via hot-line $hl_{mn}(cs_m \rightarrow cs_n)$. Note that taxis need very little effort to travel from cs_j to s_m . Although the efficiency of delivering packages via the edge $(cs_i \rightarrow cs_n)$ *directly* is low, by adding one more relay ride, it can be increased dramatically. Therefore, we can simply consider that there are also *virtual* frequent passenger flows on the edge from cs_i to cs_n. We define such edges as *virtual hot-lines*. The edge value of *virtual hot-lines* is the sum of that of parent hot-lines (t_{ij} and t_{mn} in the example), with additional time cost on transshipment. As a matter of fact, not all hot-lines can form virtual hot-lines, we thus propose several criteria to construct the virtual hot-lines.

Criterion 12.1: Spatially close

$$\mathrm{Ddist}(hl_{ij}.d, hl_{mn}.o) < \beta$$

where Ddist($hl_{ij}.\ d$, $hl_{mn}.\ o$) computes the *driving distance* from the destination node of hot-line hl_{ij} to the original node of hot-line hl_{mn}. β is a user-specified parameter, as shown in the highlighted circle in Fig. 12.7a. The criterion indicates that it only requires very little extra effort from taxi drivers (i.e., driving a short extra distance of β) to ship packages from $hl_{ij}.\ d$ to $hl_{mn}.\ o$ after dropping off passengers near $hl_{ij}.\ d$. We set β to 500 m when constructing virtual hot-lines.

Criterion 12.2: Move forward

$$\text{dist}(h_{ij}.o, h_{ij}.d) < \text{dist}(h_{ij}.o, h_{mn}.d)$$

where dist() gets the Euclidean distance of two given points. This criterion guarantees the package will always move forward and further closer to the destination after adding one more transshipment. The two hot-lines in Fig. 12.7b cannot be connected to build virtual hot-line because it violates the *move forward* criterion. We can see dist(h_{ij}. o, h_{ij}. d) > dist(h_{ij}. o, h_{mn}. d).

Criterion 12.3: Limited number of transshipments

$$|\text{Parent}(vhl)| - 1 \leq c$$

where|Parent(·)|gets the number of the parent hot-lines for the resulted virtual hot-lines; c is a user-specified constant. This criterion ensures there are no more than c transshipments for the virtual hot-lines, because considering more transshipments may increase the risk of package security issues. As for the example in Fig. 12.7c, the virtual hot-line (vhl_{ip}) is valid, while vhl_{it} is invalid, since the number of its parent hot-lines (hl_{ij}, hl_{mn} and hl_{kp}) is 3 if we set $c = 2$.

Criterion 12.4: Minimum time cost on virtual hot-line

$$\arg\min \mathcal{TC}(vhl_{\text{in}})$$

where $\mathcal{TC}()$ computes the time cost on the corresponding virtual hot-line. As illustrated in Fig. 12.7d, virtual hot-line (vhl_{in}) can be constructed by hl_{ij} and hl_{mn}, and also can be formed by hl_{ik} and hl_{pn}. This criterion ensures we choose the virtual edge with the minimum time cost, resulting from the parent hot-lines. Furthermore, the time cost of resulted virtual edge should be less than the time cost on the edge $edge_{in}$ itself, which can be retrieved from the built link graph. Mathematically, $\mathcal{TC}(vhl_{\text{in}}) < \mathcal{TC}(\text{edge}_{\text{in}})$ for the example in Fig. 12.7d.

The final routing graph consists of a set of interchange stations as nodes and both hot-lines and virtual hot-lines as edges.

3. ***Time-Dependent Routing Table Building:*** Obviously, for OD pairs having hot-lines or virtual hot-lines, the shortest paths are just the hot-lines or virtual hot-lines from origins to destinations. However, the above procedure of routing graph building cannot guarantee a *complete* graph. Thus, we need to build the time-dependent routing tables, with the objective of finding the time-dependent shortest paths for any pair of origin and destination nodes. The algorithm proposed in [18] was applied. The shortest paths can contain both hot-lines and virtual hot-lines. For paths which contain virtual hot-lines, we further retrieve their parent hot-lines and transshipment stations, and the shortest paths can be represented by the *sequence of interchange stations*. We denote the collection of the shortest paths by Path; one element shortest path for a given OD pair (e.g. Path_{ij} represents the shortest path from cs_i to cs_j, denoted by $cs_i \rightsquigarrow cs_j$) has four main attributes: the original station Path_{ij}. o, the destination station Path_{ij}.

d, the sequence of interchange stations $Path_{ij}$. seq, the expected delivery time $Path_{ij}$. et. It should be noted that there may be no valid path between some OD pairs, based on the routing graphs. For those OD pairs, we apply the following simple four-step top-*k* algorithm (i.e. Algorithm 12.1) to get their routing paths. The first two steps (i.e. Steps 1 and 2) get the top-*k* linked interchange stations to the origin and destination respectively. The last two steps (i.e. Steps 3 and 4) find the shortest path between the top-*k* linked interchange station pairs. Here, we assume cs_o is the original station, and cs_d is the destination station. Finally, for any OD pairs, we have built the time-dependent routing tables based on trajectory data in history, with the minimum expected package delivery time.

Algorithm 12.1: The Simple Top-*k* Algorithm

• Step 1: Based on the link graph (LG), we get the set of top-*k* connected interchange stations of cs_o (i.e. with the top-*k* minimum outgoing edge values), denoted by $CS_{O*} = \{cs_{o1}, cs_{o2}, \ldots, cs_{ok}\}$.

• Step 2: With the same idea, we get the set of top-*k* connected interchange stations of cs_d (i.e. with the top-*k* minimum incoming edge values), denoted by $CS_{*d} = \{cs_{1d}, cs_{2d}, \ldots, cs_{kd}\}$.

• Step 3: From CS_{O*} to CS_{*d}, we get the set of shortest paths ($\{Path_{ij}\}$), where $cs_i \in CS_{o*}$ and $cs_j \in CS_{*d}$. The set $\{Path_{ij}\}$ is a subset of $\{Path\}$.

• Step 4: If $\{Path_{ij}\}$ is empty, we can conclude that there is no delivery path from cs_o to cs_d; otherwise, the delivery path from cs_o to cs_d is the one which minimizes the delivery time cost, as shown in Eq. (12.10).

$$\arg\min_{i;j=1,\ldots,k}\left[tc(edge_{oi}) + Path_{ij}\cdot et + tc(edge_{jd})\right] \tag{12.10}$$

12.5.2 Phase II: Online Package Routing

Based on the time-dependent routing tables, we can easily obtain the shortest paths for any OD pairs. For instance, for a package to be sent from *o* to *d* the intermediate stations that it will stop can be determined by simply looking up the time-dependent routing tables. Afterwards, we can schedule *suitable* taxis to help deliver the package *segment by segment*. For example, the package delivery path in Fig. 12.8 is $o \to A \to B \to C \to d$, and for segment $o \to A$, we can just wait for a suitable taxi which will take a passenger from *o* to A (the information is contained in the real-time taxi ordering requests), and assign the package to that taxi. After the package arrives at *A*, we repeat the process, until the package reaches its final destination *d*. As discussed, these shortest paths are "optimal" in the statistical sense, completely independent of the real-time taxi ordering requests. Moreover, as indicated by the evidence provided in Fig. 12.5b, most of the delivery time is spent on waiting for the suitable taxis, which inspires us to develop a better taxi scheduling algorithm to re-plan the path dynamically.

We use the example shown in Fig. 12.8 again to illustrate our idea. Before a taxi ordering request from *o* to *A* comes in, a taxi just accepts the request to take a

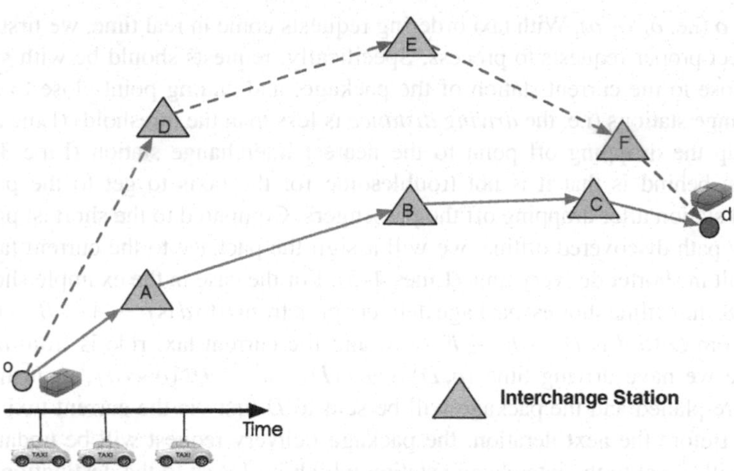
E
D
F
B
C
d
A
o
Interchange Station
Time
TAXI
TAXI
TAXI

Fig. 12.8 An illustrative example of package delivery path replanning

passenger from *o* to *d* directly. Under this circumstance, we should assign the package delivery task to the taxi heading to *d*. To handle similar situations, we thus propose an algorithm to schedule taxis adaptively according to the dynamic taxi ordering requests in real time called **AdaPlan**, to re-plan the package delivery path *iteratively*.

Algorithm 12.2 Adaptive Planning (AdaPlan) Algorithm

Input: real-time taxi ordering requests ({tor});
the package delivery request (pr);
time-dependent routing tables.
Output: the next interchange station that the package stops;
the arriving time of the package at the next stop

1 **while** $tor.t > pr.t$ **do**
2 **if** Ddist(tor. *o*, pr. *o*) $< \delta$ & Ddist(tor. d, {CS}) $< \delta$ **then**
3 $cs_{tord} \leftarrow$ closestCS(tor. d)
//closestCS(·) get the closest interchange station to the dropping off point (tor. d)
4 **if** driving time $(pr.o \rightarrow cs_{tord})$ + $\mathcal{ET}(cs_{tor} \rightsquigarrow d) < \mathcal{ET}(pr.o \rightsquigarrow d)$ **then**
5 Assign the package to the current taxi, and the next stop will be cs_{tord}
//$\mathcal{ET}(\cdot)$ gets the expected time cost for the corresponding path by looking up routing tables.
6 $pr.o \leftarrow cs_{tord}$
7 *pr. t* will be the time when the current taxi offloads the package.
8 **end if**
9 **end if**
10 **end while**

The pseudo-code of **AdaPlan** algorithm at one iteration is shown in Algorithm 12.2. Only real-time taxi ordering requests that generated after the birth time of the package delivery request can be processed (Line 1). At the first iteration, the package starts at o (i.e. $o_c \leftarrow o$). With taxi ordering requests come in real time, we first check and select proper requests to process. Specifically, requests should be with starting point close to the current station of the package, and ending point close to any of interchange stations (i.e. the *driving distance* is less than the threshold) (Line 2). We then map the dropping off point to the nearest interchange station (Line 3). The intention behind is that it is not troublesome for the taxis to get to the package off-load station after dropping off the passengers. Compared to the shortest package delivery path discovered offline, we will assign the package to the current taxi if it can result in shorter delivery time (Lines 4–5). For the case in the example shown in Fig. 12.8, the offline shortest package delivery path from o to d is $o \rightarrow A \rightarrow B \rightarrow C \rightarrow d$, while from D to d is $D \rightarrow E \rightarrow F \rightarrow d$, and the current taxi ride is from o to D. Suppose we have driving time $(o, D) + \mathcal{ET}(D \rightsquigarrow d) < \mathcal{ET}(o \rightsquigarrow d)$, then the path will be re-planed and the package will be sent to D first via the current taxi, rather than A. Before the next iteration, the package delivery request will be updated: its origin will be set to the interchange station which is closest to the destination of the previous taxi ride; its birth time will be set to the time that the schedule taxi offloaded the package to the interchange station (Lines 6–7). The information contained in the newly updated package delivery request will be used to retrieve the new shortest package delivery path and expected delivery time from the time-dependent routing tables, which would be used as the new reference for the next iteration. This iterative process terminates when the package arrives at its final destination. The rationale of the algorithm is using the offline discovered shortest paths and the associated expected time as the baseline to guide the on-line package routing process. To be more specific, if exploiting the currently available taxi ride can improve the package delivery efficiency, the package should be sent out immediately rather than waiting.

12.6 Evaluations

In this section, we empirically evaluate the performance of the **CrowdDeliver**. We first introduce the experimental setup, baseline algorithms used for comparison, evaluation metrics and results on algorithm efficiency and effectiveness. Then we conduct a case study to demonstrate the package routing trajectories for three delivery requests. We discuss several issues to be further addressed in the end.

12.6.1 Experimental Setup

1. ***Experimental Data:*** We use the real-world datasets for the evaluation, i.e. the POI data, the road network data and 1 month of taxi trajectory data generated by 7614 taxis in the city of Hangzhou, China. We determine package interchange stations according to the POI data and the road network data as discussed earlier. We split the taxi trajectory data into training and testing sets, according to the date

Table 12.1 Statistics of the Taxi Trajectory and Road Network Data Sets

Datasets	Properties	Statistics
Taxi trajectory	Number of taxis	7614
	Number of occupied rides	≈6 M
Road network	Number of road intersections	22,593
	Number of road segments	24,975

of the month. Specifically, the training set contains taxi trajectories on 1st–20th, March, 2010, which are used to build time-dependent graphs and tables. The testing trajectories were generated from 21st to 31st, March, 2010, which is used to simulate the real-time taxi ordering requests (TOR) for testing the performance of the proposed **AdaPlan** algorithm. Table 12.1 shows some statistics of the taxi trajectory and road network data.

2. ***Request Simulation:*** Since the datasets do not contain information about package delivery and real-time taxi ordering requests, we apply different mechanisms to simulate these two kinds of requests. To simulate a package delivery request (PDR), we randomly generate its birth time, origin and destination. We further eliminate requests with short-distance OD pairs since few users would request speedy shipping as ordinary delivery may be equally efficient in this case. We determine whether a loaded taxi is suitable for a given package delivery request according to the origin and destination of its upcoming passenger-delivery trip. However, such information is only available when someone orders a taxi via mobile apps. Since our taxi GPS trajectory data do not contain information about whether a taxi was requested online, in this chapter, we take a "*replay and sample*" strategy to simulate real-time taxi ordering requests. Specifically, we first "replay" all the taxi trajectories taking place during the testing time, then "sample" the taxi trajectories with a fix rate. For instance, during the next 15 min in a give time slot, if there will be 100 taxi trajectories, we randomly sample some of them, and assume they are generated by sending real-time taxi ordering requests with mobile apps of users. We will further investigate how different sampling rate values affect the performance in Sect. 12.5-D.
3. ***Evaluation Metrics:*** We adopt the following two metrics to evaluate the proposed **CrowdDeliver**.
Success Rate. The success rate is the ratio of the number of packages which can be delivered successfully within a given deadline (i.e., time duration) to the number of total packages (i.e., the number of package delivery requests simulated),

$$\mathrm{SR}(t < T) = \frac{\left|\mathcal{TC}\left(\mathrm{OptPath}\left(\mathrm{o_p} \rightsquigarrow \mathrm{d_p}\right)\right) < T\right|}{|PQ|} \tag{12.11}$$

where $\mathrm{OptPath}(\mathrm{o_p} \rightsquigarrow \mathrm{d_p})$ represents the optimal operational path generated by the proposed **AdaPlan** algorithm for a given package delivery request; T is the given dead-line. The delivery performance is better if the success rate is higher within a given shorter deadline.

Number of Transshipments. The number of transshipments ($\text{Num}_{\text{trans}}$) during a package delivery is the number of intermediate stations that the package stops at, which is equal to the number of participating taxis $\text{Num}_{p\_\text{taxis}}$ minus one (Formula 12.12). Fewer transshipments means a lower chance of package loss or damage, and perhaps less overhead cost

$$\text{Num}_{\text{trans}} = \text{Num}_{\text{relays}} = \text{Num}_{p\_\text{taxis}} - 1 \tag{12.12}$$

12.6.2 Baseline Algorithms

To show the superior performance of our proposed algorithm, we compare it with the following two baseline algorithms.

1. **FCFS**—This method adopts the *First Come First Service* strategy. Specifically, the package will be assigned to the *first* taxi that will pick up a passenger near the interchange station that the package locates, *regardless of its destination*. In fact, this algorithm is an extension of the simple and well-known *flooding* strategy [11].
2. **DesCloser**—This method assigns the package to the first taxi heading to somewhere closer to the destination of the package, compared to the current station of the package. This algorithm implements a distance-based geo-cast scheme that is commonly seen in other domains [19].

Remark: Each relay in **DesCloser** is effective as it ensures that the package would move towards its destination step by step; while some relays in **FCFS** can be ineffective as the package moves further away from its destination.

12.6.3 Experimental Objectives

We design the experiment to address the following questions.

Question 1: How much computational resource is required to generate the response for a package delivery request?

Question 2: What is the success rate of the package delivery under different time constraints?

Question 3: How many transshipments/relays are needed to take the package to its destination?

Question 4: How far the solution obtained by the proposed algorithm is from the optimal one?

Question 5: How many packages can be delivered per hour on average (i.e., the transport capacity) with the proposed system?

The first question concerns the efficiency of **CrowdDeliver**, and *Questions* 2–3 are related to its effectiveness. To answer the first question, we compute the *response time* of the algorithms with respect to different taxi trajectory sampling rates. Since passenger flows are both time-dependent and space-dependent, to assess the effectiveness of the different algorithms, we calculate their success rates and the number of transshipments with respect to packages to be dispatched to different parts of the city at different time of the day. To validate of our proposed **AdaPlan** algorithm, we compare its efficiency and effectiveness to the **Optimal** algorithm (*Question 4*). We test the transport capacity of the proposed system under different number of package requests generated per hour, and also examine the system capacity given different number (density) of interchange stations in the designate city (*Question 5*).

12.6.4 Experimental Results

1. ***Results of Response Time:*** We first analyze the main *operations* involved in the three algorithms respectively. For a given package delivery request (pr), when a new real-time taxi ordering request (tor) comes in, all three algorithms need to determine whether tor starts near the origin of the package and stops at some interchange station. Mathematically, they all perform two distance comparison operations: Ddist(tor. o, pr. o) $< \delta$ and Ddist(tor. d, {CS}) $< \delta$. **FCFS** will recruit the first taxi that satisfies the criteria, but for **DesCloser**, it needs to further determine whether the heading destination of the taxi is closer to the destination of the package, compared to its current location, and thus one more *comparison* operation is required. For **AdaPlan**, the procedure is even more complicated, requiring additional *comparison and table look-up* operations, as shown in Lines 2–9 of Algorithm 12.2. Each algorithm needs to repeat its own operation procedure at each intermediate station and thus the total response time is the *accumulated* computational time over all iterations.
   We show the comparison of response time of the three different algorithms under different sampling rates in Fig. 12.9. Overall, more response time is needed as the sampling rate increases. This is very straightforward, since a bigger sampling rate means more candidate taxis to compare, resulting in greater latency to give a response. The response time of **FCFS** is at least five times larger than that of the other two algorithms (i.e. **DesCloser**, **AdaPlan**), and **DesCloser** is always in the middle under different sampling rates. The response time of **AdaPlan** is less than 1 s even if the sampling rate is high (i.e. 50%), indicating that our proposed algorithm is quite efficient, and can plan and adjust the shipping routes in real time. We also observe an interesting phenomenon: although **FCFS** only involves two operations for each candidate taxi, it is the most time-consuming method accumulatively. In comparison, **AdaPlan** which contains the most complex operations needs the shortest response time. We argue that this is because:

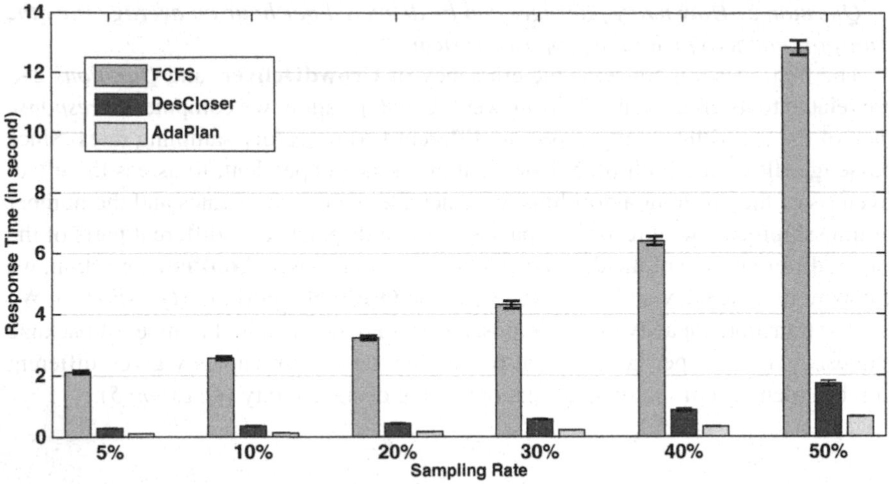
FCFS
DesCloser
AdaPlan
Response Time (in second)
Sampling Rate
5%
10%
20%
30%
40%
50%

Fig. 12.9 Evaluation results of response time for three algorithms under different sampling rates

Table 12.2 Some statistical information about the genera package delivery requests

Experiments	#of requests	Driving distance (km) [min,max)	Driving distance (km) mean ± std
Response Time w.r.t Sampling Rate I	15,000	[3.230,78.453]	21.832±12.409
Success Rate w.r.t Sampling Rate	same to the study of Response Time w.r.t Sampling Rate		
Success Rate w.r.t Region	5000 (Category I)	[3.328,52.738]	12.998±6.275
	5000 (Category II)	[3.230,66.685]	25.453±11.025
	5000 (Category III)	[3.349,78.453]	29.075±12.878
Success Rate w.r.t. Time	1000 (Time 1)	[3.934,35.514]	12.817±5.948
	1000 (Time II)		
	1000 (Time III)		
	1000 (Time IV)		
# of Transshipments w.r.t Region	same to the study of Success Rate w.r. Region		
#of Transshipments w.r.t Time	same to the study of Success Rate w.r.t Time		
AdaPlan vs. Optimal	1000	[3.110,7.980]	6.253±1.110

Time I, II, III, IV represent the package delivery requests occurring during the day time in work days, day time in rest days, night time in work days and night time in rest days respectively. Refer to Fig. 12.4 for the details of time partitions

FCFS requires much more transshipments—in other words, more rounds of computation—than the **AdaPlan** algorithm, as verified in Sect. V-D3. The detailed information about the simulated package delivery requests for this study is shown in Table 12.2. Among the 15,000 requests, the average driving distance of the package OD pairs is 21.832 km, and the driving distance can be up to 78.453 km. We randomly generated the birth time of the package delivery

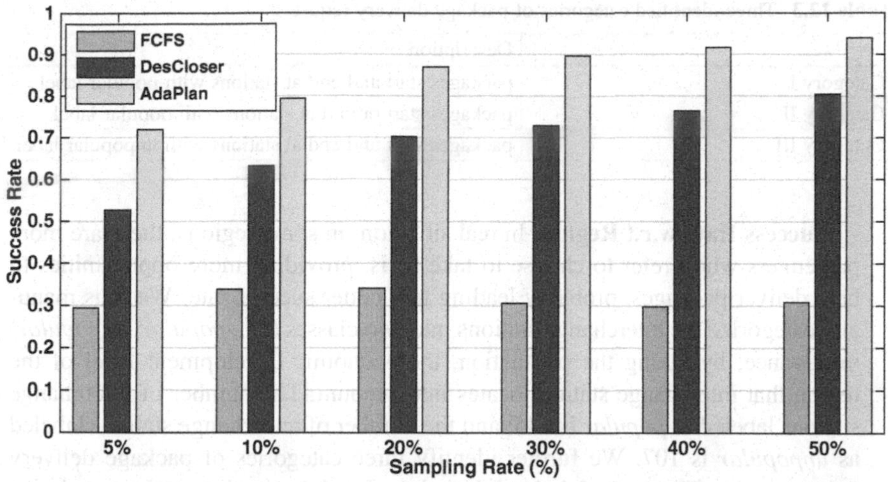
FCFS
DesCloser
AdaPlan
Success Rate
1
0.9
0.8
0.7
0.6
0.5
0.4
0.3
0.2
0.1
0
5%
10%
20%
30%
40%
50%
Sampling Rate (%)

Fig. 12.10 Evaluation results of success rate for three algorithms under different sampling rates

requests, and thus they were distributed rather evenly at the given time slots. For brevity, we do not show the request distributions on time for all experiments.

2. ***Results of Success Rate:*** As our proposed **CrowdDeliver** leverages taxis already reserved to transport passengers of which the availability is time and space dependent, we test its success rate under different sampling rates, different regions of the city, different time of the day.

   Success Rate w.r.t Sampling Rate. We report the success rate results under different sampling rates in Fig. 12.10. For **DesCloser** and **AdaPlan** algorithms in which each iteration gets the package closer to its destination, a higher sampling rate means more candidate taxis, shorter waiting time, and consequently a better success rate. However, for **FCFS**, the success rate fluctuates with the increase of sampling rate, indicating more candidate taxis cannot guarantee better performance. This is because, for FCFS, a higher sampling rate just means a shorter waiting time, but not necessarily a shorter driving time or fewer relays. **AdaPlan** algorithm performs consistently better than the other two algorithms under all sampling rates. Specifically, for **AdaPlan**, it achieves good success rate (i.e., over 70%) even if the sample rate is small (i.e. 5%); its success rate is over 85% when the sampling rate is 20%. For **FCFS**, the success rate is only around 30% under all sampling rates. Similar to the response time results, the performance of **DesCloser** is better than **FCFS**, but worse than **AdaPlan**. We set the sampling rate to 20% for the rest of the experiments. We choose relatively small sampling rate considering the fact that only a relatively small proportion of passenger's order taxis via mobile apps in real life. Note that we fix the deadline (i.e. $T = 8$ h) when computing the success rate for this study. As shown in Table 12.2, the simulated package delivery requests are same to that in the *response time study*.

Table 12.3 Three identified categories of package delivery requsets

	Description
Category I	packages start and end at stations with popular label
Category II	packages start or end at stations with popular label
Category III	packages start and end at stations with unpopular label

Success Rate w.r.t Region. In real situation, in some regions, there are more passengers who prefer to choose to take taxis, providing more opportunities to help deliver packages, probably leading to a better success rate. We thus manually categorize the interchange stations into two classes (i.e., *popular, unpopular*) in advance, by taking the population, the economic development level of the region that interchange station locates into account. The number of interchange stations labeled as *popular* is 645; and the number of interchange stations labeled as *unpopular* is 107. We further identify three categories of package delivery requests, according to the labels of the original and destination station, as shown in Table 12.3. Then, we test the performance on success rate for each category.

Figure 12.11 shows the comparison of success rate of the three algorithms under different deadlines (i.e. $T \in [1, 24]$ with an equal interval of 1 h) for the three categories. **AdaPlan** achieves the best performance for three categories under all deadlines, while the performance of **FCFS** is the worst. For the same algorithm, the performance is also different for different categories of package delivery requests. The performance for the Category I packages is the best; the performance is the worst for the Category III packages. Particularly, **AdaPlan** ensures that around 90% of the Category I packages (75% and 60% for Category II and III respectively) can be delivered within 6 h. While for **FCFS**, only less than 10% of Category II and III packages can be delivered within 6 h; only half of them can be successfully shipped within 13 h. Similar to the previous results, **DesCloser** performs better than **FCFS** and worse than **AdaPlan** for all three categories. The number of package delivery requests is the same (i.e. 5000) for all three categories. As shown in Table 12.2, the average driving distance of the Category I packages is the smallest (i.e., 12.998 km). In comparison, the average driving distance of the Category II and III is 25.453 and 29.075 km, respectively.

Success Rate w.r.t Time. We report the success rate and delivery time (mean and standard deviation) with respect to the time of the day (i.e., day-time and night-time) and day of week (i.e. work day and rest day) for the proposed **AdaPlan** algorithm in Fig. 12.12. During the day time, the packages are easier to be delivered in work days (with a higher success rate and smaller package delivery time) than in the rest days. The result is the opposite during the night time. One possible reason is that most people tend to stay at home after work for a good rest. In contrast, they are more likely to go out for fun in the evening during the rest days. Results show that our proposed **CrowdDeliver** achieves a promising efficiency. Specifically, the average package delivery time at night is 2.86 and 2.74 h in work and rest days respectively; the average package delivery time in the day-time is 2.48 and 2.30 h in work and rest days respectively. We generated

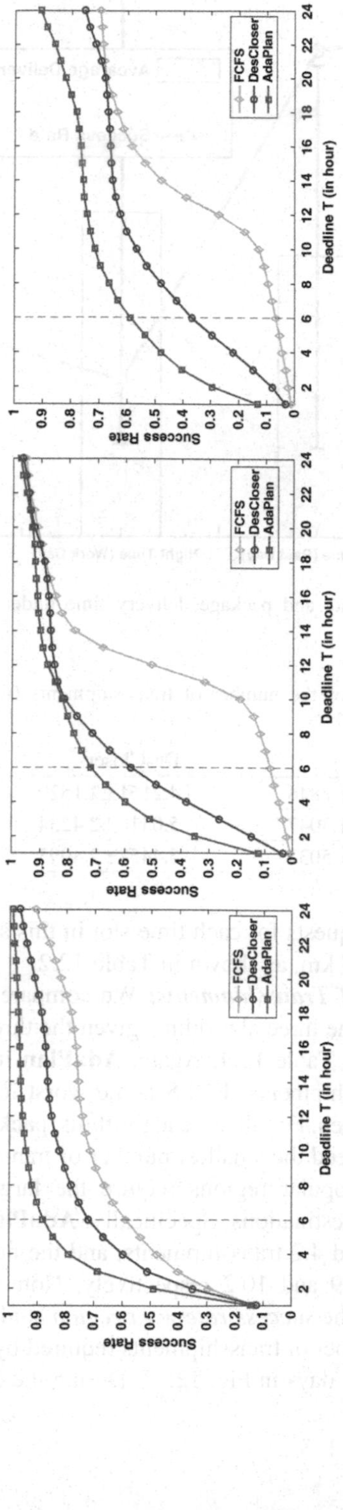

Success Rate
Deadline T (in hour)
FCFS
DesCloser
AdaPlan

Fig. 12.11 Evaluation results of success rate under different deadlines for three different categories for all three algorithms. (Left) Category I. (Middle) Category II (Right) Category III

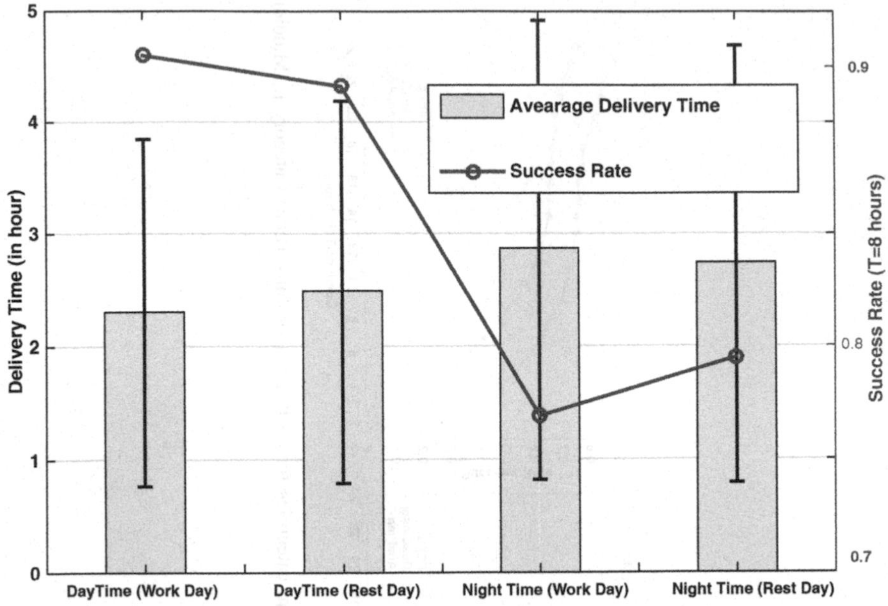
Avearage Delivery Time
Success Rate
Delivery Time (in hour)
Success Rate (T=8 hours)
DayTime (Work Day)
DayTime (Rest Day)
Night Time (Work Day)
Night Time (Rest Day)
0
1
2
3
4
5
0.7
0.8
0.9

Fig. 12.12 Results of success rate and package delivery time under different time slots (the deadline T is fixed to 8 h)

Table 12.4 Comparable results of the number of trans-shipments (mean ± *std*) for different categories of packages

	AdaPlan	DesCloser	FCFS
Category I	4.0713±1.2846	4.2161±2.1529	7.9003±5.4120
Category II	4.2379±1.3047	5.0410±2.4234	11.1874±9.5981
Category III	4.3399±1.5033	5.4458±2.4993	11.4545±7.6079

1000 package delivery requests for each time slot in this study, with an average driving distance of 12.817 km, as shown in Table 12.2.

3. ***Results of the Number of Transshipments:*** We compare the number of transshipments resulted from the three algorithms given the three categories of delivery requests, as shown in Table 12.4. Again, **AdaPlan** in general requires the smallest number of transshipments; **FCFS** is the worst for all three categories; and **DesCloser** is in-between. For all three algorithms, packages that start and end both at popular districts need the smallest number of transshipments, in contrast transported between unpopular regions require the largest number of transshipments to reach their destinations. Specifically, **AdaPlan** allows packages to be delivered within around 4.2 transshipments, and the numbers for **DesCloser** and **FCFS** are around 4.9 and 10.2 respectively. Note that the experimental setting is same to that in the *success rate w.r.t region* study.

We also show the number of transshipments required by **AdaPlan** at different hours in the work and rest days in Fig. 12.13. During the day time, **AdaPlan** on

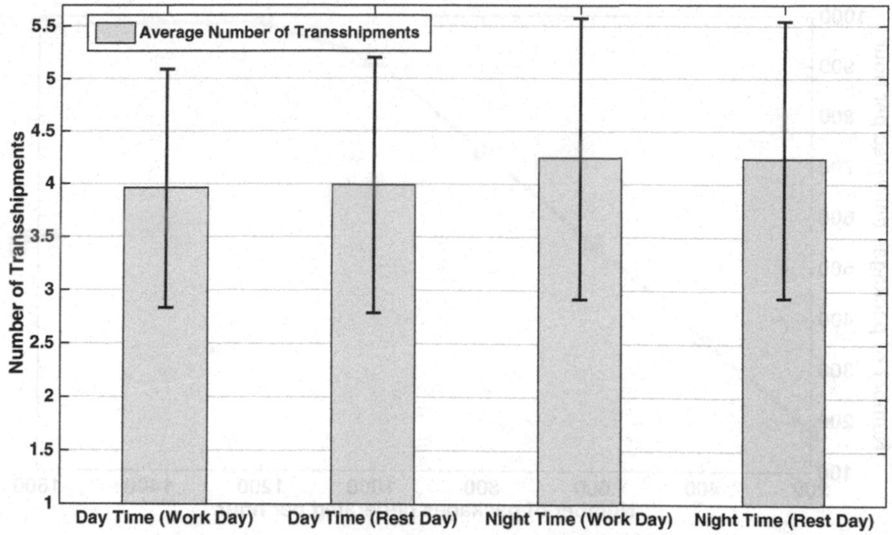
Average Number of Transshipments
Number of Transshipments
5.5
5
4.5
4
3.5
3
2.5
2
1.5
1
Day Time (Work Day)
Day Time (Rest Day)
Night Time (Work Day)
Night Time (Rest Day)

Fig. 12.13 Results of the number of transshipments (mean ± *std*) under different time slots (the deadline *T* is fixed to 8 h)

Table 12.5 Comparable results of results of the average response time and delivery time

	Optimal	AdaPlan	Statistically Optimal
Ave.Response Time (in sec.)	14280.9	0.80	0.65
Ave.Delivery Time (in hour)	1.164	2.337	4.796

average involves four transshipments whether on work days or rest days. During the night time, it requires slightly more transshipment on a work day (4.25) than on a rest day (4.23). Note that the experimental setting is same to that in the *success rate w.r.t time* study.

4. ***AdaPlan v.s. Optimal:*** To show the quality and efficiency of our proposed **AdaPlan**, we compare the distance of our solution to the ideal result of the **Optimal** algorithm w.r.t the average delivery time and average response time. The **Optimal** algorithm assumes that the spatial and temporal information of all the passenger-sending rides is foreknown, based on which it is feasible to calculate the true delivery time for a given route. Specifically, for a package delivery request, to get its optimal delivery route, we enumerate all the package delivery routes from the origin to the destination and calculate the corresponding delivery time, then pick the one with the least delivery time as the optimal result. Considering the fact that it is really time-consuming to get the optimal delivery route through an enumeration process, we use part of the network which consists of 20 interchange stations in the downtown area to conduct the experiment.

   In terms of the average response time, the **Optimal** algorithm takes almost 4 h on average to find the optimal delivery route for a package query even with just 20 interchange stations, as shown in Table 12.5. The **Statistically Optimal**

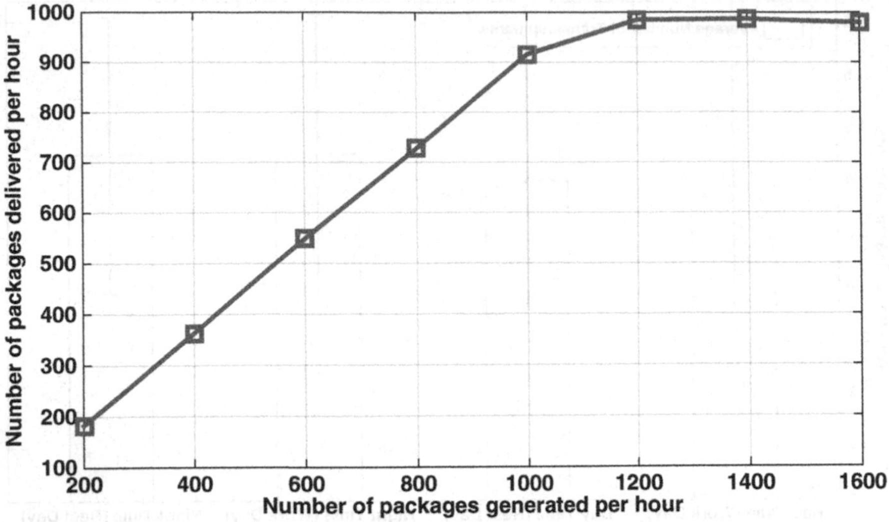
Number of packages delivered per hour
1000
900
800
700
600
500
400
300
200
100
200
400
600
800
1000
1200
1400
1600
Number of packages generated per hour

Fig. 12.14 Results of transport capacity under different numbers of package requests per hour

algorithm, which rigidly schedules taxis according to the routes stored in the routing table (i.e., without the online adaptive replanning process), takes less than a second (0.65 s) to immense a response. **AdaPlan** takes a little bit more time (0.80 s) in comparison. In other words, the last two algorithms ensure a quick response and can work in real time. Furthermore, we examine how close the average delivery time generated by the last two algorithms is to the result of the **Optimal** algorithm. As shown in Table 12.5, optimally it takes 1.164 h to complete he package delivery on average. In contrast, the route discovered by **AdaPlan** takes 2.337 h, which is almost twice of the optimal value. According to the **Statistically Optimal**, the route takes 4.796 h. This suggests that a considerable improvement has been achieved by applying the online adaptive re-planning process. However, *a noticeable gap still exists between* **AdaPlan** *and the* **Optimal**, motivating us to further improve **AdaPlan** in the future. Note that the average driving distance of the 1000 package delivery requests generated in this study is 6.253 km.

5. ***Transport Capacity:*** It is necessary to evaluate the transport capacity of the system, since it is a primary consideration when applied to real-life scenarios. Figure 12.14 shows that the number of packages that the proposed system can transport per hour increases linearly with the number of hourly package requests when it is smaller than 1000. In other words, this suggests that the system mains a 90% success rate. However, the success rate gradually declines when the number of hourly package requests increases from 1000 to 1200, and cannot handle over 1200 requests in an hour. This result indicates that the transport capacity of our system is around 985 packages per hour on average. It should be noted that, to approximate the potential maximum transport capacity of the system, we further

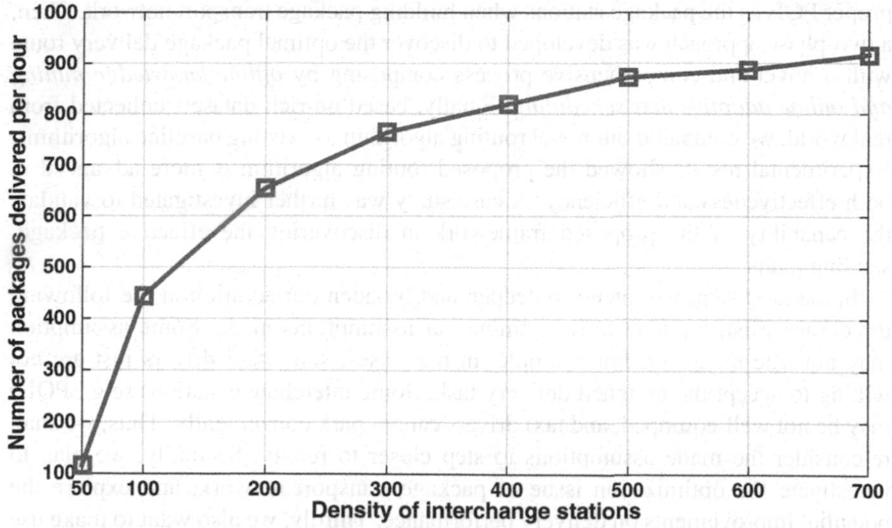
Number of packages delivered per hour
1000
900
800
700
600
500
400
300
200
100
50
100
200
300
400
500
600
700
Density of interchange stations

Fig. 12.15 Results of system capacity under different densities of interchange stations

relax a couple of constraints a bit in this study. More specifically, (1) we assume that *all real-time taxi ordering requests are generated by dedicated mobile apps* (i.e., the sampling rate is equal to 1); (2) we further assume *a taxi can be recruited multiple times for multiple package deliveries* before picking up passengers (i.e., a recruited taxi is allowed to deliver a maximum of k packages, 9 where *k* is set to 3); (3) we also *exclude interchange stations near transportation hubs* (e.g., railway/long-coach stations, popular metro stations and airport) since taxis head to those destinations probably have little extra room for hitchhiking packages. As a result, we keep 700 interchange stations for this study. We further examine the transport capacity under different density of interchange stations, as shown in Fig. 12.15. As expected, the more interchange stations our system has, the higher transport capacity it can achieve. The transport capacity is extremely low—around 115.6 packages per hour on average—with 50 interchange stations. In this study, at each condition, we randomly sample the given number of interchange stations from the entire pool. We are aware that the choice of interchange stations may affect the transport capacity result, but the overall trend is the same. Note that we set the number of hourly package requests to 1200 in the second study, and the deadline to 12 h in both capacity studies.

12.7 Conclusions and Future Work

In this chapter, we introduced a new framework named **CrowdDeliver** which is able to send citywide packages leveraging the unintendedly cooperation from participants. Using packages as the new taxi hitchhikers, initially, **CrowdDeliver** selected proper POIs as the package stations when building package transport network. Then, a two-phase approach was developed to discover the optimal package delivery route with a novel and comprehensive process composing by *offline knowledge mining and online adaptive taxi scheduling*. Finally, based on rich datasets collected from real world, we compared our novel routing algorithm to existing baseline algorithms. Experimental results showed the proposed routing algorithm is more advanced in both effectiveness and efficiency. A case study was further investigated to validate the capability of the proposed framework in discovering the effective package-sending paths.

In the next step, we intend to deepen and broaden our solution in the following directions. First, we plan to re-examine our assumptions made. Some assumption may not true in reality. For example, in real cases, some taxi drivers just are not willing to accept the assigned delivery task. Some interchange stations (e.g., POIs) may be not well-equipped, and taxi drivers cannot park conveniently. Thus, we shall re-consider the made assumptions to step closer to reality. Secondly, we plan to investigate the optimization issue of package transport network, and explore the potential improvements on delivery performance. Thirdly, we also want to make use more hitchhiking resources, such as opportunities provided by other transportation means. Last but not least, we intend to develop mobile apps for both online shoppers and taxi drivers, and test the performance in real settings.

References

1. Rohm AJ, Swaminathan V. A typology of online shoppers based on shopping motivations. J Bus Res. 2004;57(7):748–57.
2. "Same-Day Dreamer," Economist, London, England, 2014. [Online]. http://www.economist.com/news/business/
3. Bast H, Delling D, Goldberg A, et al. Route planning in transportation networks[M]//Algorithm engineering. Cham: Springer; 2016. p. 19–80.
4. Zheng X, Liang X, Xu K. Where to wait for a taxi? New York: Proceedings of the ACM SIGKDD International Workshop on Urban Computing; 2012. p. 149–56.
5. Chen P, Chankov S. Crowdsourced delivery for last-mile distribution: an agent-based modelling and simulation approach. In: 2017 IEEE International Conference on Industrial Engineering and Engineering Management (IEEM). New York: IEEE; 2017. p. 1271–5.
6. Du J, Guo B, Liu Y, Wang L, Han Q, Chen C, CrowDNet ZY. Enabling a crowdsourced object delivery network based on modern portfolio theory. IEEE Internet Things J. 2019;6 (5):9030–41.
7. Sadilek A, Krumm J, Horvitz E. Crowdphysics: planned and opportunistic crowdsourcing for physical tasks. Atlanta: Proceedings of ICWSM; 2013.
8. Sadilek A, Kautz H, Bigham JP. Finding your friends and following them to where you are. New York: Proceedings of the fifth ACM International Conference on Web Search and Data Mining; 2012. p. 723–32.
9. McInerney J, Rogers A, Jennings NR. Crowdsourcing physical package delivery using the existing routine mobility of a local population. New York: The Orange D4D Challenge; 2014.
10. Arslan AM, Agatz N, Kroon L, Zuidwijk R. Crowdsourced delivery—a dynamic pickup and delivery problem with ad hoc drivers. Transp Sci. 2019;53(1):222–35.

11. Devari A. Crowdsourced last mile delivery using social networks, M.S. thesis. State University, New York: Buffalo, NY, USA; 2016.
12. Chen C, Wang Z, Zhang D. Sending more with less: crowdsourcing integrated transportation as a new form of citywide passenger–package delivery system. IT Profess. 2020;22(1):56–62.
13. Chen C, et al. CrowdDeliver: planning city-wide package delivery paths leveraging the crowd of taxis. IEEE Trans Intell Transp Syst. 2017;18(6):1478–96.
14. Chen W, Mes M, Schutten M. Multi-hop driver-parcel matching problem with time windows. Flex Serv Manuf J. 2017;30:517–53.
15. Hu Z, Askin RG, Hu G. Hub relay network design for daily driver routes[J]. Int J Prod Res. 2019;57(19):6130–45.
16. Castro PS, Zhang D, Li S. Urban traffic modelling and prediction using large scale taxi GPS traces. New York: Proceedings of International Conference on Pervasive Computing; 2012. p. 57–72.
17. Yue Y, Zhuang Y, Li Q, Mao Q. Mining time-dependent attractive areas and movement patterns from taxi trajectory data. New York: Proc. IEEE International Conference on Geoinformatics; 2009. p. 1–6.
18. Dean BC. Continuous-time dynamic shortest path algorithms, M.S. thesis. Dept. Comput. Sci., MIT: Cambridge, MA, USA; 1999.
19. Zhang L, Yu B, Pan J. Geomob: a mobility-aware geocast scheme in metropolitans via taxicabs and buses. New York: Proceedings of IEEE INFOCOM 2014-IEEE Conference on Computer Communications; 2014. p. 1279–787.

Chapter 13
CrowdExpress: Making Citywide Packages Arrive as many as Possible

13.1 Introduction

People have put forward more and more diverse requirements for the urban express services, e.g., the express service as a requirement of the fast speed, the contactless one as a requirement of the epidemic prevention during the current coronavirus crisis [1]. For online shoppers, speed and cost are still two major issues, and these two issues are usually conflicting in nature. Urban crowdsourced logistics (referred as crowdshipping), which completes the delivery of packages with the concept of delivering from little or no effort from the crowd, is considered to be an effective way to alleviate this contradiction. Therefore, in this chapter, we propose a similar idea, using the collaborative efforts of taxi crowds to send packages with passengers in a shared room and transportation network. Based on previous studies that paid special attention to taxis, we propose having packages cooperate with existing taxis that carry passengers on the street, i.e., the existing mobility of taxi drivers [2, 3], let packages take hitchhiking rides. Specifically, our goal is to design a specific crowdsourced logistics, **CrowdExpress**, which can reduce costs and ensure that packages arrive on time [4, 5]. We believe that compared to the system in Chap. 12 aimed at finding the optimal package-sending route with the least delivery time cost, it is more common and reasonable in actual situations. Online shoppers are actually more concerned about whether their package can arrive at homes within the deadline, and there is not much interest in the specific arrival time [6]. Considering that the quality of passenger service should be the top-priority of taxi drivers, the proposed **CrowdExpress** only needs the participating taxi drivers to pay little extra effort and time, and will not significantly reduce the quality of service or cause any interference to passengers. As discussed in the Chap. 12, we can take the

---

Part of this chapter is based on a previous work: C. Chen, S. Yang, Y. Wang, B. Guo and D. Zhang, "CrowdExpress: A Probabilistic Framework for On-Time Crowdsourced Package Deliveries," in IEEE Transactions on Big Data, doi: https://doi.org/10.1109/TBDATA.2020.2991152.

C. Chen et al., *Enabling Smart Urban Services with GPS Trajectory Data*,
https://doi.org/10.1007/978-981-16-0178-1_13

proposed taxi-based package delivery problem as a route planning problem, but there is a completely different goal, that is, to deliver the package to the destination on time (within the time limit specified by the user). In order to make the idea of organizing passenger and package flows seamlessly through the taxi transportation network feasible, we need to address the following two main research challenges:

1. Package flow and passenger flow in both time and space are incompatible. In more detail, (1) compared with the package flow, the passenger flow presents an obvious peak-hour pattern; (2) for economic reasons, most passengers only take a taxi when the destination is close, for example, within 4 km [7]. While for package deliveries, the destination is generally far away from the departure place (e.g., longer than 5 km) [2]. Therefore, we believe that the routing algorithm based on the direct query-matching framework may not work [8, 9], because a single free ride may not be able to deliver the package to the destination; on the contrary, we need the cooperative relay of the taxi.
2. The requests for demands of packages and passengers have high uncertainties. Although the regular spatial and temporal patterns of passenger flow have been unveiled from coarse-grained taxi GPS trajectory data, it is still challenging to accurately predict passenger demands in a quite fine granularity (e.g., the passenger demands in the next 5 min for a given road section). At the same time, the same situation also occurs in packaging demands [10]. In short, there are uncertainties in both requests, making the time cost estimation and comparison of the packet routing path challenging.

With the aforementioned research challenges and goals, the main contributions of this chapter are:

1. We proposed an innovative modification to crowdsourcing logistics, that is, a mixed transportation mode of passengers and packages, using *unconscious cooperation* between a crowd of passenger-occupied taxis to meet the punctual delivery of packages across the city, in order to reduce transportation costs and improve transportation efficiency. In addition, the taxi driver performs package transfer (uploading and offloading) before picking up (after dropping off) passengers, which can ensure the quality of service to passengers.
2. We formulate the package routing problem as a punctual arrival problem [11], in order to resolve the uncertainty of package and passengers requests. We propose a two-stage probabilistic framework named as **CrowdExpress** to solve it. In the first stage, we build a package transport network by mining the historical GPS trajectory data of taxi offline. In the second stage, for each package delivery request generated in real time, we propose an online adaptive taxi scheduling algorithm based on the maxProb probability model to iteratively determine the next stop of the package "on-the-fly".
3. We conducted extensive evaluations using road network data and taxi GPS trajectory data to verify the efficiency and effectiveness of **CrowdExpress**. These data are generated from more than 19,000 taxis in a month in New York (NYC). The results showed that it responds within 25 ms. About 9500 packages

can be passed daily, with a success rate of over 94%, which is always better than the benchmark method.

13.2 Preliminary, Problem Statement and System Overview

In this section, we provide definitions of some basic concepts, elicit assumptions we have made, and give a formal problem statement. Finally, we give an overview of the proposed **CrowdExpress** system.

13.2.1 Preliminary

13.2.1.1 Basic Concepts

We define the basic concepts used in this chapter as follows:

Definition 13.1 (Time Slot Slicing) We divide a whole day into different time slots (periods) according to the day type, since the traffic conditions are changing in different time slots, resulting in large variance in travel time. A work day is divided into three time slots and a rest day is divided into two time slots, as detailed in Table 13.1.

Definition 13.2 (Package Transport Network) A package transport network is a graph $G(S, E)$, consisting of a node set S and an edge set E. Each element s in S is an interchange station which is responsible for package collections and storage. Each element $e(i, j)$ in E is a non-stop directional transport route from node s_i to node s_j, implying that there is an abundant passenger flow for hitchhiking packages. It should be noted that the edge in the package transport network has different meaning from the edge defined over the road network. There can be multiple driving paths over the road network that connect two interchange stations, associating with different travel time at different time slots.

Definition 13.3 (Real-Time Taxi Ordering Request) A taxi ordering request (tor) is defined as a triplet $\langle o_t, d_t, r_t \rangle$, where o_t and d_t refer to the passenger's origin and

Table 13.1 Time slot slicing

	Work days	Rest days
Night-time hours	00:00–06:59	00:00–07:59
	19:00–23:59	19:00–23:59
Day-time hours	09:00–16:59	08:00–18:59
Rush hours	07:00–08:59	NG
	17:00–18:59	

intended destination, respectively. r_t refers to the time that the passenger submits the request. The time when the passenger is picked up is usually after r_t.

Definition 13.4 (Travel Time Probability Function) Each edge in the package transport network (G) is associated with an independent random travel time (cost) t_{ij} whose probability density function is denoted by $p_{ij}(t) \cdot p_{ij}(t)$ varies at different time slots since traffic conditions vary.

For instance, the probability that a package spends a time in the interval $[0, t_0]$ from node s_i to node s_j directly can be computed by the definition, as shown in Eq. (13.1).

$$P_{ij}(t \leq t_0) = \int_0^{t_0} p_{ij}(t)dt \tag{13.1}$$

The travel time probability function in each time slot can be obtained separately, according to Eq. (13.1). What is more, as can be observed, the travel time probability is a monotonic increasing function of the time t.

Definition 13.5 (Travel Time Discretization) To simplify the calculation of travel time probability along an edge, we consider the travel time in a discrete manner. More precisely, we use a piecewise constant function with equal step width t to discretize different travel times.

In the discrete case, the integral shown in Eq. (13.1) can be replaced using the formula shown in Eq. (13.2).

$$P_{ij}(t \leq t_0) = \frac{\#\text{traveltimes} < \alpha\tau}{\#\text{traveltimes}} \tag{13.2}$$

$$t_0 = (\alpha - 1)\tau + \delta \tag{13.3}$$

where #traveltimes refers to the number of all the possible travel times from node i to node j which are recorded in the taxi trajectory in history, while #traveltimes < at refers to the number of travel times less than at after time discretization as defined; $\alpha \in N^{+}$ and $0 \leq \delta < \tau$. As can be seen in Eq. (13.3), α and δ are the quotient and reminder of $\frac{t_0}{\tau}$, respectively. For the edge from s_o to s_1 shown in Fig. 13.1, $P_{o1}(t \leq 5) = 0.3; P_{o1}(5 < t \leq 10) = 0.7$.

Definition 13.6 (Arriving-on-Time Probability) The arriving on time probability of a package-delivering path within a given time duration (i.e., deadline, t_0) is defined as the ratio of the number of travel times less than t_0 to the number of all possible travel times (suppose that the package is shipped from s_i to s_j via s_k), as follows

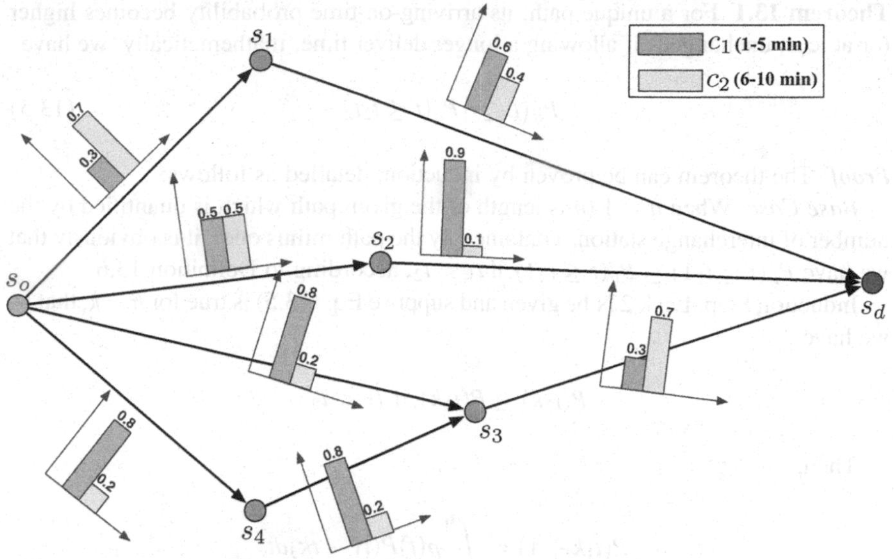

C_1 (1-5 min)
C_2 (6-10 min)
s_1
s_2
s_3
s_4
s_o
s_d
0.7
0.3
0.6
0.4
0.5 0.5
0.9
0.1
0.8
0.2
0.7
0.3
0.8
0.2
0.8
0.2

Fig. 13.1 An illustrative example of package delivery paths from s_o to s_d, as well as the distribution of discrete travel times on each edge

$$P_{ij}\left(t \leq t_0 \int_{t_0-t_1}^{t_0} dt_2 \int_0^{t_1} p_{ik}(t_1)p_{kj}(t_2 - t_1)dt_1 \qquad (13.4)$$

It is obviously that the integral computation becomes more complicated if the given path contains more interchange stations, since more travel time combinations can be generated.

Similarly, to ease the computation of integral in Eq. (13.4), we first let the travel time be considered in the discrete manner, then the integral can be degraded to the sum computation. Taking the path ($path_1 : s_o \rightarrow s_1 \rightarrow s_d$) shown in Fig. 13.1 as an example, its arriving-on-time probability within 15 min can be computed as

$$P_{ij}\left(t \leq 15 \underbrace{0.3 \times (0.6 + 0.4)}_{\text{Case I}} + \underbrace{0.7 \times 0.6}_{\text{Case II}} = 0.72\right.$$

For the given path, two cases in total can lead to the successful arrival of packages by the deadline. Case I: If the first recruited taxi driver spent no more than 5 min in the first segment (i.e., $s_o \rightarrow s_1$), due to the sufficient time margin left, then the second recruited taxi driver can arrive at s_d by the deadline at a hundred percent. Case II: If the first recruited taxi driver took more than 5 min in the first segment, then the package can be arrived on time only if the second recruited taxi driver spent no more than 5 min to accomplish the second segment (i.e., $s_1 \rightarrow s_d$)

Theorem 13.1 For a unique path, its arriving-on-time probability becomes higher (or at least unchanged) if allowing a longer deliver time, mathematically, we have

$$P_{ij}(t \leq t_1 P_{ij}(t \leq t_2 t_2 \tag{13.5}$$

Proof The theorem can be proven by induction, detailed as follows:

Base Case. When $n = 1$ (n is length of the given path which is quantified by the number of interchange stations contained by the path minus one), it is obviously that we have $P_{ij}(t \leq t_1 1) \leq P_{ij}(t \leq t_2 1)$, if $t_1 < t_2$, according to Definition 13.6.

Induction Step. Let k 2 N be given and suppose Eq. (13.5) is true for $n = k$, that is, we have

$$P(t_2 k) \geq P(t_1 k), \text{if } t_1 < t_2$$

Then,

$$P(t_1 k + 1) = \int_0^{t_1} p(t)P(t_1 - tk)dt$$

where *p*(*t*) is the probability density function on the edge from the origin to the first stop.

$$\begin{aligned} P(t_2 k + 1) &= \int_0^{t_2} p(t)P(t_2 - tk)dt \\ &= \int_0^{t_1} p(t)P(t_2 - tk)dt \\ &\quad + \int_{t_1}^{t_2} p(t)P(t_2 - tk)dt \\ &\geq \int_0^{t_1} p(t)P(t_1 - tk)dt \\ &\quad + \int_{t_1}^{t_2} p(t)P(t_2 - tk)dt \end{aligned}$$

$$= P(t_1 k + 1) + \int_{t_1}^{t_2} p(t)P(t_2 - tk)dt$$

$$\geq P(t_1 k + 1)$$

Conclusion. By the principle of induction, Eq. (13.5) is true for all $n \in N$.

Definition 13.7 (Maximum Probability of Arriving-on-Time) For an OD pair, the maximum probability of arriving-on-time (with time cost no greater than t_0) is defined as the maximal one among all probabilities on all possible *N* paths from the origin (s_i) to the destination (s_j), denoted by $u_{ij}(t \leq t_0)$. Inanotherword, $u_{ij}(t \leq t_0)$ serves as the upper bound of $P_{ij}(t \leq t_0)$.

If a package at node s_i is sent to s_k in the next step, the probability that the package spends a time in the interval $[\omega, \omega + d\omega]$ on edge s_i, s_k is $p_{ik}(\omega)d\omega$, thus the time margin at node s_k is $t_0 - w$ On the basis of Bellman's principle of optimality [10, 12], no matter which node s_j that the package is elected to send next, the package must follow the optimal routing strategy in shipping from node s_k to the destination s_j within the remaining time $t_0 - w$. Therefore, the maximum probability of arriving-on-time can be formally defined recursively as follows:

$$u_{ij}(t \leq t_0) = \max_{k \neq j} \int_0^{t_0} p_{ik}(\omega) u_{kj}(t \leq t_0 - \omega) d\omega \tag{13.6}$$

Intuitively, according to Definition 13.6, to compute the maximum probability of arriving-on-time for an OD pair, one needs to find all possible paths from the origin to the destination, which is well-known as an NP-hard problem [18]. To make this point clear, we use the example shown in Fig. 13.1 again. It is easy to find all paths from s_0 to s_d in this example, i.e., $path_1 : s_o \rightarrow s_1 \rightarrow s_d$, $path_2 : s_o \rightarrow s_2 \rightarrow s_d$, $path_3 : s_o \rightarrow s_3 \rightarrow s_d$ and $path_4 : s_o \rightarrow s_4 \rightarrow s_d$, respectively. For each path, similar to the computation in Eq. (13.4), it is not difficult to obtain the corresponding arriving-on-time probability within a given deadline (e.g., t_0=15 min). As a result, $u_{ij}(t \leq 15) = 0.95$ for the example when delivering the package via path 2. It is obvious that $u_{ii}(t) = 1$ given any deadline, since no travel time is needed if the package stays still. We exclude the round trip in the study.

13.2.1.2 Assumptions

In this chapter, we make the exactly same assumptions to [13], regarding taxi driver participation constraints and willingness, as well as the package traceability. To avoid overlapping and save room, we do not elaborate the details here. Readers can refer to [13] for the assumption details.

13.2.2 Problem Statement

The collaborative crowdsourced package deliveries leveraging the relays of passenger-occupied taxis can be viewed as the problem of finding arriving-on-time paths, and thus can be formulated as follows:

Given:

1. A historical set of taxi trajectory records Tr, such as from the past month in the designated city,

2. A set of real-time taxi ordering requests *TOR* from mobile phone apps, and a set of real-time package delivery requests *PR*. Note that these two requests come in stream,
3. A given deadline specified by the user for each package delivery request.

Objectives:

1. Build a package transport network (i.e., the identification of interchange stations and estimation of edge values) based on the historical taxi trajectory data.
2. Find a package delivery path for each package request (*pr*), which can make the package arrive at the destination by the deadline. However, such package delivery path may not be unique or existed. To migrate the issue, we thus transform the problem to the arriving-on-time problem, i.e., finding the optimal one that is expected to have the maximum probability of arriving-on-time.

Constraints:

1. Only taxis that accept the real-time taxi ordering requests after the package delivery request is posted can be scheduled, i.e., tor. $r_t > pr.\ t_p$
2. A recruited taxi can be available to participate again only after completing the current task (i.e., dropping off the package at the predefined interchange station).

13.2.3 System Overview

We develop a two-phase system called **CrowdExpress**, i.e., offline package transport network building and online taxi scheduling and package routing to find the optimal route with the maximum probability of arriving-on-time for each package delivery request within a given deadline, by collaboratively recruiting taxi drivers

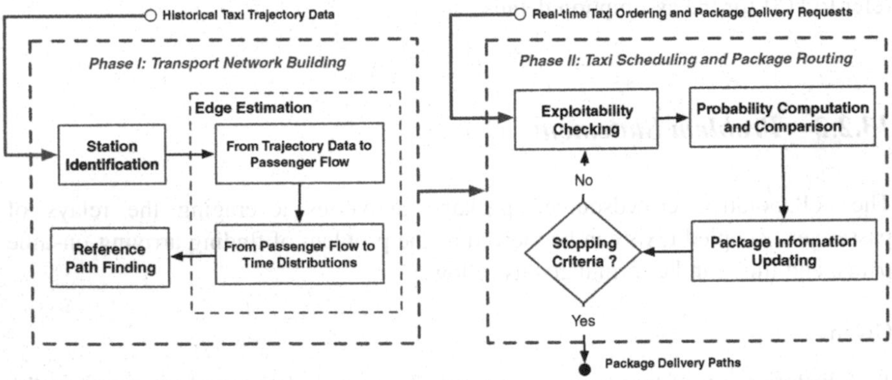
Historical Taxi Trajectory Data
Phase I: Transport Network Building
Edge Estimation
Station Identification
From Trajectory Data to Passenger Flow
From Passenger Flow to Time Distributions
Reference Path Finding
Real-time Taxi Ordering and Package Delivery Requests
Phase II: Taxi Scheduling and Package Routing
Exploitability Checking
Probability Computation and Comparison
No
Stopping Criteria ?
Package Information Updating
Yes
Package Delivery Paths

Fig. 13.2 The overview of the CrowdExpress system

that have been reserved to passengers (occupied by passengers), as shown in Fig. 13.2.

Phase I is an offline process, with the historical taxi trajectory data as input, aiming to identify the package interchange stations, estimate the edge values, as well as find the reference paths for any given OD package pairs. Based on the constructed package transport network, for a real-time incoming individual package delivery request, Phase II mainly takes four online steps to tackle with, namely, Exploitability Checking, Probability Computation and Comparison, Package Information Updating and Stopping Criteria Checking, with the streaming taxi ordering requests as input. The system finally outputs the corresponding package delivery paths. The technical details will be presented in the next two sections.

13.3 Offline Package Transport Network Building

The task of offline package transport network building is to identify interchange stations (i.e., node locations) as well as the estimation of travel time distributions (i.e., edge values). Here, we mainly take a three-step procedure to achieve the objectives, detailed as follows.

13.3.1 Package Interchange Station Identification

One basic principle for the identification of interchange stations is that they should be located where passengers are frequently picked-up or dropped-off, to take as full advantage of passenger-sending rides for the package hitchhiking as possible, and probably to minimize the extra efforts imposed to the taxi drivers as well. Fortunately, with the taxi trajectory data left, such information (i.e., pick-up and drop-off points) can be easily extracted. Then, we cluster them using DBSCAN algorithm as it is capable of merging closer data points with arbitrary distributions [14]. Finally, locations near the point centroids of each cluster and the road sides are identified as the locations of the interchange stations to serve the purposes of package collection, storage and receiving.

13.3.2 Edge Estimation

13.3.2.1 From Trajectory Data to Passenger Flow

It is straightforward to infer the passenger flow between any two interchange stations during a given time slot from the trajectory data [13]. Specifically, we first group the trajectories according to their starting time (t_s). Second, to compute the passenger flow from s_i to s_j, we count the number of the trajectories satisfying Eqs. (13.7) and (13.8). It should be noted that there could be no passenger flow between some interchange station pairs.

$$\mathrm{Ddist}(\mathrm{Tr}_i.o, loc(s_i)) \leq \epsilon \tag{13.7}$$

$$\mathrm{Ddist}(\mathrm{Tr}_i.\mathrm{d}, loc(s_j)) \leq \epsilon \tag{13.8}$$

where $\mathrm{Tr}_i.\ o$ and $\mathrm{Tr}_i.\ \mathrm{d}$ are the original and destination points of Tr_i, respectively; loc $(\cdot)$ gets the latitude and longitude location of the given interchange station; Ddist $(a \cdot b)$ calculates the driving distance from point a to point b; ϵ is a user-specified parameter. The physical meaning of ϵ is that any passenger-delivery ride which starts and ends near a pair of interchange stations (i.e., with driving distance less than ϵ) can be hitchhiked for the package delivery between this pair. Hence, for a given OD pair, a bigger ϵ would result in a bigger number of passenger flow. It is worth noting that, for a specific trajectory, there could be multiple interchange station pairs that satisfy Eqs. (13.7) and (13.8), in other words, can provide package hitchhiking ride between all these pairs. Therefore, a bigger ϵ also leads to a bigger number of interchange station pairs, suggesting that the corresponding trajectory can be more capable of providing hitchhiking rides. However, for passengers, a bigger ϵ may mean a longer waiting time for the reserved taxis, since the taxi driver might have to travel farther to collect the package before picking up passengers. To control for the additional waiting time, we set ϵ to 500 m.

13.3.2.2 From Passenger Flow to Time Distributions

To estimate an edge value, we need to estimate two parts, i.e., the waiting time and the driving time [13]. The driving time is simply the travel time of each taxi trajectory. The waiting time on the edge is defined as the time required to wait for a suitable hitchhiking ride that can transport a package from s_i to s_j directly. To address this problem, we employ the Non-Homogeneous Poisson Process (NHPP) to model the behavior of passenger taking taxis [15]. According to the statistical frequency of passenger taking taxis from s_i to s_j in history (i.e., passenger flow), we can estimate the waiting time of packages at different time slots at the interchange stations. Specifically, the waiting time on the edge from s_i to s_j is

$$t_w = \text{waiting time} = \frac{\Delta T}{\bar{N}} \tag{13.9}$$

where $\bar{N}$ is the average number of passengers taking taxis from s_i to s_j during the given time slots; ΔT is the length of that time slot. Note that the waiting time obtained by Eq. (13.9) is in the statistical sense, and it could be much smaller in the real case due to the timely availability of right passenger sending trips.

The waiting time component is substituted in advance when computing the arriving-on-time probability of a given path at a given time duration, thus the corresponding time distributions is obtained by discretizing the driving time component only according to Definition 13.7.

13.3.3 Reference Path Finding

On the basis of the constructed package transport network, we find two reference paths for each given OD pair, i.e., the shortest path when assuming value on each edge is the minimum travel time (*SPath_min*), and the shortest one when assuming value on each edge is the maximum travel time (*SPath_max*), respectively. More specifically, given an OD pair, when choosing all minimum travel times on edges, we can find the shortest path, i.e. *SPath_min*, using *Dijkstra's* algorithm [16]. *SPath_min* refers to the case that the package can be delivered in the most efficient manner if sent via that path. Similarly, when choosing all maximum travel times on edges, we can also discover the shortest path, i.e. *SPath_max*. *SPath_max*, as a comparison, refers to the case that the package can be delivered in the lowest efficient manner if sent via that path. *Inaddition*, the corresponding total travel times of those two paths are also recorded, which can be used to guide the online taxi scheduling and package routing upon the real-time taxi ordering requests in the second phase, with details would be further addressed in the next section. It should be noted that, although it is a time-consuming procedure to find two shortest paths for each OD pair with a computation complexity of $O(k^2)$, it can be operated offline

13.4 Online Taxi Scheduling and Package Routing

Given an OD pair of the package, the task of online taxi scheduling is essentially a sequential decision-making one. To be more specific, the phase should make a decision whether utilizing the current available hitchhiking ride for the package delivery or waiting for the future rides, according to the upcoming taxi ordering requests generated in real-time and the remaining time margin. The package would be hitchhiked immediately if the corresponding maximum arriving-on-time probability is greater than the expected value if waiting for the future opportunities. The upper bound value of maximum arriving-on-time probability if hitchhiking the current ride is actually related to the case of sending the package via the highest efficient path to the final destination after reaching the intermediate stop (same to the destination of current ride) with the help of current ride. Refer to Sect. 13.3.2.1 for the details. While the expected maximum arriving-on-time probability if hitchhiking the future rides corresponds the case of sending to the other intermediate stops before sending the package to the final destination from that intermediate stop via the most efficient path. Refer to Sect. 13.3.2.2 for the details. We thus propose maxProb algorithm to compute and compare these two cases iteratively until reaching the package's destination to determine whether the waiting is worthy.

While the task of online package routing is much simpler, i.e., just assigning the delivery task to the scheduled taxi. As a result, the package will be delivered to the next stop, which is same to the destination of the taxi. The two steps impact each other mutually. On one hand, both the potential hitchhiking rides for the package delivery and the time margin is highly related to the current stop where the package locates; on the other hand, which the next stop that the package will head to is determined by the scheduled taxi. In the following, we mainly focus on the step of online taxi scheduling, which includes the following operations.

13.4.1 Reference Path Finding

Before triggering this phase, we first need to conduct the exploitability checking, i.e., to determine: (1) whether the origin of an upcoming taxi ordering request is close to the location of the package; and (2) whether it ends at one of the interchange stations. If both conditions are met, then we further need to compare the maximum arriving-on-time probability of sending the package via s_k (hitchhiking the current) to the maximum arriving-on-time probability of sending the package via other potential next stations (the other neighbors of s_o in the transport network) (waiting for the future). If the former value is greater, then the package will be sent out immediately by hitchhiking the current taxi ride; otherwise, the system will wait for the new future taxi ordering requests that may lead to a higher arriving-on-time probability and the decision will made again, given the new time margin. Thus, the core of the online taxi scheduling is the maximum arriving-on-time probabilities computation and comparison.

13.4.2 Probability Computation and Comparison

13.4.2.1 Probability Computation if Hitchhiking the Current

According to the definition, the maximum arriving-on-time probability for case if hitchhiking the current (suppose that the recruited taxi will go to s_k) can be computed as follows:

$$P_{od}\left(t \leq t_0 - \Delta t \int_0^{t_0-\Delta t} P_{\mathrm{ok}}(\omega) u_{kd}(t \leq t_0 - \Delta t) d\omega \right. \tag{13.10}$$

where $\mathrm{Path}_{\mathrm{okd}}$ refers to the path from s_o to s_d while stopping at s_k in the next; Δt is the time difference between the occurring time of the taxi ordering request and the birth time of the package delivery request, i.e. $\Delta t = tor.\, t - pr.\, t$; t_0 is the given time duration of the deadline; u_{kd} is the maximum arriving-on-time probability from s_k to s_d, as defined in Definition 13.6.

By definition, it is easy to compute the arriving-on-time probability of a determined path under a given deadline time, such as $\int_0^{t_0-\Delta t} P_{ok}(t)dt$. By contrast, it is rather challenging to get the value of the latter part in Eq. (13.10). As discussed, one naive and straightforward way is first to enumerate all the possible paths from s_k to s_d, then compute the value of arriving-on-time probability for each of them, finally pick up the maximum one as the final value. It is easy to understand that the trivial method cannot work in real cases as the problem of finding all possible path for a given OD pair is NP-hard. Actually, it is also no need and some branches in the transport network can be trimmed recursively. We propose a novel algorithm named maxProb to compute the probability, which mainly consists of two operations, i.e., initialization and deep-first searching.

Initialization: From s_k to s_d, it will be easy to find two shortest paths, SPath_min (s_k, s_d), SPath_max (s_k, s_d). We further obtain the two corresponding reference paths from s_o to s_d via s_k, and compute their arriving-on-time probabilities given the remaining time, which are two boundaries and used to guide the process of branch trimming. For brevity, we use $minP_{o \to k \rightsquigarrow d}$ and $maxP_{o \to k \rightsquigarrow d}$ to represent the arriving-on-time probabilities of $s_o \to$ SPath_min (s_k, s_d) and $s_o \to$ SPath_max (s_k, s_d), respectively.

Depth-First-Searching: From s_k to s_d, we mainly apply the Depth-First-Search (DFS) method to recursively get each possible path [19], and compute the maximum probability of arriving-on-time. One exception is that the user specifies an extremely long deadline, mathematically, $maxP_{o \to k \rightsquigarrow d} = 1$, implying that the package can be delivered on time for sure via the reference path. Therefore, no DFS is needed and a simple taxi scheduling can be enough under such circumstance. The overall procedure of DFS starting from s_k can be summarized as follows.

Algorithm 13.1 FindNextStop(s_k, s_d, TN)

1 ngb = Neigh(s_k, TN); // get the neighbouring stations of s_k based on the transport network topology.
2 ns = ∅;
3 refP = $minP_{o \to k \sim d}$;
4 **for** i == 1 to |ngb| **do**
5 $\quad s_{ki} = ngb(i)$;
6 $\quad$ **if** *refP*≤ $P(t \le t_0 - \Delta t - t\_\min(s_{ki}, s_d) k \to ki)$ **then**
7 $\quad\quad$ ns = ns ∪ ngb(i);
8 $\quad$ **end if**
9 **end for**

The core function is to find the next package stop of s_k, with the pseudocode shown in Algorithm 13.1. The very beginning task is to get the neighouring stations of s_k, given the topology of the built package transport network (Line 1). DFS starts to find the next stop from one of the neighbouring nodes of s_k (e.g., s_{ki}) in the loop (Lines 4–9). Whether s_{ki} can be the next package stop is determined by the inequation shown in Line 6. In the in-equation, the reference probability (refP) is first set to $minP_{o \to k \leadsto d}$ (Line 3). t_min (s_{ki}, s_d) is the time cost of the reference path SPath_min (s_{ki}, s_d) estimated by the historical taxi trajectory data; the right part of the in-equation is the maximum arriving-on-time probability which corresponds to the ideal case that the package can be shipped from s_{ki} to s_d in the most-efficient way (i.e., the time cost on each edge in the package transport network is the minimal). If the in-equation satisfies, it indicates that there exists a potential path which can lead to a higher maximum arriving-on-time probability than the reference path, thus DFS will continue to search with a new start from s_{ki} recursively, with the same procedure to the DFS starting from s_k; otherwise, DFS will be terminated and the related branches will be trimmed at the same time. Thus, a recursion may be stopped either at some intermediate node or generates a successful path reaching the given destination. If a valid path is resulted (Path_valid), refP will be updated using its corresponding probability if and only if it is greater than the previous value.

The whole DFS ends when all neighhours of s_k are checked by repeatedly calling the above recursive DFS. Finally, the maximum arriving-on-time probability if assigning the package delivery task to the current available taxi shall be the final value of refP.

13.4.2.2 Probability Computation if Waiting for the Future

If the future taxi rides heading to any one of the other neighouring nodes of s_o except for s_k (marked as $\{ngb(s_o)\} - s_k$) could lead to a higher arriving-on-time probability, compared to the case if hitchhiking the current, a better decision should be the waiting. The maximum arriving-on-time probability if waiting for the future can be computed as follows:

$$P_{od}\left(t \le t_0 - \Delta t' \quad \max_{s_j \in ngb(s_o) - s_k} \int_0^{t_0 - \Delta t'} P_{oj}(\omega) u_{jd}(t \le t_0 - \Delta t') d\omega \right. \tag{13.11}$$

where $\Delta t' = \Delta t + t_w(s_o, s_j)$ and $t_w(s_o, s_j)$ refers to the edge value component of waiting time from s_o to s_j. As can be seen, the major difference between Eq. (13.10) and Eq. (13.11) is the time margin. More specifically, less time margin is left for the package deliveries as an additional time cost would be induced while waiting for the future taxi rides. Here, we simply use the average waiting time to approximate the additional time cost.

Similarly, all maximum arriving-on-time probabilities of waiting for the future exploitable taxi rides from s_o can be computed, and the maximal one among them will be chosen to represent the maximum arriving-on-time probability if assigning the package delivery task to the future taxis.

13.4.2.3 Probability Comparison

As discussed, once receiving a real-time taxi ordering request, on-line taxi scheduling and package routing will be activated, and the package may be shipped to some intermediate stop by hitchhiking the current ride or stands still at the current stop by comparing those two maximum arriving-on-time probabilities. Note that the remaining time margin shrinks as time goes by, the two probabilities computed in Eqs. (13.10) and (13.11) are dynamically changed, thus the better decision (hitchhiking the current or waiting for the future) can be also adjusted adaptively "on-the-fly".

13.4.3 Package Information Updating

After the package is sent to the next station whether by hitchhiking the current or future rides, the information about the package delivery request should be updated. To be more specific, the origin of the package should be set as the updated station that the package locates; the birth time should be set as the time when the package arrives at the current stop. The newly updated package delivery request will be used as the input of Phase II.

13.4.4 Stopping Criteria Check

For a package delivery, the previous three operations will be iteratively conducted until one of the following two stopping criteria is satisfied: (1) the package has arrived at its destination; (2) the time is running out (the package cannot be delivered by the deadline), in that case, the system would report failure. For those failure package deliveries, empty taxis can be recruited dedicatedly to send them to the destinations. However, the topic is out of the scope of the chapter.

13.5 Evaluations

In this section, we empirically evaluate the performance of the proposed maxProb algorithm. We first introduce the experimental setup, baseline algorithms used for comparison, evaluation metrics and results on algorithm efficiency and effectiveness. We discuss some open research issues to be further addressed in the end.

13.5.1 Experimental Setup

13.5.1.1 Experiment Data

We use the real-world datasets for the evaluation, i.e. the road network data which is extracted from OpenStreetMap, and 1 month of taxi trajectory data generated by over 19,000 taxis in the city of New York (NYC), US. Readers can refer to [17] for the details on how to extract the road network from the crowd-sourced open platform (i.e., OpenStreetMap) correctly. We determine package interchange stations according to the algorithm discussed earlier.

For the taxi trajectory data, we split it into training and testing sets, according to the date of the month. Specifically, the training set contains taxi trajectories on 1st–20th, January, 2013, which are used to build package transport network. It should be noted that for the taxi trajectory data in NYC, no detailed travel routes between the pick-up and drop-off points are provided due to the privacy considerations. The testing trajectories were generated from 21st to 31st, January, 2013, which are used as the real-time taxi ordering requests (TOR) for testing the performance of the proposed maxProb algorithm. Table 13.2 shows some basic statistics of the taxi trajectory and road network data. It should be noted that we do not differentiate the trajectory data collected from work days and rest days. We simply exclude trajectory data collected in rush hours in work days since the passenger flow pattern is quite different at that time and the operation of **CrowdExpress** at that time may worsen the traffic conditions.

13.5.1.2 Package Delivery Request

Since the data sets do not contain information about package delivery requests, we apply simple mechanism to simulate it. The novel package delivery system targets the city-wide person-to-person service. Hence, to simulate a package delivery request (PDR), we randomly generate its birth time, origin and destination. Regarding the origin and destination, any package should be originated and ended at the interchange stations. We further eliminate requests with short-distance OD pairs (i.e., with driving distance less than 3 km) since few users would request speedy shipping as ordinary delivery may be equally efficient in this case. Besides, based on the total number of pick-ups and drop-offs of taxi passengers in the station, we categorize the stations into two kinds, i.e., the popular and the unpopular.

Table 13.2 Statistics of the taxi trajectory and road network data sets

Datasets	Properties	Statistics
Taxi trajectory	Number of taxis	>19,000
	Number of occupied rides	≈13 M
Road network	Number of road intersections	11,999
	Number of road segments	15,202

13.5.1.3 Evaluation Metrics

We adopt the following three metrics to evaluate the proposed maxProb.

Success Rate. The success rate is the ratio of the number of packages which can be delivered successfully within a given deadline (i.e. time duration) to the number of total packages (i.e. the number of package delivery requests simulated).

$$SR(t \leq deadT) = \frac{\left|\mathcal{TC}\left(\text{Path}\left(o_p \rightsquigarrow d_p\right)\right) \leq deadT\right|}{|PQ|} \tag{13.12}$$

where $\text{Path}(o_p \rightsquigarrow d_p)$ represents the optimal path generated by the proposed maxProb algorithm for a given package delivery request; deadT is the given deadline. The delivery performance is better if the success rate is higher within a given shorter deadline.

Regarding the deadline setting, we do not set an absolute value for all package deliveries since package originated (ended) at different locations would need absolutely varied time. Thus, for an individual package delivery, we set a relative deadline separately instead, according to the following equation:

$$\text{deadT} = \text{t}_{\text{avg}} + \text{extraT} \tag{13.13}$$

in which t_{avg} is the average value of the time cost by the two reference paths, which is obviously different for packages with different OD pairs. extraT is the extra time value imposed by the user; a smaller extraT indicates that the user needs the package more urgently and wants it to be arrived more timely.

Number of Relays. The number of relays ($\text{Num}_{\text{relays}}$) during a package delivery is defined as the number of participating taxis $\text{Num}_{\text{p\_taxis}}$ (Formula (13.14)).

$$\text{Num}_{\text{relays}} = \text{Num}_{\text{p\_taxis}} \tag{13.14}$$

On one hand, fewer relays generally mean a lower chance of package loss or damage, and perhaps less overhead cost. On the other hand, fewer number of participating taxis may imply requiring less reward cost to taxi drivers. Thus, the performance is better if the number of relays is smaller.

Package Throughput. The average number of package deliveries that the system can complete successfully per day. The system achieves better performance if the package throughput is bigger.

13.5.2 Baseline Algorithms

To show the superior performance of our proposed algorithm, we compare it with the following three baseline algorithms.

1. **FCFS**—This method adopts the First Come First Service strategy. Specifically, the package will be assigned to the first taxi that will pick up a passenger near the interchange station that the package locates, regardless of its destination. In fact, this algorithm always favors the strategy of hitchhiking the current, which is also known as an extension of the simple and well-known flooding strategy.
2. **DesCloser**—This method assigns the package to the first taxi that will head to somewhere closer to the destination of the package, compared to the current station of the package. This algorithm implements a distance-based geo-cast scheme that is commonly seen in other domains.
3. **Direct**—This method waits for the taxi heading to the destination near the interchange stations that the package will be delivered directly, without any intermediate stops. Specifically, the package will be assigned to the taxi that will pick up and drop off a passenger near the interchange stations that the package locates and heads, respectively. Thus, no relays are needed.

Remark. Each relay in DesCloser is effective as it ensures that the package would move towards its destination step by step; while some relays in FCFS can be ineffective as the package moves further away from its destination. For Direct, it may be inefficient for package deliveries where there is little passenger flow in-between. However, all three baseline algorithms do not take the arriving-on-time probability of package deliveries into account, thus probably resulting in a high failure rate.

13.5.3 Experimental Objectives

We plan the experiments to address the following questions.

Question 1: How much computational resource is required to generate the response for a package delivery request?

Question 2: How does maxProb perform under different given deadlines?

Question 3: How does maxProb perform w.r.t the birth time of package deliveries?

Question 4: How does maxProb perform w.r.t the locations (both origins and destinations) of package deliveries?

Question 5: How many packages can be delivered daily on average (i.e., throughput) with the proposed system?

The first question concerns the efficiency of maxProb, and *Questions 2–4* are related to its effectiveness. To answer the first question, we compute the response time of the algorithms. Since passenger flows are both time- and space-dependent, to assess the effectiveness of the different algorithms, we calculate their success rates and the number of relays with respect to packages to be dispatched to different parts of the city at different time of the day. We test the throughput of the proposed system and the success rate under different number of package requests generated per hour, and also examine the system throughput given different number (density) of interchange stations in the designate city (*Question 5*).

13.5.4 Experimental Results

13.5.4.1 Results of Response Time

We first analyze the main operations involved in the four algorithms respectively. For a given package delivery request (pr), when a new real-time taxi ordering request (tor) comes in, all four algorithms need to determine whether tor starts near the origin of the package and stops at some interchange station, i.e., exploitability checking. FCFS will recruit the first taxi that satisfies the criteria, but for DesCloser, it needs to further determine whether the heading destination of the taxi is closer to the destination of the package, compared to its current location. For Direct, it also needs to determine whether the taxi would head to the destination of the package. Thus one more comparison operation is required for both DesCloser and Direct algorithms. For maxProb, the procedure is even more complicated, mainly requiring additional probability computation and comparison} operations, as discussed previously. Each algorithm needs to repeat its own operation procedure at each intermediate station (except for Direct) and thus the total response time is the accumulated computational time over all iterations.

We show the comparison results of average response time of the four different algorithms in Fig. 13.3. The average response time of FCFS is the biggest while that of Direct is the smallest among all algorithms. The average response times of DesCloser and maxProb are in-between and DesCloser costs slightly more time than maxProb. More precisely, the average response time of Direct is within 7 ms;

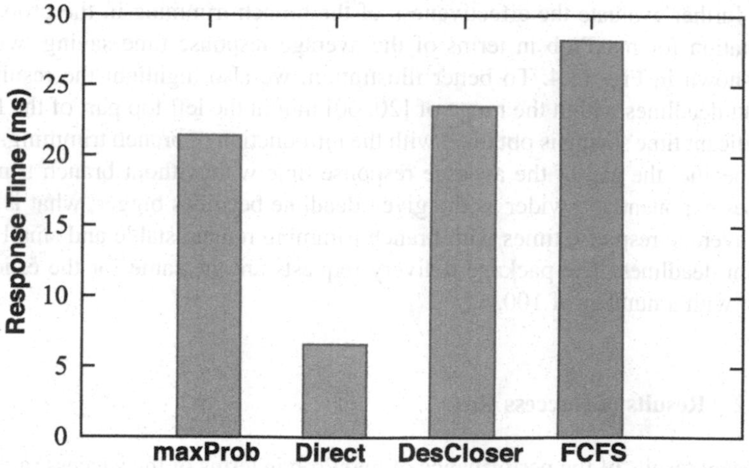
Response Time (ms)
30
25
20
15
10
5
0
maxProb
Direct
DesCloser
FCFS

Fig. 13.3 Evaluation results of response time for four different algorithms

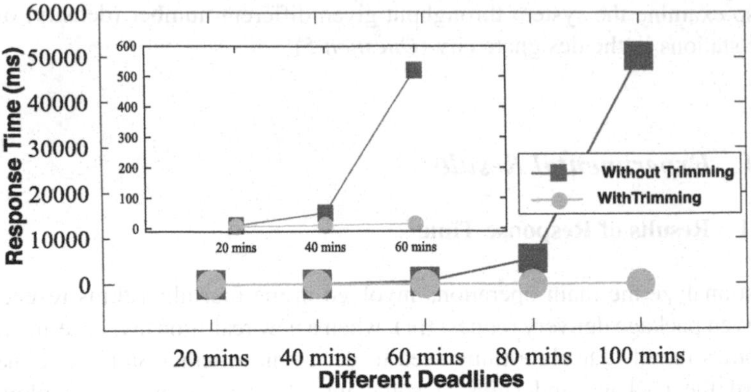
60000
50000
40000
30000
20000
10000
0
Response Time (ms)
600
500
400
300
200
100
0
20 mins
40 mins
60 mins
Without Trimming
WithTrimming
20 mins
40 mins
60 mins
80 mins
100 mins
Different Deadlines

Fig. 13.4 Comparison results of response time of maxProb with/without branch trimming w.r.t different deadlines

the average response time of maxProb is around 22 ms; the average response time of FCFS is no more than 30 ms. All results indicate that all four algorithms are quite efficient, and can plan and adjust the shipping routes in real time. We also observe an interesting phenomenon: FCFS only involves two simple comparisons for each candidate taxi, but it is the most time-consuming method accumulatively. In comparison, although maxProb algorithm contains the most sophisticated operations (i.e., DFS), it needs a shorter response time than FCFS and DesCloser. We argue that this is because: FCFS and DesCloser require more rounds of computation (i.e., more relays) than the maxProb algorithm. In another word, both FCFS and DesCloser generate more unnecessary package relays. The efficiency of maxProb can be guaranteed because: (1) the number of connected intermediate stops to the package's origin is limited, and (2) all the paths with the highest efficiency between any two stops in the transport network are pre-computed and stored.

We further evaluate the effectiveness of the branch trimming in the probability computation for maxProb in terms of the average response time saving, with the result shown in Fig. 13.4. To better illustration, we also highlight the result w.r.t different deadlines within the range of [20, 60] min in the left-top part of the figure. A significant time saving is obtained with the introduction of branch trimming. To be more specific, the gap of the average response time with/without branch trimming increases exponentially wider as the given deadline becomes bigger, what is more, all the average response times with branch trimming remain stable and small under all given deadlines. The package delivery requests are the same for the efficiency studies, with a number of 100.

13.5.4.2 Results of Success Rate

We present results of the performance of maxProb in terms of the success rate under different deadlines in Fig. 13.5. More specifically, extraT is set in a range from 20 to

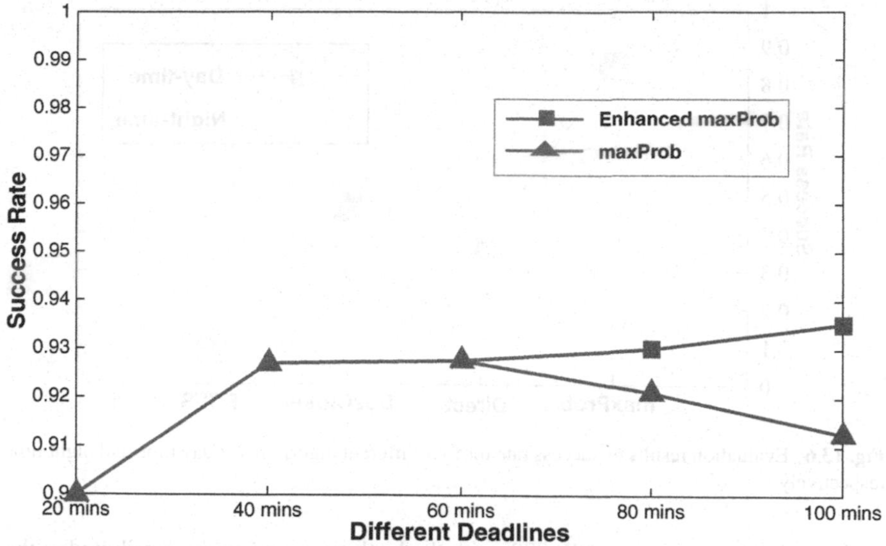
Success Rate
1
0.99
0.98
0.97
0.96
0.95
0.94
0.93
0.92
0.91
0.9
Enhanced maxProb
maxProb
20 mins
40 mins
60 mins
80 mins
100 mins
Different Deadlines

Fig. 13.5 Evaluation results of success rate under different deadlines

100 min, with an equal interval of 20 min. As one can see, the success rate under all deadlines is high, with a value above 90%. The success rate firstly becomes slightly higher then decreases gradually as users allow a longer deadline. The highest success rate appears when the extraT is set 60 min, with the value of around 93%. The observed phenomena seems somehow counterfactual at the first glance as the success rate should be higher when setting a longer deadline in intuition. The root cause is that: when giving a bigger deadline, the arriving-on-time probability if hitching the current becomes greater, as guaranteed by Theorem 13.1. An extreme case is that the probability would always equal one and dominates the other possibilities, as a consequence, the maxProb algorithm tends to select the hitchhiking the current strategy while routing packages. Under such circumstance, maxProb degrades to the FCFS algorithm to some extent, causing the negligible decrease of the success rate. To avoid such degradation, the key issue is to lower the value computed by Eq. (13.10). Thus, one potential solution can be the reduction of remaining time margin during package routing when applying maxProb algorithm. Specifically, if hitchhiking the current strategy is always preferred at first during package routing, the remaining time margin can be manually reduced to 90% of the true one that imposed by the user. We call such variant as the enhanced maxProb algorithm (i.e., introducing the modification of the time margin). The effectiveness of the enhanced maxProb algorithm is evaluated, with the results shown in Fig. 13.5. With the enhanced version, the success rate increases when the deadline gets longer. It should be noted that the success rate of FCFS is much lower, compared to maxProb, which will be verified in the following experiments. The number of

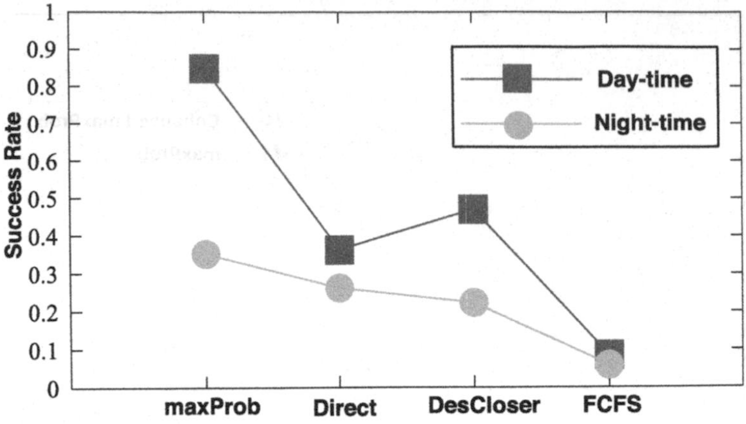
Success Rate
1
0.9
0.8
0.7
0.6
0.5
0.4
0.3
0.2
0.1
0
Day-time
Night-time
maxProb
Direct
DesCloser
FCFS

Fig. 13.6 Evaluation results of success rate for four different algorithms at day time and night time respectively

package delivery requests is 10,000, with the birth time uniformly distributed at the day-time (i.e., from 8:00 to 18:00).

We also report the result of success rate for four different algorithms w.r.t the time of the day. For simplicity, we do not provide the success rate of the four algorithms under every hour of a day, and just split a day into two time slots, i.e., day time from 8:00 to 18:00 and the rest is the night time, respectively. In the time dimension, as shown in Fig. 13.6, for all four algorithms, the success rate is higher at day time than night time, except for FCFS which only achieves the similarly low success rate at both time slots. Direct algorithm returns quite close performance at day time and night time. As predicted, the success rate of FCFS algorithm is the lowest, i.e., below 10%. Compared to the other three baseline algorithms, maxProb achieves the best success rate at both day time and night time. In this experiment, the number of package delivery requests is 10,000, with the birth time uniformly distributed at the day time and nigh time, respectively; the extraT is fixed to 60 min.

In real situation, there are more passengers who prefer to take taxis at some interchange stations, providing more hitchhiking opportunities for package deliveries, probably leading to a better success rate. We thus manually categorize the interchange stations into two classes (i.e. popular, unpopular) in advance, by taking its total number of pick-ups and drop-offs in history into account. The number of interchange stations labeled as popular is 19; and the number of interchange stations labeled as unpopular is 15. We further identify three categories of package delivery requests, according to the labels of the original and destination station, as shown in Table 13.3. Then, we test the performance on success rate for each category. The number of package delivery requests for each category is same, with a value of 10,000.

Figure 13.7 shows the comparison of success rate of the four algorithms under a given deadline (the extraT is fixed to 60 min in this study) for the three categories. maxProb achieves the best performance for three categories, while the performance

Table 13.3 Three identified categories of package delivery requests

	Description
Category I	packages start and end at popular stations (p2p)
Category II	packages start or end at popular stations (p2u or u2p)
Category III	packages start and end at unpopular stations (u2u)

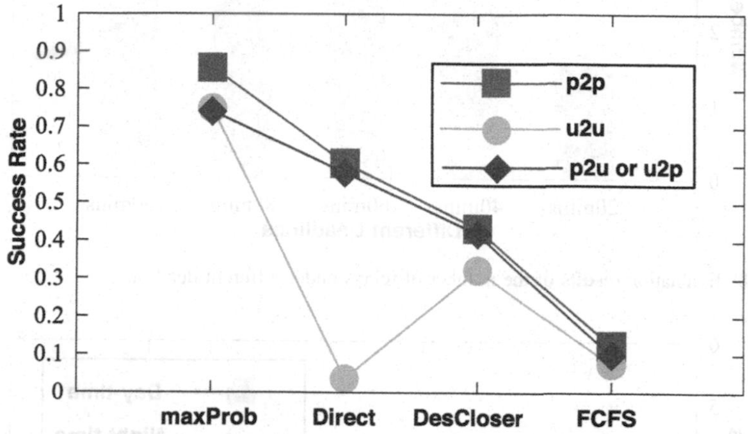
Success Rate
1
0.9
0.8
0.7
0.6
0.5
0.4
0.3
0.2
0.1
0
p2p
u2u
p2u or u2p
maxProb
Direct
DesCloser
FCFS

Fig. 13.7 Evaluation results of success rate for four different algorithms of different package categories

of FCFS is the worst. One exception is the success rate of Direct for the Category III packages. In such case, without any relay at the intermediate stations, Direct obtains an extremely low success rate (i.e., under 5%), which is due to the fact that there is insufficient passenger flow between two unpopular stations. For the same algorithm, the performance is also different for different categories of package delivery requests. The performance for the Category I packages is the best; the performance is the worst for the Category III packages. Particularly, maxProb ensures that around 90% of the Category I packages (75% for Category II and III) can be arrived-on-time successfully. While for FCFS, only less than 10% of all Category packages can be delivered by the deadline. Similar to the previous results, DesCloser performs better than FCFS and worse than maxProb for all three categories.

13.5.4.3 Results of Number of Relays

We compare the number of relays of maxProb under different given deadlines, with the results shown in Fig. 13.8. One can observe that the number of relays presents an ascending tendency with the increase of the given deadline. One possible reason is that more ineffective relays (i.e., the package moves back and forth towards the destination commonly) are resulted since maxProb is inclined to hitchhike the coming taxi immediately, as discussed.

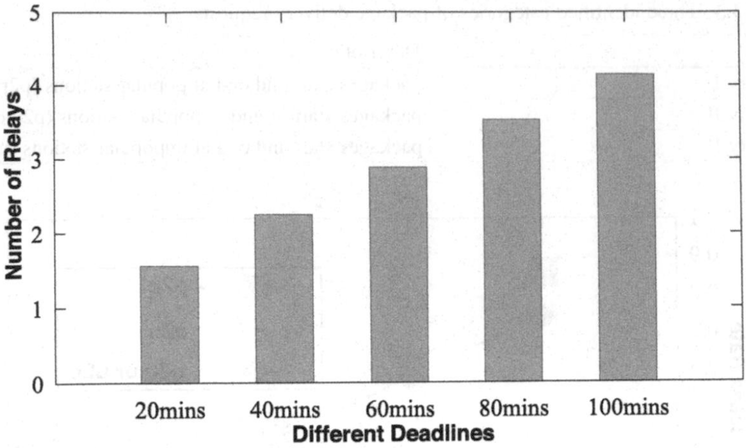
Number of Relays
0
1
2
3
4
5
20mins
40mins
60mins
80mins
100mins
Different Deadlines

Fig. 13.8 Evaluation results of the number of relays under different deadlines

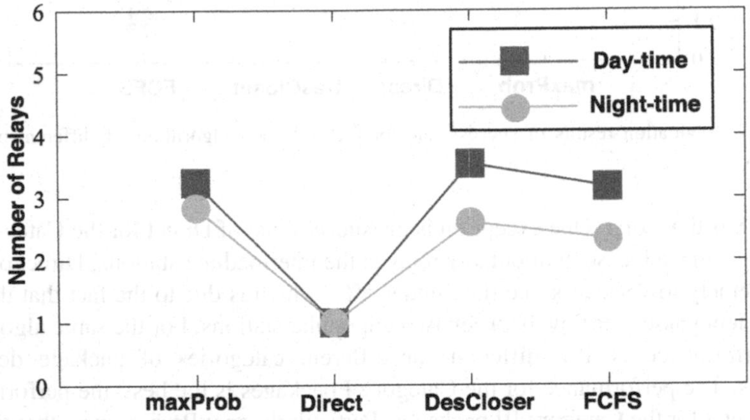
Number of Relays
0
1
2
3
4
5
6
Day-time
Night-time
maxProb
Direct
DesCloser
FCFS

Fig. 13.9 Evaluation results of the number of relays for four algorithms at day time and night time respectively

We also report the results of the number of relays for four algorithms w.r.t the day time and nigh time respectively, as shown in Fig. 13.9. As can be predicted, the number of relays for Direct shall equal one. For the other three algorithms, slightly more relays are generally required at the day time than the night time. Moreover, a little surprisingly, the number of relays resulted from DesCloser and FCFS are quite close to that obtained by maxProb, implying that the number of relays is somehow independent of the adopted algorithms for a successful package delivery. We will investigate deeper about the potential causes qualitatively and quantitatively such as the geographical and temporal distributions of the successful package deliveries in the future work.

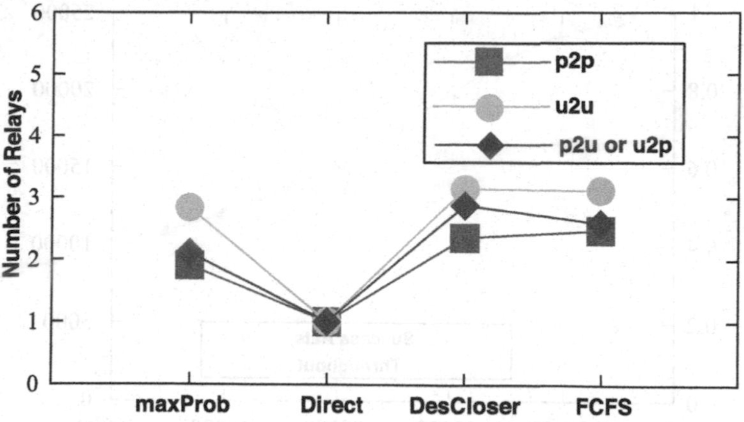
Number of Relays
0
1
2
3
4
5
6
maxProb
Direct
DesCloser
FCFS
p2p
u2u
p2u or u2p

Fig. 13.10 Evaluation results of the number of relays for four algorithms for different package categories

We further report the results of the number of relays for four algorithms w.r.t the package categories, as shown in Fig. 13.10. Similarly, the number of relays for Direct is one, regardless of the package categories. Compared to the other two algorithms (i.e., DesCloser and FCFS), maxProb requires slightly fewer relays for all three package categories. Similar to the results of the success rate w.r.t the package categories, for all algorithms except for Direct, the performance in terms of the number of relays is the best for the Category I packages; the performance for the Category III packages is the worst.

13.5.4.4 Results of Package Throughput

It is necessary to evaluate the package throughput of the system, since it is a primary consideration when applied to real-life scenarios. Figure 13.11 shows the results on the number of packages that the proposed system can transport successfully w.r.t the number of total generated package delivery requests per day, together with the success rate. More specifically, package throughput increases gradually with the number of generated package delivery requests before approaching a stable value. On the contrary, the success rate declines almost linearly with the number of generated package delivery requests. As can be seen, the maximum package throughput is around 20,000 per day, however, the corresponding success rate is quite low (i.e., around 40%) and might be not applicable in real life. To make the proposed system practical in real situations, the package throughput should be around 9500 per day while maintaining a relatively promising success rate.

We further examine the package throughput under different density (number) of interchange stations, as shown in Fig. 13.12. As expected, the more interchange stations our system has, the higher throughput it can achieve (the number of package delivery requests for this study is 10,000). We argue that the root cause is the

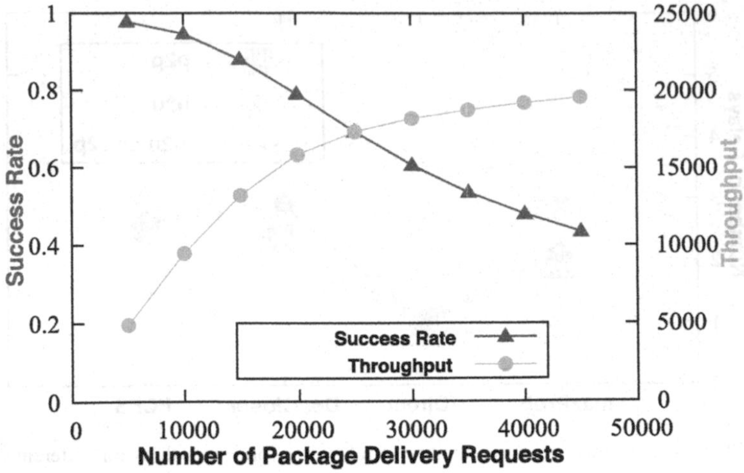
Success Rate
Throughput
Number of Package Delivery Requests
Success Rate
Throughput
0
0.2
0.4
0.6
0.8
1
0
5000
10000
15000
20000
25000
0
10000
20000
30000
40000
50000

Fig. 13.11 Evaluation results of success rate and throughput w.r.t. package delivery requests

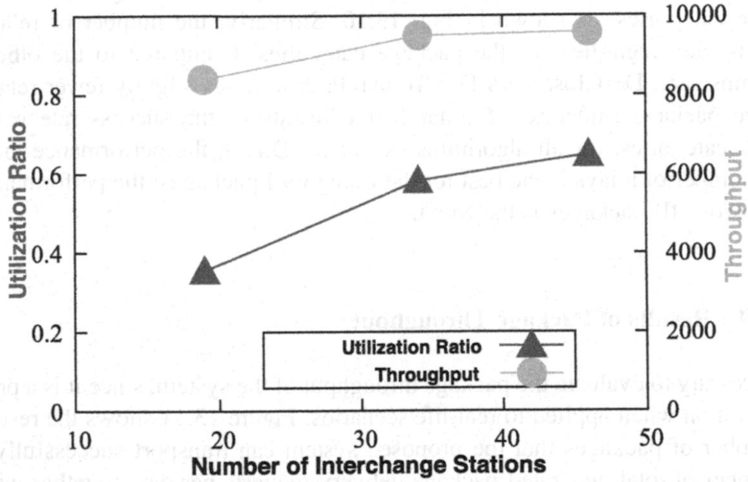
Utilization Ratio
Throughput
Number of Interchange Stations
Utilization Ratio
Throughput
0
0.2
0.4
0.6
0.8
1
0
2000
4000
6000
8000
10000
10
20
30
40
50

Fig. 13.12 Evaluation results of utilization ratio and throughput w.r.t. number of interchange stations

improvement on utilization ratio of taxi trips for package deliveries, which is defined as the ratio between the number of taxi trips involved in package deliveries and the total number of taxi trips. We thus plot the utilization ratio under different density of interchange stations in Fig. 13.12 as well. As evidenced, an increasing utilization ratio (35.5%, 58.3% and 64.9% respectively) is achieved as the interchange stations in the city becomes denser (18, 34 and 46 respectively). Moreover, there is still a considerable room to increase the package throughput further if placing more interchange stations to better utilize the taxi trips. We are aware that the choice of interchange stations (different locations but with the same number) may affect the throughput result, but the overall trend shall be the same. %However, quite limited improvement on throughput can be achieved when the number of interchange station increases from 34 to 46. is the number of hitchhiking rides provided by the taxis than can be utilized in the city.

13.6 Conclusions and Future Work

In this chapter, we propose a new package delivery route planning framework, named as **CrowdExpress**. The framework uses hitchhiking ride services provided by occupied taxis to transport packages in a timely manner without reducing the quality of passenger service. Specifically, we first build a package transportation network by mining taxi GPS trajectory data offline. On this basis, a novel and comprehensive two-stage method of package delivery path planning is proposed. Finally, we compared the proposed method in this chapter with three baseline algorithms using the real-world datasets, and the results show that the proposed method is more effective and efficient.

In the next step, we plan to deepen and broaden our solutions in several directions. First, we plan to associate the packaging deadline with the packaging price. Currently, we have set a general and relative deadline for the delivery of all packages and deal with them equally. In the near future, we plan to set different priorities for packages based on the user's expected arrival time. For example, if users want their packages to arrive earlier, they will have to pay higher fees. Also new package routing algorithms should be developed accordingly. Second, we intend to take measures to improve system coverage and throughput. For example, optimizing switching stations (number and location) and grouping packets with close destinations. Finally, we plan to implement and test our system in real settings with real users to collect feedback on how to further improve the service.

References

1. Tirachini A, Cats O. COVID-19 and public transportation: current assessment, prospects, and research needs[J]. J Publ Transport. 2020;22(1):1.
2. Arslan AM, Agatz N, Kroon L, Zuidwijk R. Crowdsourced delivery—a dynamic pickup and delivery problem with ad hoc drivers. Transp Sci. 2019;53(1):222–35.
3. Liu Y, et al. FooDNet: toward an optimized food delivery network based on spatial crowdsourcing. IEEE Trans Mob Comput. 2019;18(6):1288–301.
4. Benjaafar S, Li Y, Daskin M. Carbon footprint and the management of supply chains: Insights from simple models. IEEE Trans Autom Sci Eng. 2013;10(1):99–116.
5. Rohm AJ, Swaminathan V. A typology of online shoppers based on shopping motivations. J Bus Res. 2004;57(7):748–57.
6. Chen C, Yang S, Wang Y, et al. CrowdExpress: a probabilistic framework for on-time crowdsourced package deliveries[J]. IEEE Trans Big Data. 2020;

7. Chen C, Zhang D, Li N, Zhou Z-H. B-Planner: Planning bidirectional night bus routes using large-scale taxi GPS traces. IEEE Trans Intelli Transport Syst Trans Syst. 2014;15(4):1451–65.
8. Devari A. Crowdsourced last mile delivery using social networks, Master's thesis. Buffalo, NY: The State University of New York at Buffalo; 2016.
9. Li B, Krushinsky D, Reijers HA, Van Woensel T. The share-a-ride problem: People and parcels sharing taxis. Eur J Oper Res. 2014;238(1):31–40.
10. Tan X, Shu Y, Lu X, Cheng P, Chen J. Characterizing and modeling package dynamics in express shipping service network. New York: Proceedings of IEEE Congress on Big Data; 2014. p. 144–51.
11. Nie Y, Fan Y. Arriving-on-time problem: Discrete algorithm that ensures convergence. Transp Res Rec. 2006;1964:193–200.
12. Archetti C, Savelsbergh M, Speranza MG. The vehicle routing problem with occasional drivers. Eur J Oper Res. 2016;254(2):472–80.
13. Chen C, et al. CrowdDeliver: planning city-wide package delivery paths leveraging the crowd of taxis. IEEE Trans Intell Transport Syst. 2017;18(6):1478–96.
14. Ester M, et al. A density-based algorithm for discovering clusters in large spatial databases with noise. Proc 2nd Int Conf Knowl Discov Data Mining. 1996;96:226–31.
15. Zheng X, Liang X, Xu K. Where to wait for a taxi? New York: Proceedings of the ACM SIGKDD International Workshop on Urban Computing; 2012. p. 149–56.
16. Skiena S. Dijkstra's algorithm. In: Implementing discrete mathematics: combinatorics and graph theory with mathematica. Reading, MA, USA: Addison-Wesley; 1990. p. 225–7.
17. Alarabi L, Eldawy A, Alghamdi R, Mokbel MF. TAREEG: A MapReduce-based web service for extracting spatial data from OpenStreetMap. New York: Proceedings of the 2014 ACM SIGMOD International Conference on Management of Data; 2014. p. 897–900.
18. Golden BL, Levy L, Vohra R. The orienteering problem. Naval Res Logist (NRL). 1987;34 (3):307–18.
19. Tarjan R. Depth-first search and linear graph algorithms. SIAM J Comput. 1972;1(2):146–60.

Part VII
Open Issues, Future Directions and Conclusions

Chapter 14
Open Issues and Conclusions

14.1 Open Issues

14.1.1 Data Issue: The More or the Less?

In this monography, trajectory data mining techniques is the key enablers to smart urban services. The intelligence related to those smart urban services relies on the human-centered knowledge that is extracted from GPS trajectories in a data-driven manner. Intuitively, more human-centered data sources could convey more valuable knowledge, such as the body-sensing devices [1] and social media [2]. However, the import of new data sources would easily cause the human privacy issue which is a very important aspect in the field of urban computing. The GPS trajectory data employed in this monography is relatively safe, since it is more like a ready-made data source with the wide equipment of GPS devices in vehicles. Although there are many kinds of data sources associate with human behaviours, which dataset can be employed and how to safely use it are still served as considerable open issues in the research on enabling smart urban services.

14.1.2 Model Issue: Effectiveness or Lightweight?

The currently impressive development of intelligent transportation benefits from big data techniques and the advanced AI algorithms. Although the performance is getting better, the complexity of models is also gradually growing. To make the situation worse, these models are heavy and not easy to be deployed to real-world scenarios, especially for the deep learning models with a huge number of parameters and calculations. Hence, the balance between effectiveness and lightweight is very worth considering when developing smart urban services.

C. Chen et al., *Enabling Smart Urban Services with GPS Trajectory Data*,
https://doi.org/10.1007/978-981-16-0178-1_14

14.1.3 Application Issue: Ground Truth

GPS trajectory data is highly related to human behaviour, thus we are able to understand different behaviors through mining the trajectory data. However, the fact is that many implicit aspects of human behaviours cannot be accurately recorded or labeled with trajectory data, such as passenger's trip purposes [3], driver's driving styles. In other words, we know a trajectory relates to one kind of behaviours but we cannot know what exactly it is. Hence, even with a plenty of data and advanced data mining technologies, it is still difficult to find the exact relationship between trajectories and the target human behavior, because of the lack of ground truth. With regard to this point, we mostly employ the unsupervised approaches like clustering algorithms and interpretation to obtain a rough target distribution [4]. Nevertheless, to some extent, this fact limits the confidence of some applications.

14.2 Future Directions

14.2.1 Data Foundations and Embedded Semantics

With the development of mobile computing, smart urban services will impose increasingly complex requirements on the data foundation techniques [5]. For example, algorithms like map-matching and compression are required to run efficiently and online in mobile devices while retaining ample spatiotemporal properties. On the other hand, the latent knowledge in trajectories would gradually appear to be inefficient for diverse requirements. To narrow the gap, it is emerging to enrich the semantics by fusing GPS trajectories with the multi-source urban data [6–8]. For example, using POI data to augment the activity semantics of movements, so as to reveal passengers' trip purposes [4]. However, it is notable that the basic premise of using multi-source data is to avoid privacy problems.

14.2.2 Lane-Level Trajectory-Based Services

Due to the uncertainty of collected GPS trajectories (e.g., positioning errors, unstable sampling), most studies merely utilize the coarse-grained movement nature within GPS trajectories. Even with the correction techniques like map-matching, trajectories would still lose vehicles' real position information on roads. As a result, it is extremely difficult to develop those smart services which depend on vehicles' fine-grained positions on roads, such as the real-time driving assistance and collaboration on the lane level. Such a situation can be changed with the development of high-precision positioning technology [9]. By then, the reported trajectory could be a new indicator of vehicles' real movements on roads. Compared to visual images or videos, the high-precision trajectory is much cheaper and lighter to enable the real-position-dependent smart urban services in several scenarios.

14.2.3 Vehicle Collaboration with IoV, 5G and Edge Computing

The evolution from the single-agent intelligence to the multi-agent intelligence is an emerging trend of smart city development. In transportation systems, the real-time collaboration of vehicles on roads can be of great significance to the driving safety, city perception, and intelligent dispatching [10, 11]. However, such services are hard to be achieved by mining GPS trajectory data. On the one hand, very few trajectories were collected at the same time and roads, which means no sufficient data in the collaborative scenario for the data mining approach. On the other hand, dealing with a group of vehicles' GPS trajectories will take unsustainable communication and computing cost. Fortunately, the gaps could be gradually narrowed with the development of the latest ICT technologies like the Internet of Vehicles (IoV) [12], 5G Communication [13] and Edge Computing [14]. Specifically, the implementation environment of collaboration services would be significantly improved in terms of sensing, communication and computing. Hence, in the near future, mining trajectory data will become a promising approach to enable the multi-vehicle collaboration services in transportation systems.

14.2.4 Promoting Smart City with Smart Mobility

Urban services in different fields of smart city are interrelated but with unbalanced developments. In recent years, data sharing and communication between different urban fields are getting easier and quicker. As an essential part of urban cities, smart mobility is expected to be a good impulse and carrier to promote the intelligence of other fields of smart city. Chapters 12 and 13 present one kind of typical cases, in which the Crowdshipping service utilizes the taxi-based mobility to resolve the speed and cost conflict of the goods delivery. In the future, we believe that more "beyond" would benefit from smart mobility with the development of ICT technologies.

14.3 Conclusions

In the increasingly complex transportation systems, problems (e.g., overload, low efficiency, and environmental pollution) are getting worse. Additionally, the mobility needs of different parties (i.e., people and goods) have also increased due to the growing concerns of *efficiency, flexibility, affordability*. These facts put forward higher requirements for the intelligentization of urban mobility. Hence, how to enable smart urban services for a spectrum of parties becomes a significant and urgent issue.

In general, the in-depth understanding of human behaviours is the key to enabling smart urban services, so that it is crucial to obtain the majority/minority human mobility knowledge. Fortunately, GPS trajectory data that records where and when people move in the city is accumulatively gathering and readily available on a large scale, providing us a time-evolving view to understand the mobility behaviors in the city from a data-driven perspective. What's more, recent years also witness the superior performance of AI algorithms in data mining tasks, making the data-driven knowledge extraction more in-depth and comprehensive. Hence, the GPS trajectory data mining brings us extensive opportunities to enable smart urban services. Specifically, we summarize the opportunities as follows:

- Majority pattern identification: spatiotemporal traffic flow evolving patterns, urban area correlations, taxi demand evolving patterns.
- Minority anomaly detection: anomalous driving behaviors, abnormal events.
- Personalized/individual analysis: preference modelling, driving strategy modelling, intention inference.
- Extra utilization besides human mobility: crowdshipping, crowdsensing, and etc.

The general framework of developing smart urban service based on the GPS trajectory data mining, consists of four components as illustrated in Fig. 14.1. First of all, different parties participate the human mobility with different mobility behaviors, continuously reporting GPS trajectory data to data centers. On top of that, to make the mobility smarter, an initial and essential work is to conduct a systematical

Fig. 14.1 Framework of developing smart urban service with trajectory data

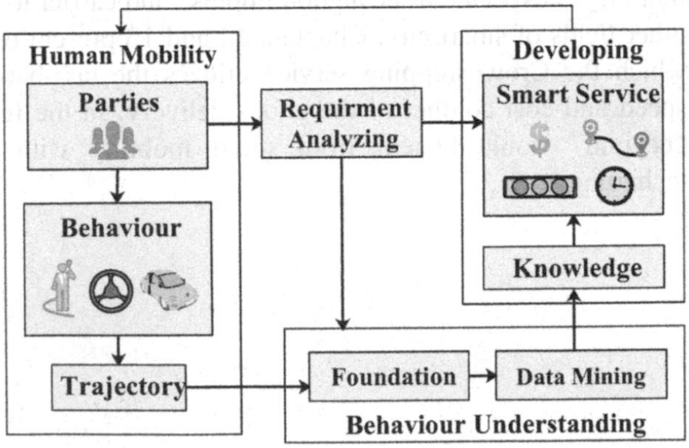
Human Mobility
Parties
Behaviour
Trajectory
Requirment
Analyzing
Developing
Smart Service
Knowledge
Foundation
Data Mining
Behaviour Understanding

requirement analysis for different mobility-parties, which would be used in the behavior understanding and service developing. Specifically, we summarize the requirements of different mobility-parties for smart services as follows:

- Passengers: As the most important group in human mobility, passengers' requirements can be summarized as providing services provisions covering the whole life circle of trips. Before trips, passengers usually want to know more about the forthcoming trip, such as travel time and fare. During trips, it is expected to reduce passengers' anxiety which comes from the potential fraud and safety concern of driver's driving. Generally, after trips, passengers would continue to perform activities. Hence, it is expected to receive personalized recommendations or advertisements regarding their intended activities.
- Drivers: Drivers participate in human mobility with more complex behaviors, like customer searching, route selection, and driving. Generally, their requirements can be summarized as "4 more". Know more indicates that drivers hope to receive helpful information for driving efficiency, such as the real-time/future traffic conditions of the roads they are heading for. The most concern of drivers is about their income. Earn more relates to good taxi service strategies in the situations of passenger delivering and searching, which strongly determine the high or low income. Save more relates to the concerns about time and money. The time saving is also a driving efficiency issue associates with know more and driving strategy. The money saving requirement mainly comes from the fuel economy concern. Drivers always expect to take the most fuel-efficient driving route according to the personal driving style and the real-time traffic conditions. At last, although the current vehicles are equipped with rich driving assistance devices, the driving safety still remains inherent correlation with drivers' driving style. Hence, drive more safely is about the need of a driving-style-aware service to warn aggressive driving behaviors.
- Urban planners: Even though urban planners are able to gain access to a large amount of urban data, it is still challenging for them to obtain insights from observations to help make decisions. In addition, urban planners also desire to reduce dedicated vehicles deployed for specific tasks (such as urban sensing), which is meaningful for alleviating the burden of transportation systems. Hence, smart dispatching strategies are required to effectively transform those specific tasks to other carriers such as buses and taxis.
- Travellers: How to enjoyably tour in an unfamiliar city is quite important to travellers. In this sense, the natural requirement of travellers is using taxi as a real-time probe. As a matter of fact, such requirement can be regarded as a route planning issue, and coupled with two essential aspects, namely the limited travelling time of travelers and personalized setting.
- Goods: Intactness, speed and cost are three major concerns of the goods delivery. Speed and cost are two optimization objects, but they generally conflict with each other. With regard to this point, how to balance the delivering speed and cost is identified as an urgent requirement for smart urban services for goods.

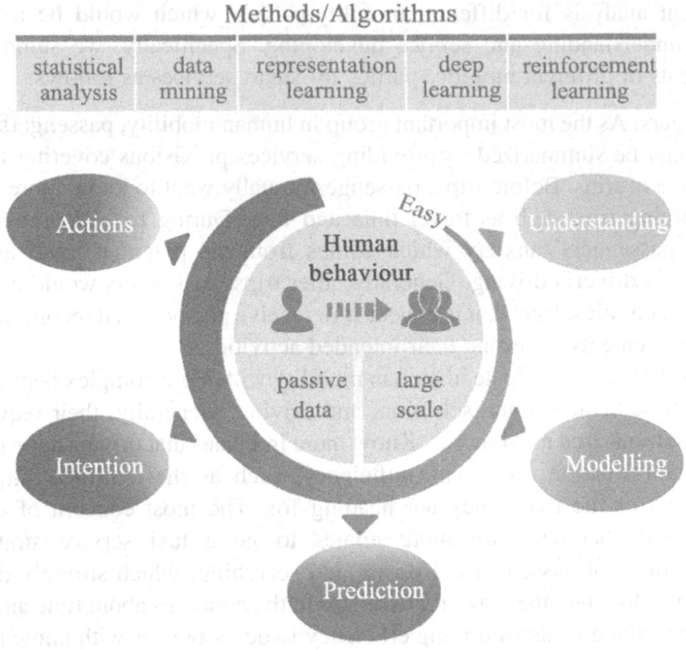
Methods/Algorithms
statistical analysis
data mining
representation learning
deep learning
reinforcement learning
Easy
Actions
Understanding
Human behaviour
passive data
large scale
Intention
Modelling
Prediction

Fig. 14.2 Human behaviour understanding: methods and usages

The behaviour understanding is the core component of this framework, which aims to extract the requirement-related mobility knowledge from trajectory data. Within this component, the data foundation techniques include data selection, cleaning, map matching, compression, privacy protection, and so on. The data mining methods include statistical analysis, basic data mining algorithms, machine learning, representation learning, deep learning, reinforce learning, and so on. The use of data foundation and data mining methods depends on the requirements, since they indirectly determine which kind of data format should be used and what knowledge should be extracted. In general, the usages of the extracted human behavior knowledge, from easy to difficult, can be categorized into five levels as shown in Fig. 14.2, namely understanding, modelling, prediction, intention and actions.

On the stage of service developing, parties' requirements are employed to specify what to do (i.e., functions), while the corresponding knowledge tells how to do. In addition, it is necessary to consider the computing capability and system limitations in the real-world scenarios when transforming the developing process into engineering tasks. Finally, smart services would be applied to different parties, then improve or even change the mobility in cities.

The advanced ICT technologies are changing people's daily life, and making the development of smart urban services full of opportunities and challenges. This monograph shows how to develop smart urban services for a spectrum of parties by using one kind of pervasive and easily accessible data (i.e., GPS trajectory data).

It aims to provide feasible guidelines for the researches of GPS trajectory data mining, from the perspectives of data foundation, knowledge extraction and applications. We also discuss several open issues and future directions, aiming to inspire sustainable developments as well as interesting questions for the field.

References

1. Bharti P, De D, Chellappan S, Das SK. HuMAn: complex activity recognition with multi-modal multi-positional body sensing. IEEE Trans Mob Comput. 2018;18(4):857–70.
2. Chua AYK, Banerjee S. Customer knowledge management via social media: the case of Starbucks. J Knowl Manag. 2013;17(2):237–49.
3. Chen C, Jiao S, Zhang S, Liu W, Feng L, Wang Y. TripImputor: real-time imputing taxi trip purpose leveraging multi-sourced urban data. IEEE Trans Intell Transport Syst. 2018;19 (10):3292–304.
4. Chen C, Liao C, Xie X, Wang Y, Zhao J. Trip2vec: a deep embedding approach for clustering and profiling taxi trip purposes. Pers Ubiquit Comput. 2019;23(1):53–66.
5. Li D, Shan I, Shao Z, Zhou X, Yao Y. Geomatics for smart cities-concept, key techniques, and applications. Geo-spatial Informat Sci. 2013;16(1):13–24.
6. Zhang D, Lee K, Lee I. Semantic periodic pattern mining from spatio-temporal trajectories. Inf Sci. 2019;502:164–89.
7. Natal IDP, Cordeiro RAC, Garcia ACB. Activity recognition model based on GPS data, points of interest and user profile. In: International symposium on methodologies for intelligent systems. Cham: Springer; 2017. p. 358–67.
8. Petry LM, Silva CLD, Esuli C, Renso A, Bogorny V. MARC: a robust method for multiple-aspect trajectory classification via space, time, and semantic embeddings. Int J Geogr Inf Sci. 2020;34(7):1428–50.
9. Li W, Fu Q, Liu B, Li J. Review of low-cost and high-precision positioning technology. Bull Survey Mapp. 2020;2:1–8.
10. Li P, Wu X, Shen W, Tong W, Guo S. Collaboration of heterogeneous unmanned vehicles for smart cities. IEEE Netw. 2019;33(4):133–7.
11. Lv C, Wang H, Cao D, Zhao Y, Auger DJ, Sullman M, Mouzakitis A. Characterization of driver neuromuscular dynamics for human–automation collaboration design of automated vehicles. IEEE/ASME Trans Mechatron. 2018;23(6):2558–67.
12. Contreras-Castillo J, Zeadally S, Guerrero-Ibañez JA. Internet of vehicles: architecture, protocols, and security. IEEE Int Thing J. 2018;5(5):3701–9.
13. Panwar N, Sharma S, Singh AK. A survey on 5G: the next generation of mobile communication. Phys Commun. 2016;18:64–84.
14. Shi W, Cao J, Zhang Q, Li Y, Xu L. Edge computing: vision and challenges. IEEE Internet Things J. 2016;3(5):637–46.

Printed in the United States
by Baker & Taylor Publisher Services